INTRODUCTION TO
CIRCUIT
ANALYSIS

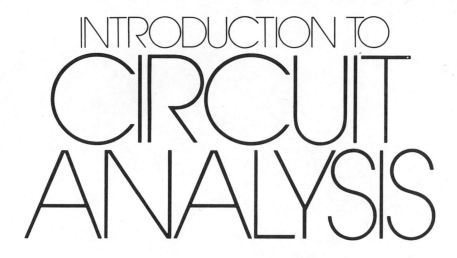

INTRODUCTION TO CIRCUIT ANALYSIS

TIMOTHY N. TRICK
University of Illinois
Urbana, Illinois

JOHN WILEY & SONS

New York   Santa Barbara   Chichester   Brisbane   Toronto

Library of Congress Cataloging in Publication Data:
Trick, Timothy N   1939–
    Introduction to circuit analysis.

    Includes bibliographical references and index.
    1. Electric circuits.   2. Electric network analysis.
I. Title.
TK454.T74   621.319′2   77-10843
ISBN 0-471-88850-8

Printed in the United States of America

10 9 8 7 6 5 4 3 2 1

To Dorothe

Preface

This book was written for a first course in electrical engineering circuit analysis. It has been taught at the University of Illinois to sophomores in a three-hour, one-semester course that is accompanied by a discussion–laboratory session that meets four hours every week. Typically, the students have completed a physics course on electricity and magnetism, and are currently registered in a course on computer programming and a course on differential equations. If the course is taught at a slower pace over a two-quarter or two-semester span, the physics and mathematics requirements could be relaxed. Also, the book uses no computer algorithms so that a course in computer programming is not required. Computer algorithms are omitted for the following three reasons. It is difficult to teach both the principles of circuit analysis and numerical algorithms for circuit analysis in the first one-semester circuit analysis course for engineers. Second, the examples necessary to demonstrate the circuit analysis principles can always be made simple enough so that computer aids are not required. Finally, the first step in circuit design is the conception of a circuit in the designer's mind and a quick oversimplified pencil-and-paper analysis to see if the conceived circuit can be expected to do the job. The computer is a valuable tool in circuit design, but it has never made a poor circuit designer into a circuit design genius. Good circuit designers understand well both circuit theory and the devices with which they must work. This does not mean that the computer cannot be used as a teaching aid—merely that numerical algorithms are not taught in our first course. For example, one could have available an interactive graphics terminal that would allow a student to input the topology and parameter values of a circuit and obtain a graphical display of the response. At the University of Illinois we have this capability on the PLATO system. A course on numerical algorithms for circuit analysis and design is available to our more advanced undergraduate students.

This book was written because I felt that most first course circuit analysis books currently on the market have one or more of the following defects. They are too old and treat only three basic components, the resistor, the capacitor and the inductor. They are very shallow surveys that fail to teach the student the mathematical and physical principles of circuits. They are so abstract that students see no connection between the material they are studying and electrical circuits in the real world. To overcome these defects we introduce electronic devices in our first course and include an introductory treatment of their physical characteristics. The principles for formulating and solving circuit equations are carefully explained without using abstract mathematical notation. Numerous practical examples are included in the text.

The book consists of 15 chapters and two appendices. Chapter One reviews electrostatic fields and conduction in various media. A final section on integrated circuit diffused resistors is included to show the student how many elements can be simultaneously fabricated on a single silicon wafer and interconnected by means of photolithographic techniques. This chapter is intended for review and physical motivation of the student before we begin to postulate the mathematical models and laws of circuit theory in the following chapters. Its omission will have a negligible effect on the presentation of the material in the succeeding chapters. In fact, this chapter should be omitted or left as a reading assignment in a one-quarter or one-semester course on dc, transient, and ac analysis.

Chapter Two introduces signals. The important topics are the average and rms values of a periodic signal and the addition of sinusoids of the same frequency using phasors. One should review Appendix B, Complex Numbers, before introducing the phasor concept. Also, the students should be introduced to the mathematical characterization of some nonperiodic signals, such as the step function, pulse function, and exponential function. A section is included on the Fourier series and frequency spectra of periodic signals, but it is strictly optional. Usually this material is covered in a more advanced course on linear circuit and system analysis. It is included here to introduce the student briefly to the concept of the frequency content of a signal. The examples in this section are a half-wave rectified signal and a pulse. These examples are referred to in Chapter Six in the sections on diode waveshaping and diode communication circuits to illustrate that the frequency content of an input signal can be changed by means of nonlinear elements. Again, Section 2.1.3 can be deleted in a one term course.

The ideal resistor, ideal current source, ideal voltage source and Kirchhoff's laws are the topics of Chapter Three. In addition, it is shown how to model, scale, and analyze practical circuits, including the linear ladder network.

The mesh and node method of analysis are covered in Chapter Four. In the node method, voltage sources often cause difficulty. Frequently, a student writes the node equations with the current in the voltage source neglected. In the mesh method, often the voltage across the current source is neglected. I hope that the approach used in this chapter will eliminate these mistakes. The concept of tree branches and link branches is introduced in the final section to point out to the student that for a given tree every other voltage in the circuit is dependent on the voltages of the branches of the tree. Similarly, every branch current in a circuit is dependent on the link branch currents. This gives us another technique for writing an independent set of Kirchhoff equations. Again, it may be necessary to skip this section in a one-term course.

Equivalent circuits is the subject of Chapter Five. The substitution principle is presented and it is shown how to reduce a circuit by series/parallel operations. The Thévenin equivalent, source transformations, and superposition are also discussed and the D/A converter is used as one of the examples. A section on symmetrical circuits is included, but probably should be omitted in a one-term course. The final section discusses the important concept of voltage and current source splitting, which has numerous applications in modeling and analysis.

The diode is the topic of Chapter Six. In a more conventional circuit analysis course, this chapter can be omitted or it can be taught in an accompanying laboratory course. Another option would be to introduce the students to the diode rectifier and clipper in Section 6.1, the semiconductor diode characteristics in Section 6.4 and the concept of small-signal analysis in Section 6.5. This small-signal section is a good preparation for the concept of two-port modeling with dependent sources in Chapter Seven.

Chapter Seven introduces the concept of modeling two-ports with dependent sources. Examples include the operational amplifier. Also, the analysis of practical circuits with dependent sources is discussed. This material is contained in Sections 7.1 to 7.4. The remainder of this chapter discusses transistor modeling and circuit analysis. In a one-term course Sections 7.5 to 7.11 can be deleted or covered in an accompanying laboratory course.

The capacitor and inductor are introduced in Chapter Eight. This chapter gives a complete treatment of energy storage in linear, nonlinear, and time-varying capacitors and inductors. Again, only the linear constant capacitor and inductor characteristics should be stressed in a one-term course.

The transient analysis of linear constant circuits with zero or constant inputs is discussed in Chapter Nine. The treatment is general although all examples are restricted to first- or second-order circuits. This chapter also includes the important topics of the analysis of switched circuits and circuits with dependent states.

Chapter Ten shows that the exponential input to circuits, whose input-output relation is characterized by a linear constant coefficient differential equation, leads to the important concept of the complex frequency circuit model and network function. From this circuit model it is shown how one can determine both the transient behavior and the steady-state response to a sinusoidal input. Again, in a one-term course, the concept of a complex frequency model and network function can be introduced in a couple of lectures and then one can move into the more important chapters on ac analysis and the frequency response of circuits.

Sinusoidal steady-state analysis (ac analysis) is the topic of Chapter Eleven. Analysis, equivalent circuits, power, maximum power transfer, impedance matching, and three-phase systems are discussed in this chapter and numerous practical examples are illustrated.

The frequency response of first- and second-order circuits is discussed in Chapter Twelve including resonance, lowpass, bandpass, and highpass filters. The RC operational amplifier filter is also introduced. Sections on Bode plots, scaling, and complex loci are also included.

The transformer is introduced in Chapter Thirteen. The first four sections develop the transformer models and Section 13.5 discusses the practical application of the transformer in impedance matching and power systems. The frequency response of the nonideal transformer is described in Section 13.6. The last two sections present more advanced transformer models and the modeling of three or more coupled coils. Only the first four or five sections can reasonably be covered in a one-term course..

Chapter Fourteen introduces the important topic of sensitivity analysis. Practical examples include the sensitivity of Wheatstone bridge measurements and the accuracy of a D/A converter due to parameter tolerances. Also included is an active filter example.

Chapter Fifteen concludes the book, and it introduces the circuit properties of reciprocity and interreciprocity and their application in analysis.

Appendix A gives a very thorough review of the independence of equations and the solution of liner algebraic equations by means of the elimination of variables and determinants. It is included for students who feel that they need a review of these fundamentals.

There are a variety of options available for teaching the material in this book. The discussion in the first seven chapters was deliberately restricted to signals and resistive circuits so that Chapters One to Seven and Appendices A and B can be taught in a three-lecture hour/week one-semester, or one-quarter course at the beginning sophomore level. Chapters Eight to Thirteen can be taught in a similar course the following term. In fact, in the two-semester sequence there should be enough time for the interesting and practical material in Chapter Fourteen on sensitivity analysis and the material in Chapter Fifteen on reciprocity and interreciprocity. A second option would be to cover the material in Chapters One to Thirteen in a five-lecture hour/week, one-quarter, or one-semester course.

Because of the great diversity of subject matter that now exists in electrical engineering schools, many curricula require an introductory electrical and electronic circuit analysis course that covers dc, transient, and ac analysis in a three-lecture hour/week, one-semester course. The outline used in this course is listed below.

Chapter	Topics Discussed	Lecture Hours
1	Optional—brief review of conduction in materials and the fabrication of a diffused resistor.	1
2	Review of complex numbers in Appendix B, discussion of average and rms values of periodic signals, addition of sinusoids of the same frequency, optional-introduction to signal spectra and the Fourier series.	3
3	Lumped circuits, ideal elements, Kirchhoff's Laws, review of Appendix A, independence of the Kirchhoff equations, circuit analysis, scaling, and ladder networks.	4
4	Node and mesh methods, optional discussion of tree and fundamental loop methods.	4
5	Entire chapter on equivalents circuits with Section 5.4.4 on symmetrical circuits optional.	4
6	This chapter on diodes is optional. One could cover Sections 6.1, 6.4, and 6.5.	2

Chapter	Topics Discussed	Lecture Hours
7	Sections 7.1 to 7.4 on the modeling and analysis of two ports with dependent sources including the operational amplifier	3
8	Emphasis on energy storage and response of only the linear constant inductor and capacitor.	2
9	All sections on transient analysis.	5
10	Brief introduction to the network function and how it can be used for transient analysis and ac analysis.	2
11	ac analysis, all sections.	5
12	First-order and second-order frequency response plots, Bode plots, and scaling.	3
13	First four or five sections on transformers.	2

$$\text{Quizzes} \qquad \frac{4}{44} \text{ hours}$$

Simultaneously most students also enroll in an introductory laboratory that meets four hours a week, two hours are discussion, and two hours are spent in the laboratory on experimentation. The laboratory introduces the students to signals, instrumentation, diodes, operational amplifiers, transistors, transient responses, frequency response measurements, transformers, and ac power measurements. This laboratory gives the student additional exposure to the signal concepts discussed in Chapter Two. Also, the material on diodes can be omitted from the lecture and covered in the laboratory. Similarly, the material on transistors in Sections 7.5 to 7.11 is omitted from the lecture and selected topics are covered in the laboratory. Thus, students who enroll in both the introductory course on circuit analysis and the introductory electrical engineering laboratory have three lecture hours, two laboratory discussion hours, and two laboratory experimentation hours during the week on the above material for one semester. At the University of Illinois students take these two courses as second semester sophomores. These courses are typically followed by introductory semester courses on fields and waves, semiconductor devices, electronic circuits, digital systems, and a course on signals, circuits, and systems that covers Fourier and Laplace transforms and state variables.

In conclusion, I wish to express my deepest gratitude to Professor M. E. Van Valkenburg for his encouragement to undertake this venture and for his numerous suggestions for improving the quality of many of the presentations. Also, special thanks go to Professors E. C. Jones, Jr. of Iowa State University and S. K. Mitra of the University of California at Davis for their valuable criticism as reviewers, to James S. Lehnert and my daughter Patricia Ann Trick for their careful reading of the manuscript and many helpful suggestions from a student's viewpoint, and to Professors M. L. Babcock, M. S. Helm, S. W. Lee, W. D. O'Brien, Jr., W. H. Olson, R. B. Uribe, and P. E. Weston who, as instructors of the course at the University of Illinois, made valuable comments on the content and presentation of the material. Also, I sincerely thank Mrs. Rose Harris and Mrs. Dee Wrather for their uncanny ability to decipher my notes to produce the typewritten manuscript.

Timothy N. Trick

Contents

Three Circuit Analysis: Models and Kirchhoff's Laws

Four Node and Mesh Methods for Analysis

Eight **Energy Storing Elements**

Nine **Response of Circuits with Energy Storage Elements**

Ten **Network Functions**

Eleven AC Analysis

Twelve Frequency Response

Thirteen The Transformer

Fourteen Sensitivity Analysis

Fifteen Reciprocity and Interreciprocity

Appendix A Solution of Linear Algebraic Equations

Appendix B
Complex Numbers

Index

Chapter One
Introduction

Electrical systems transmit information, process data, make measurements, and transfer energy by means of electrical signals. The system
could be a radio transmitter or receiver, a measuring device such as an
oscilloscope, ohmmeter, ammeter, or voltmeter, a large computer, a small
calculator, a power station, or a small electronic test generator. These
systems are composed of a variety of components, such as batteries,
generators, resistors, capacitors, inductors, transistors, motors, transformers, antennas, and the like. With the great technological advances in
large-scale integrated circuits (LSI), many complex electrical systems can
be made inexpensively, and, as a result, electronics will play an even more
significant role in our lives in the future.

The heart of all these systems is the electrical circuit or network, which
is an interconnection of electrical elements to form an electrical system.
The circuit can be an interconnection of physical electrical elements or it
can be a "pencil-and-paper" circuit in which the actual electrical components are modeled by idealized electrical elements for the purpose of
mathematical analysis. This book introduces the reader to some of the
devices and the fundamental laws and concepts necessary for the understanding and design of electrical circuits. To do this we give an
elementary introduction to such electrical components as batteries, electric
generators, resistors, capacitors, and transistors, for example, and we

1

develop mathematical or graphical models for the components in our circuit based on the characteristics of these devices that we can calculate or measure at a set of terminals. This knowledge of the basic components and the analysis of the circuit model will develop our intuitive knowledge of electrical circuits and allow us to design circuits. To analyze the circuit model we need to know *circuit theory,* the formulation of the underlying principles for the analysis and design of electrical circuits. Over the years, devices may change, but it is unlikely that the basic laws and theorems for the analysis of electrical circuits will change.

Electrical phenomena are described by the laws of electromagnetism as formulated by Coulomb, Oersted, Ampere, Faraday, and Maxwell. None of these laws deal with electrical circuits, but they are the basis of field theory. Whereas the basic laws of field theory are Maxwell's equations, the basic laws of circuit theory are Kirchhoff's equations. In order to study circuit theory we need only to understand what is meant by voltage, current, power, and energy. This is not to say that field theory is nonessential to the electrical engineer. On the contrary, field theory is very essential to the understanding of devices. The radiation of energy from an antenna can only be understood by means of field theory. The operation of a cathode ray tube, a diode, or a motor can only be explained by the laws of field theory. In this book our interest is circuit theory, although some basic principles from field theory will occasionally be introduced to explain the electrical behavior of a device.

In this first chapter we introduce basic notation and discuss the fundamental laws of static electric field theory. The terms voltage, current, and energy are defined, and current flow in various media is discussed. The chapter concludes with a discussion of the fabrication processes for integrated circuits.

1.1 NOTATION AND PHYSICAL CONSTANTS

In this book the metric system of measurement will be used unless practical use dictates otherwise. In particular, the International System of Units (abbreviated SI in all languages), adopted by the National Bureau of Standards in 1964 and by the Institute of Electrical and Electronics Engineers (see *IEEE Spectrum,* August 1965 and March 1966, and *ASTM/IEEE Standard Metric Practice,* IEEE Std, 268-1976), will be followed. This system is based on the meter, kilogram, second, ampere, and Kelvin. In the semiconductor device and integrated circuit area circuit size measurements are typically in terms of the mil ($\frac{1}{1000}$ of an inch), and device distance measurements are in terms of micrometers (1 μm = 10^{-6} meter) because of the smallness of these circuits and devices. The standard symbols for some of the terms and units with which we will be working are

Table 1.1 Terminology: Notation and Unit of Measurement

Quantity	Symbol	Unit
Force	F or f	newton (N) = kg-m/s²
Charge	Q or q	coulombs (C)
Energy	W or w	joules (J) = N − m
Power	P or p	watts (W) = J/s
Current	I or i	amperes (A) = C/s
Voltage	V or v	volts (V) = J/C
Electric field	$\mathscr{E}$	volts/meter (V/m)
Flux linkages	Ψ	webers (Wb) = V − s
Magnetic flux density	B	teslas (T) = Wb/m²
Frequency	f	hertz (Hz) = cycles/second
Frequency	ω	radians/second = $2\pi f$
Temperature	T	Kelvin (K)
Time	t	second(s)
Resistance	R	ohm (Ω) = volt/ampere
Conductance	G	siemens (S) or mho (℧) = ampere/volt
Inductance	L	henry (H) = weber/meter
Capacitance	C	farad (F) = coulomb/volt
Elastance	S	(farad)⁻¹, $S = 1/C$
Reciprocal inductance	Γ	(henry)⁻¹, $\Gamma = 1/L$

Table 1.2 Prefixes

Prefix	Factor	Symbol	Example
atto	10^{-18}	a	aC = 10^{-18} coulombs
femto	10^{-15}	f	fF = 10^{-15} farads
pico	10^{-12}	p	ps = 10^{-12} seconds
nano	10^{-9}	n	nA = 10^{-9} amperes
micro	10^{-6}	μ	μV = 10^{-6} volts
milli	10^{-3}	m	mW = 10^{-3} watts
centi	10^{-2}	c	cm = 10^{-2} meters = 1 centimeter
deci	10^{-1}	d	dW = 10^{-1} watts
deka	10	da	daW = 10 watts
hecto	10^{2}	h	hV = 10^{2} volts
kilo	10^{3}	k	kHz = 10^{3} cycles/s
mega or meg	10^{6}	M	MW = 10^{6} watts
giga	10^{9}	G	GHz = 10^{9} cycles/s
tera	10^{12}	T	THz = 10^{12} cycles/s

given on the inside front cover of the book along with some of the physical constants used in the laws established to explain experimental observations of electrical phenomena.

Table 1.1 lists some of the quantities that will be defined later, along with the symbols associated with these quantities and their unit of measurement. Usually the capital symbol is used to denote a constant value, whereas the small symbol will denote time dependence. Also, we will frequently prefix these quantities with terms such as mega or micro, for example, megahertz or microamps, which mean multiply the frequency by 10^6 and multiply the current by 10^{-6} amperes, respectively. These decimal prefixes, their meaning, abbreviation, and an example are given in Table 1.2.

1.2 STATIC ELECTRIC FIELDS

Matter is commonly modeled by atoms. Basically the atom consists of a nucleus containing neutrons and protons with electrons orbiting outside the nucleus. The electron and proton have an electrical charge e of 1.6019×10^{-19} coulomb; however, the charge of the proton is $+e$ and the charge of the electron is $-e$. The electron, which is the fundamental particle of all models of atomic structure has a rest mass of 9.1066×10^{-31} kg. It is known that like charges repel and unlike charges attract. If a charge is brought into the vicinity of another charged body it experiences a force. Thus, we say that a charged body is surrounded by an electric field $\mathscr{E}$. This electric field is calculated from Gauss's law. The integral form of Gauss's law is expressed below.[1]

$$\oint_S \epsilon \mathscr{E} \cos \theta \, dS = \int_V \rho \, dv \tag{1}$$

Gauss's law (1) states that the total electric flux $\epsilon \mathscr{E}$ leaving a surface S of volume V is equal to the charge enclosed by that surface. In Equation 1 ϵ is the permittivity of the medium in which the field is located ($\epsilon_o = 8.854 \times 10^{-12}$ farad/m in free space), ρ is the charge density, dS is the incremental surface area, $\cos \theta$ is the cosine of the angle between the vector $\mathscr{E}$ and the normal $\mathscr{E}_n$ to the surface as shown in Figure 1.1, and dv is the incremental volume. The differential form of Gauss's law in rectangular coordinates is[1]

$$\frac{\partial \mathscr{E}_x}{\partial x} + \frac{\partial \mathscr{E}_y}{\partial y} + \frac{\partial \mathscr{E}_z}{\partial z} = \frac{\rho}{\epsilon} \tag{2}$$

where $\mathscr{E}_x$ is the x component of the $\mathscr{E}$ vector, and so on.

For example, in Figure 1.2 a surface S surrounds a uniformly charged

[1] W. H. Hayt, Jr., *Engineering Electromagnetics*, 3rd edition, McGraw-Hill, 1974, pp. 63–77.

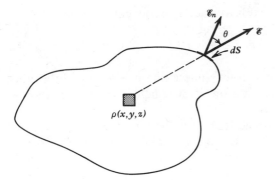

Figure 1.1 Electric field surrounding a charged body.

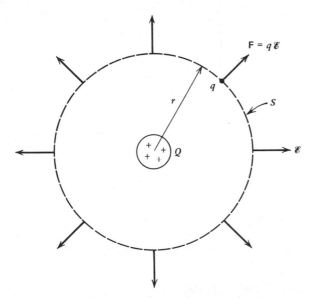

Figure 1.2 Electric field surrounding a uniformly charged sphere.

sphere with total charge Q. Due to the symmetry of the example it follows that $\mathscr{E}$ is normal to the surface S and uniform on the surface so that (1) yields

$$e\mathscr{E}4\pi r^2 = Q\mathbf{a}_r$$

or

$$\mathscr{E} = \frac{Q}{4\pi r^2 \epsilon}\mathbf{a}_r \tag{3}$$

An incremental test charge q on this surface experiences a force

$$\mathbf{F} = q\mathscr{E} = \frac{qQ}{4\pi r^2 \epsilon}\,\mathbf{a}_r \tag{4}$$

where $\mathbf{a}_r$ is the unit vector normal to the sphere and it is assumed that the test charge q has a negligible effect on the electric field $\mathscr{E}$ in (3). Equation 4 is known as Coulomb's law.

A second and very useful example is the calculation of the electric field between the pair of equal but oppositely charged metal plates in Figure 1.3.

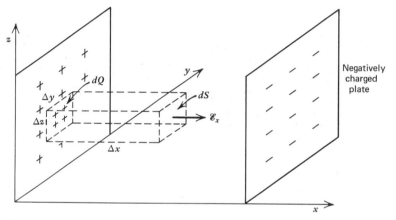

Figure 1.3 Calculation of the electric field between oppositely charged metal plates.

If we assume that the parallel plates are infinite sheets, then the electric field is confined between the plates. Furthermore, the y and z components of the electric field, $\mathscr{E}_y$ and $\mathscr{E}_z$, are zero. Then, (1) becomes

$$\int \epsilon \mathscr{E}_x \, dS = \int dQ \tag{5}$$

where in Figure 1.3 the total charge contained in the box $\Delta x \, \Delta y \, \Delta z$ is the surface charge dQ contained in the incremental area dS on the surface of the plate. Since the electric field is zero on all of the surfaces of the box except the right face, then between the plates,

$$\mathscr{E}_x = \frac{Q_A}{\epsilon} \tag{6}$$

Note that $\mathscr{E}_x$ is constant between the plates and proportional to the charge per unit area, Q_A.

In the case when the area A of the plates is finite and each plate has a charge $+Q$ and $-Q$, respectively, then

$$\mathcal{E}_x \approx \frac{Q}{\epsilon A} \tag{7}$$

This approximation is very good in the center midway between the two plates but becomes less accurate as one moves toward the edges of the plates.

If an electron lies between two oppositely charged metal plates as shown in Figure 1.4, it experiences a force

$$F_x = -e\mathcal{E}_x \tag{8}$$

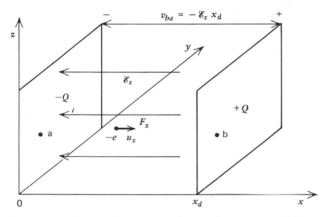

Figure 1.4   Motion of an electron in a uniform electric field.

In this discussion it is assumed that $\mathcal{E}_x$ is negative since it is directed in the negative x direction. From Newton's law,

$$m\frac{du_x}{dt} = -e\mathcal{E}_x \tag{9}$$

where u_x denotes the velocity of the electron in the x direction. Thus, at time t_1, the velocity

$$u_x(t_1) = u_x(t_o) - \frac{e\mathcal{E}_x}{m}\int_{t_0}^{t_1} dt \tag{10}$$

where $u_x(t_o)$ denotes the velocity of the electron at time t_o and the negative sign in the second term simply implies that the electron moves in the opposite direction of the electric field.

The *work done by the electric field* in moving an incremental test charge q through an electric field $\mathcal{E}$ between two points is the integral of the force over the distance, that is,

$$w_{ba} = \int_a^b q\mathscr{E} \cos \theta \, dl \tag{11}$$

as illustrated in Figure 1.5. Note that when the incremental distance dl along the path is perpendicular to the electric field $\mathscr{E}$ then no energy is required since $\cos 90° = 0$.

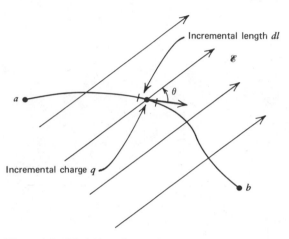

Figure 1.5 Work done in moving an incremental charge through an electric field.

For example, the work done by the field in moving an electron from point a to point b in Figure 1.4 is

$$w_{ba} = \int_0^{x_d} (-e)\mathscr{E}_x \, dx = -e\mathscr{E}_x x_d \tag{12}$$

In (11) the charge is negative but $\theta = 180°$ since the electron is moving in the opposite direction of the field. Thus, the energy is positive [$\mathscr{E}_x$ is negative in (12)], that is, the field is doing the work. Note also that

$$w_{ab} = \int_{x_d}^0 (-e)\mathscr{E}_x \, dx = +e\mathscr{E}_x x_d \tag{13}$$

that is, the total energy required to move a charge in a closed path is zero. Also, (11) depends only on the endpoints a and b and is independent of the path taken between these points.

Finally, the *voltage* v or potential difference between two points is defined as the work done per unit charge by an external source in going from one point to the other, that is, from (11)

$$v_{ba} = v_b - v_a = -\int_a^b \mathscr{E} \cos \theta \, dl \qquad (14)$$

This is the integral relation between voltage and the electric field. The differential form is

$$\frac{\partial v}{\partial x} \mathbf{a}_x + \frac{\partial v}{\partial y} \mathbf{a}_y + \frac{\partial v}{\partial z} \mathbf{a}_z = -\mathscr{E} \qquad (15)$$

where $\mathbf{a}_x$ is a unit vector in the x direction so that

$$\frac{\partial v}{\partial x} = -\mathscr{E}_x \qquad (16)$$

and so on.

Upon the application of either (14) or (16) to the parallel plate example in Figure 1.4 we find that the voltage between the plates is

$$v_{ba} = -\int_0^{x_d} \mathscr{E}_x \, dx = -\mathscr{E}_x x_d \qquad (17)$$

Thus v_{ba} is positive in Figure 1.4 since $\mathscr{E}_x$ is negative. Note that from our definition the positively charged plate is said to be at a higher or positive potential with respect to the negatively charged plate. Also, since the electric field in the y and z directions is zero, then the voltage between any two points on the same plane perpendicular to the x axis is zero. Thus, all points on a metal plate are at the same potential. Next we calculate the velocity of the electron at any point x between the plates.

In Figure 1.4 let us suppose that an electron lies between the two plates, hence it experiences a constant force as given by (8). Assuming that the electron is initially at rest at point a, then, from (9),

$$u_x(t) = -\frac{e\mathscr{E}_x}{m} t \qquad (18)$$

Since $u_x(t) = dx/dt$, then

$$x = -\frac{e\mathscr{E}_x}{2m} t^2 \qquad (19)$$

Upon substitution of (19) into (18) we obtain

$$u_x = \sqrt{\frac{2e}{m}} \sqrt{-\mathscr{E}_x x}$$

But the voltage between a point x and the negatively charged plate is

$$v_x = -\mathscr{E}_x x$$

so that

$$u_x = \sqrt{\frac{2ev_x}{m}} \qquad (20)$$

Thus, the velocity u_x of an electron at a distance x from the negative plate is proportional to the square root of the voltage v_x where v_x is the potential difference between the parallel plane a distance x from the negative plate and the negative plate. Equation 20 also assumes that the velocity of the electron is zero initially at $x = 0$ in Figure 1.4.

1.3. VOLTAGE GENERATION

There are two ways in which one can create a potential difference between two points, either by a separation of charges or by varying the intensity of a magnetic field that couples a wire. In about 600 B.C. the early Greeks first generated electric fields knowingly by rubbing a piece of silk or fur on an amber (elektron, in Greek) rod. This friction technique was used by Coulomb in France and Cavendish in England to arrive at Coulomb's law expressed by Equation 4. The disadvantage of these electrostatic methods is that it is difficult to maintain a continuous current flow. A decade before 1800 a big breakthrough in energy conversion occurred with the discovery by the Italian physicist Alessandro Volta that when two different electrodes (e.g., zinc and copper) are immersed in an electrolyte, a potential difference results. This is the principle of the battery. A chemical reaction maintains the charge differential between the electrodes essentially constant for reasonable values of current flow until certain chemicals are depleted—we say the battery is dead. In solids a net potential difference exists in a closed circuit when the junctions of the dissimilar materials are maintained at different temperature levels or one of the junctions is radiated with light. This is the basis for the thermocouple and the photoelectric device, respectively.

The second method of generating electric energy is the rotating generator. This source of voltage is governed by Faraday's law, which states that "the voltage induced in a coil is proportional to the time-rate-of-change of the flux linkages encircling the coil." Figure 1.6 illustrates a coil being rotated with angular velocity ω radians/s in a magnetic field. The strength of the magnetic field (flux density $\mathbf{B}$) is measured in teslas (webers/m²), and the flux ϕ in webers is calculated by the equation

$$\phi = \int_{\substack{\text{area of} \\ \text{the coil}}} B \cos \theta \, dS \qquad (21)$$

where θ is the angle between $\mathbf{B}$ and the incremental area dS. If, instead of

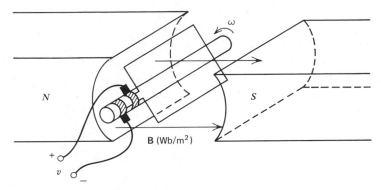

Figure 1.6   ac electric generator.

one turn, we have two turns of wire in the field, the voltage v doubles. Thus, Faraday's law states that the induced voltage is equal to the rate of change of flux linkages Ψ, where $\Psi = N\phi$ and N is the number of turns. This law is expressed by the equation

$$v = d\Psi/dt \tag{22}$$

The polarity of v alternates, since the rate of change in the magnetic flux linkages surrounding the coil increases and decreases, and so on, as the coil rotates. With careful design the voltage v can be made sinusoidal. For example, in Figure 1.6 at any time instant

$$\theta = \theta_o + \omega t \tag{23}$$

where θ_o is the reference angle at $t = 0$ and ω is the angular velocity of the rotating coil in radian/s. Assuming that B is constant, from (21)–(23) we obtain

$$v(t) = N \frac{d}{dt}\left[B \cos\left(\theta_o + \omega t\right) \int dS \right]$$

or

$$v(t) = NB\omega A \sin(\theta_1 + \omega t) = V \sin\left(\theta_1 + \omega t\right) \tag{24}$$

where N is the number of turns on the coil, A is the area of the coil, V represents the amplitude of the sinusoid, and $\theta_1 = \theta_o + 180°$ is referred to as the phase angle of the sinusoid. This waveform is illustrated in Figure 1.7. The factor ω is the angular frequency of the sinusoidal signal in radians/s. The period T of the sinusoid is $2\pi/\omega$ seconds, and the quantity $f = 1/T = \omega/2\pi$ is called the frequency of the sinusoid and is measured in units of Hertz (Hz) in honor of the German physicist Heinrich Rudolf Hertz, whose experiments in electromagnetic radiation led to the development of radio-telegraphy and the beginning of electronics as a significant branch of

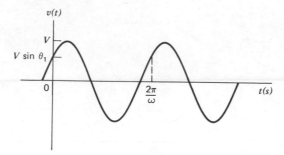

Figure 1.7 Alternating voltage.

science and technology. In North American power generating plants the frequency of the voltage waveform is 60 Hz, while in most of Europe, Africa, Australia, and Asia it is 50 Hz; in South America both 50 Hz and 60 Hz systems are found.

Historically the terms alternating current (ac) and direct current (dc) refer to circuits in which the source voltage is sinusoidal (ac) or constant (dc), for example, a battery. A variety of electronic generators are also available that produce sinusoidal, square, triangular, and pulse waveforms with adjustable amplitude and period. Thus, whenever we write the voltage v, it is understood that the voltage is possibly a function of time, that is, $v(t)$. In fact, *when a quantity is not time dependent it is usually capitalized.* The MKS unit of measurement of potential difference or voltage is the volt, in honor of Volta.

1.4 CURRENT

In order to obtain a flow of electricity we need a potential difference (voltage) and mobile charges. Analogously, in order to obtain fluid flow we need a pressure differential and a fluid. In electrical circuits we are not interested in how each charge moves but in the rate of flow of charge with respect to time. We define this rate of flow as

$$i = \frac{dq}{dt} \tag{25}$$

where i is called the current and is measured in the MKS units coulombs per second or amperes in honor of the French physicist André M. Ampere. *The convention is to call current flow positive in the direction that positive charges move.* Current flow in various media is discussed below.

1.4.1 CONDUCTION IN METAL

Let us consider the segment of wire illustrated in Figure 1.8. We shall assume that a uniform electric field exists in the wire. One of the properties

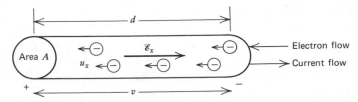

Figure 1.8   Wire segment.

of metals is that one or more of the electrons of an atom are relatively free to move under the influence of an electric field. In a vacuum a free electron would be accelerated to the left in Figure 1.8. However, in metal the atoms are so dense that collisions with other atoms result. Under these conditions the electrons actually drift to the left with an average velocity proportional to the electric field. The total current flow to the right in Figure 1.8 is

$$i = e\frac{nAd}{T} \tag{26}$$

where n is the free electron density (cm^{-3}), nAd is the total number of free electrons in the wire segment, and T is the time it takes for these electrons to move through the wire segment. Assuming that the electrons move with an average drift velocity $u_x = -d/T$, then

$$i = -enAu_x \tag{27}$$

The drift velocity of the electrons is proportional to the electric field.

$$u_x = -\mu_e\mathscr{E}_x \tag{28}$$

where the constant of proportionality μ_e (cm^2/V-s) is called the mobility of the electrons. The mobility is reasonably constant if the electric field is not too large and the temperature of the material is constant.

From the above equations the current flow per unit area is

$$J = ne\mu_e\mathscr{E}_x = \sigma\mathscr{E}_x \tag{29}$$

where $J = i/A$ and σ is called the *conductivity* of the material. Also,

$$\rho = 1/\sigma$$

is called the *resistivity* of the material and is measured in ohm-meters in honor of the German physicist Ohm who first proposed Equation 26. The total current in the wire is

$$i = AJ = \frac{A\sigma}{d}\mathscr{E}_xd = \frac{A\sigma}{d}v \tag{30}$$

where A is the cross-sectional area of the wire in square meters, d is the length of the wire in meters, and v is the voltage drop across the wire. The resistance of the wire in ohms is defined as

$$R = \frac{\rho d}{A} \tag{31}$$

Note that resistance is a function of temperature since the mobility of the free charges is temperature dependent. The temperature of the wire is determined not only by the ambient temperature but also by the energy dissipated in the wire, which is determined by the current flow over a period of time. If the current flow is too great a sufficient amount of heat will be generated to destroy the wire.

1.4.2 CONDUCTION IN SEMICONDUCTORS

In a semiconductor crystal the atoms are held together by highly localized electronic bonds between their neighbors. For example, in silicon each atom has four valence electrons and four nearest neighbors. Each valence electron forms a *covalent bond* with an electron from one of its nearest neighbors as illustrated in Figure 1.9.

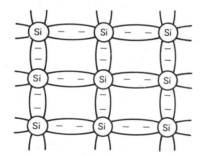

Figure 1.9   Silicon crystal covalent bonds.

A silicon crystal has 5×10^{22} atoms/cm^3. At absolute zero these bonds are complete. In order for an electron to leave one of these covalent bonds it must gain an energy of approximately 1.1 electron volts (1 eV = 1.6×10^{-19} joule). At room temperature approximately 10^{10} electrons/cm^3 gain enough energy to leave this covalent bond (one free electron for every 10^{12} atoms). When an electron is free of the covalent bond it leaves the silicon ion positively charged. This vacancy in the covalent bond, Figure 1.10, is called a hole. Experiments have shown that the hole behaves as a positively charged particle with a charge of 1.6×10^{-19} coulombs and with a certain effective mass that is not necessarily equal to the mass of the electron. *When the crystal is placed in an electric field both electrons and holes drift (electrons to the left and holes to the right in Figure 1.10), and the current is nearly twice the current due only to the free electrons.* Since

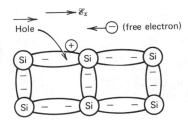

Figure 1.10   Vacancy in a silicon crystal covalent bond.

there are effectively two different charged particles (electrons and holes) that move in response to an electric field, the conductivity of the material is

$$\sigma = e(n\mu_n + p\mu_p) \tag{32}$$

where n is the number of electrons per unit volume, μ_n is the mobility of the electrons, p is the number of holes per unit volume and μ_p is the mobility of the holes.[2] If the crystal is a pure (intrinsic) semiconductor, then there is one hole for each free electron ($p = n$).

1.4.3 CONDUCTION IN DOPED SEMICONDUCTORS

The conductivity of semiconductors can be increased by adding impurities to the crystal. For example, the addition of a donor impurity atom (an atom with five valence electrons such as arsenic, antimony, and phosphorus) to the crystalline material, results in one electron outside the covalent bond for each donor atom (see Figure 1.11). At room temperature this extra electron has a high probability of being a free electron. If

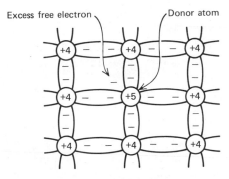

Figure 1.11 n-type semiconductor.

<hr />

[2] The Hall effect experiment conclusively proves the existence of both positive and negative charge carriers in semiconductors. See P. E. Gray and C. L. Searle, *Electronic Principles: Physics, Models and Circuits*, Wiley, Chapter 2, 1969.

approximately one donor atom for every 10^6 Si atom is added to the crystal, then

$$N_d \approx 10^{17} \text{ atoms/cm}^3 \qquad (33)$$

where N_d is the number of donor atoms per unit volume. Thus, the conductivity has been increased by a factor of 10^7. In this case the electrons are the majority carriers, and the semiconductor is called an n type. In computing the conductivity of the material we can neglect the number of holes, since in equilibrium one can say that the recombination of electron-hole pairs must equal the thermal generation of electron pairs. This hypothesis leads to the equation[3]

$$np = n_i^2 \qquad (34)$$

where n_i is the number of free electrons per unit volume in the intrinsic (pure) semiconductor. Thus if $n = N_d = 10^{17}$, then

$$p \approx \frac{(10^{10})^2}{10^{17}} = 10^3 \text{ holes/cm}^3 \qquad (35)$$

Clearly, the contribution of the holes can be neglected in computing the conductivity of the n-type semiconductor.

The equilibrium equation (34) can be upset $(np > n_i^2)$ by supplying additional energy to the crystal. For example, light illuminating the crystal can cause excess electron-hole pairs to be generated. This is the principle of the photoresistor or photo cell. The more intense the light the lower the resistivity of the material. When the light is turned off the excess carriers recombine and the crystal returns to equilibrium. The average time that it takes for the excess charges to recombine is called the *lifetime* of the excess carriers.

One can also improve the conductivity of an intrinsic semiconductor by adding acceptor atoms (atoms with only three valence electrons) to the crystal, such as boron gallium, indium and aluminum, Figure 1.12. In this crystal holes are the majority carriers, and the crystal is referred to as a p-type.

Let us conclude this section by calculating the conductivity of the intrinsic semiconductor and an n-type semiconductor. In intrinsic silicon at room temperature $n = p = 10^{10} \text{ cm}^{-3}$. Given this density of electrons and holes, the approximate mobility of electrons in silicon is $1600 \text{ cm}^2/\text{V-s}$ and the mobility of holes in silicon is $400 \text{ cm}^2/\text{V-s}$. These values are obtained from the graph in Figure P1.14 of the problem set at the end of this chapter. Note that the mobility decreases with increasing impurity concentration.

$$\sigma = 1.6 \times 10^{-19} (10^{10} \times 1600 + 10^{10} \times 400) (\Omega\text{-cm})^{-1}$$

[3] A. S. Grove, *Physics and Technology of Semiconductor Devices*, Wiley, 1967, pp. 100–105.

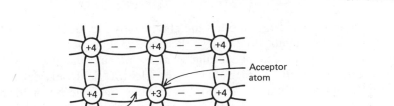

Figure 1.12   *p*-type semiconductor.

or

$$\sigma = 3.2 \times 10^{-6}(\Omega\text{-cm})^{-1}$$

In *n*-type silicon if the number of donor atoms is $10^{17}\,\text{cm}^{-3}$, then from (35) the number of holes is $10^3\,\text{cm}^{-3}$ and from Figure P1.14 in the problem set we find that $\mu_n \approx 600\,\text{cm}^2/\text{V-s}$ for a concentration of 10^{17} donor atoms/cm³. Since $p = 10^3/\text{cm}^3$ we use $\mu_p \approx 400\,\text{cm}^2/\text{V-s}$ so that

$$\sigma = 1.6 \times 10^{-19}(10^{17} \times 600 + 10^3 \times 400)$$

or

$$\sigma = 9.6\,(\Omega\text{-cm})^{-1}$$

Note that the conductivity of the doped semiconductor is almost a factor of 10^7 greater than the conductivity of the pure semiconductor, which is approximately the increase in free carriers due to doping. In the above calculation the current flow due to holes is negligible and was omitted. In the *n*-type semiconductor we say that electrons are the majority carriers, whereas in the *p*-type semiconductor holes are the majority carriers.

1.4.4 CONDUCTION IN A VACUUM

Let us conclude our discussion of current flow with a discussion of current flow in a vacuum tube. Figure 1.13 illustrates two parallel plates with a potential difference v_b between them, which sets up an electric field $\mathcal{E}_x = v_b/d$ between the plates where d is the distance between the plates.

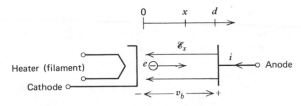

Figure 1.13   Conduction between parallel plates.

The positive plate is called the anode, and the negative plate is called the cathode. Now current cannot flow unless there are mobile electric charges in the field. Free charges can be obtained (a) by placing the plates in a sealed tube with a gas that ionizes when the electric field between the plates is of sufficient strength, or (b) by making the electric field of sufficient strength to liberate electrons from the cathode (this is called field emission and requires a very large voltage), or (c) by thermionic emission, which is the most frequently used approach and which is discussed below.

The discovery of thermionic emission by Edison made possible the development of the vacuum tube in the early part of this century. Basically two metal cylindrical shells with different diameters are mounted in a glass vacuum tube as shown in Figure 1.14. The smaller cylinder is the cathode and a heater (filament) is placed inside it which heats the cathode to a temperature in the range 1200 K to 2500 K. At these temperatures the electrons in the cathode have sufficient kinetic energy to be liberated from cathode, and under the action of the electric field the electrons flow to the anode.

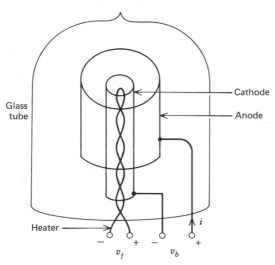

Figure 1.14   Vacuum tube construction.

The current flow in a vacuum tube can be computed as follows. An electron in the electric field experiences a constant force to the right in Figure 1.13 opposite to the direction of the electric field $\mathscr{E}_x$.

$$F_x = -e\mathscr{E}_x \tag{36}$$

Since F_x is constant, the electron experiences a constant acceleration.

Recall from (20) that the velocity of the electron with zero initial velocity at a distance x from the cathode is related to the potential difference between the point x and the cathode by the equation

$$u(x) = \sqrt{\frac{2e}{m}} \, v(x) \tag{37}$$

where u denotes velocity and v voltage. From (16) it follows that

$$\mathscr{E}_x = -\frac{dv(x)}{dx} \tag{38}$$

and from (2)

$$\frac{d\mathscr{E}_x}{dx} = \frac{\rho(x)}{\epsilon_o} \tag{39}$$

if we neglect the y and z components of the field. Therefore,

$$\frac{d^2v(x)}{dx^2} = -\frac{\rho(x)}{\epsilon_o} \tag{40}$$

The current density is

$$J = \rho(x)u(x) \, A/m^2 \tag{41}$$

Upon the elimination of $\rho(x)$ and $u(x)$ from (37), (40), and (41) we obtain

$$\frac{d^2v(x)}{dx^2} = -\frac{1}{\epsilon_o}\sqrt{\frac{m}{2e}} \frac{J}{\sqrt{v(x)}} \tag{42}$$

The solution of this equation yields[4]

$$i = 2.34 \times 10^{-6}\frac{A}{d^2} v_b^{3/2} \tag{43}$$

Equation 43 is sketched in Figure 1.15. Note that the current flow is unidirectional, that is, almost no current flows for negative values of v_b

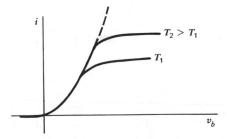

Figure 1.15   Vacuum tube i-v characteristic.

[4] W. H. Hayt, Jr. and G. W. Hughes, *Introduction to Electrical Engineering*, Appendix B, McGraw-Hill, 1968.

since few electrons are liberated from the anode. Also, for large values of v_b the electrons move to the anode faster than they are emitted from the cathode so that as v_b increases the current is eventually limited by the emission rate of electrons from the cathode. If the temperature of the cathode is increased, electrons are liberated at a faster rate and the current limit increases as shown in Figure 1.15. Note that when the current reaches this limit it departs from the $\frac{3}{2}$ power law.

A significant advance in radio communications was made with the invention of the triode amplifier. In the early 1900s L. deForest found that if a wire mesh or screen is placed between the cathode and the anode, then a small negative voltage between this screen and the cathode could control the current flow between the cathode and the anode. If the anode current is passed through a resistor, then the voltage across the resistor will have the same waveform as the voltage between the screen and the cathode but can be larger in which case we have voltage gain. The vacuum tube served as the principal electronic circuit component until a better understanding of current flow in solids was obtained and solid-state technology was developed. In 1945 Bell Laboratories launched an intensified research program to better understand the basic physics of semiconductors. This program led to the invention of the transistor by W. Shockley, W. H. Brattain, and J. Bardeen in 1948 and to the era of solid-state electronics. The semiconductor diode and transistor are introduced in Chapters Six and Seven.

1.4.5 CATHODE-RAY TUBE

There is one vacuum tube that has not been completely displaced by solid-state technology. This device is the cathode-ray tube (CRT) used in the television, oscillosope, and many other display devices. This tube, illustrated in Figure 1.16, consists of a heater to liberate electrons from the cathode, an anode with a small hole in it, a pair of vertical deflection plates, a pair of horizontal deflection plates, and a screen coated with a phosphor that emits light when struck by an electron. The voltage v_z accelerates the electrons to the anode where primarily those with velocity components in the z direction pass through the hole in the anode. Usually a focusing mechanism is added between the anode and the vertical deflection plates to correct the trajectories of electrons that do not have velocity components only in the z direction. Assuming that the field is uniform between the cathode and the anode, the electrons pass through the anode with a velocity given by

$$u_z = \sqrt{\frac{2ev_z}{m}} \qquad (44)$$

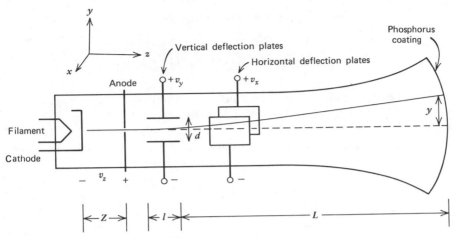

Fkgure 1.16 Cathode-ray tube (CRT).

where u_z is the z component of the velocity.

The time required for the electron to move through the vertical deflection plates is

$$t_l = l/u_z \tag{45}$$

During that time the electron experiences a force $F_y = -e\mathscr{E}_y$. The velocity of the electron in the y direction when it emerges from the vertical deflection plates is determined below.

$$m\frac{du_y}{dt} = e\frac{v_y}{d} \tag{46}$$

so that

$$u_y = \frac{e}{m}\frac{v_y}{d}t_l = -\frac{v_y l}{du_z}\cdot\frac{e}{m} \tag{47}$$

The deflection in the y direction as the electron leaves the vertical deflection plates is determined from the equation

$$\frac{dy}{dt} = u_y = \frac{e}{m}\frac{v_y}{d}t \tag{48}$$

so that

$$y = \frac{e}{m}\frac{v_y}{d}\frac{t_l^2}{2} = \frac{e}{2m}\frac{v_y}{d}\left(\frac{l}{u_z}\right)^2 \tag{49}$$

The total deflection Y at the screen is

$$Y = \frac{e}{2m}\frac{v_y}{d}\left(\frac{l}{u_z}\right)^2 + u_y\left(\frac{L}{u_z}\right) \tag{50}$$

$$= \frac{e}{2m}\frac{v_y}{d}\left(\frac{l}{u_z}\right)^2 + \frac{e}{m}\frac{v_y lL}{du_z^2} \tag{51}$$

Upon the elimination of u_z from (51) we obtain

$$Y = \frac{1}{2} \frac{v_y}{v_z} \frac{l}{d} \left(L + \frac{1}{2} l \right) \tag{52}$$

which gives the total y-axis deflection as a function of the voltages v_y, v_z, and the dimensions of the tube. A similar formula could be derived for the horizontal deflection.

Electrons can also be deflected by means of a magnetic field. In Figure 1.17 a charge q moving with a velocity u experiences a force

$$F = quB \sin \theta \tag{53}$$

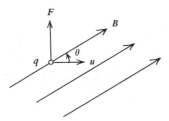

Figure 1.17   Electron in a magnetic field.

where F is perpendicular to the plane formed by the u and B vectors and in the direction that a right-hand screw would move in turning the velocity vector toward the magnetic flux density vector B. The generation of the magnetic field requires an electromagnet, which consists of a coil usually wrapped on an iron core in order to increase the magnetic flux density. Magnetic deflection systems are used in television because the CRT and associated deflection circuitry are less expensive than electrostatic deflection systems. However, in oscilloscopes and other types of displays, where writing speed and possibly weight are important, electrostatic deflection systems are used.

1.5 INTEGRATED CIRCUITS

Before 1960 electrical circuits consisted of a collection of discrete components, such as resistors, capacitors, inductors, vacuum tubes, diodes, transistors, motors, and the like, whose terminals were soldered together to form an electrical system. In 1959 it was discovered that, by combining solid-state technology and photolithographic technology used in the manufacture of printed circuit boards, many transistors, diodes, and resistors could be fabricated simultaneously on the same semiconductor wafer and interconnected by means of thin film deposition of metal on the wafer. These circuits are called monolithic integrated circuits. One basic fabrication process is the planar process (described below) that enables the simultaneous fabrication of resistors, diodes, and transistors on a silicon wafer. At this time we will describe only the fabrication of diffused resistors.

1.5.1 DIFFUSED RESISTORS

The process begins with a p-type silicon semiconductor wafer with 10^{15} acceptor atoms/cm^3. The wafer is 5 to 10 cm in diameter and 150 μm thick for mechanical rigidity. The top surface of the wafer is chemically cleaned and polished and then placed in a furnace and heated to 1275 to 1475 K. In this furnace an epitaxial (Greek word meaning "arranged upon") layer approximately 10 μm thick is grown on the substrate as shown in Figure 1.18a. This is accomplished in the furnace by passing vapors containing silicon over the surface of the wafer. Also n-type impurities are added to the vapor. The donor atom concentration is approximately 10^{16} atoms/cm^3.

After the epitaxial growth a layer of silicon dioxide (SiO$_2$) is formed on the surface of the wafer. The SiO$_2$ layer acts as an insulator and protects the surface from contamination. Next, photolithographic techniques are used to etch windows in the SiO$_2$ so that impurities can be selectively diffused into the epitaxial layer through the windows in the oxide. This is accomplished by spreading a photoresist on the surface of the wafer and aligning a *mask* on top of the wafer. The wafer is then exposed to ultraviolet light and portions of the photosensitive resist not covered by the opaque portions of the mask harden. The unexposed portions of the photoresist are washed way. Next a hydrofluoric acid solution is applied to the surface of the wafer, etching away the silicon dioxide not protected by the photoresist. This step forms a window in the oxide through which acceptor or donor atoms can be diffused into the epitaxial layer as shown in Figure 1.18b. In this step the wafer is placed into a furnace and acceptor atoms diffuse through any open windows. In this step the p-type diffusion forms rings that isolate n-type regions from the rest of the wafer. On these n-type islands the components are made. The n-type island can be electrically isolated from the p-type substrate because the pn junction forms a diode (see Chapter Six). If the n-region is floating or at a positive potential with respect to the p-region, then essentially no current flows between the n and p regions.

The next step is to grow more silicon dioxide on the surface and then apply a second mask to the wafer to expose a new diffusion window over the n-type islands. The wafer is placed in a diffusion furnace and a p-type diffusion is made, as shown in Figure 1.18c. This diffusion is 1 to 3 μm thick.

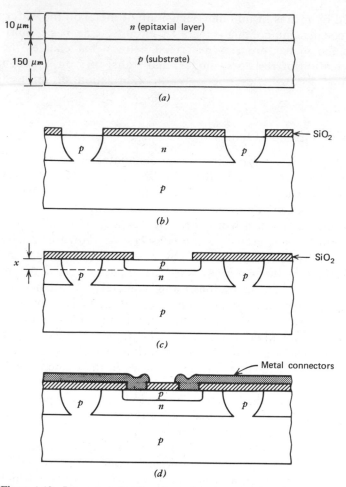

Figure 1.18 Integrated circuit resistor. (a) p-type semiconductor crystal with n-type epitaxial layer. (b) p-type isolation diffusion. (c) p-type resistor diffusion. (d) Metal connections added.

Finally another oxide layer is grown and a third mask is used to etch windows for metal contacts to the components. In this case our final p-type diffusion serves as our resistor and a metal contact is made to each end of the resistor as shown in Figure 1.18d. A top view of this resistor is shown in Figure 1.19. Transistors and diodes can be fabricated simultaneously with two additional masks and diffusion steps. These devices are discussed in Chapters Six and Seven.

After metal has been evaporated onto the surface of the wafer a final mask is used to etch the connections between the components. Typically a component occupies an area of $10^4 \, \mu m^2$ or $10^{-4} \, cm^2$ and the area of the

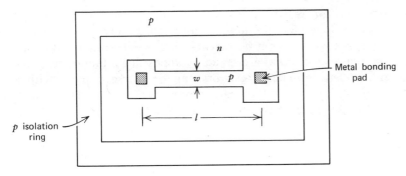

Figure 1.19   Top view of *p*-type diffused resistor.

wafer is approximately 20 cm² so that thousands of components can be made simultaneously and interconnected. Typically many identical circuits are made on the same wafer. The initial cost for tooling up to make an integrated circuit is high, but if millions of the circuits can be sold, the cost for a complex electrical circuit can be reduced to less than a dollar in many cases.

The resistance of the diffused resistor in Figure 1.19 is

$$R = \left(\frac{\rho}{x}\right)\left(\frac{l}{w}\right) = R_s \frac{l}{w} \tag{54}$$

The quantity R_s is called *sheet resistance* and is measured in ohms/square ($\Omega/\square$). In making an integrated circuit the number of steps should be kept at a minimum to maximize the yield of "good" circuits. Thus, all resistors are diffused at the same time so that x, and hence the sheet resistance, is a constant for each resistor. Thus, different values resistance are obtained by varying the ratio l/w. This ratio is referred to as the number of squares necessary to realize a given resistance.

In the *p*-type silicon let the number of acceptor atoms be $N_A = 10^{18}$ then from Figure P1.14 $\mu_p \approx 100$ cm²/V-s. Let $x = 1 \ \mu\text{m} = 10^{-4}$ cm then

$$R_s \approx \frac{1}{xqN_A\mu_p} = \frac{1}{(10^{-4} \text{ cm})(1.6 \times 10^{-19} \text{ C})(10^{18} \text{ cm}^{-3})(100 \text{ cm/V-s})} = 625 \ \Omega/\square$$

If $l = 100 \ \mu\text{m}$ and $w = 20 \ \mu\text{m}$, then $l/w = 5$ squares to that

$$R = (625 \ \Omega/\square)(5\square) = 3125 \ \Omega$$

Typically, the value of the diffused resistor has a tolerance of ±20% and cannot be trimmed to yield more accuracy. Thus, in the design of integrated circuits, it is important that critical voltage or current responses be dependent on ratios of resistors instead of on the absolute value of resistors because in integrated circuits resistors can usually be matched to within ±2%. In cases where accurate resistors are needed one must resort to thick or thin film resistors that are made by depositing a film of resistive material onto a substrate. These resistors can be trimmed (e.g., with a laser) to yield accurate values. To learn more about integrated circuits, consult the references cited at the end of this chapter.

REFERENCES

HISTORICAL PAPERS

1. Süsskind, C. "The Early History of Electronics," *IEEE Spectrum*, Vols. 5–7, Parts I–VI, August 1968, December 1968, April 1969, August 1969, April 1970, September 1970.
2. Geddes, L. A. and H. E. Hoff. "The Discovery of Bioelectricity and Current Electricity," *IEEE Spectrum*, Vol. 8, December 1971, pp. 38–46.
3. Dibner, B. "Michael Faraday—A Centennial," *IEEE Spectrum*, Vol. 4, August 1967, pp. 115–119.
4. Shiers, G. "On the Origins of Electron Devices," *IEEE Spectrum*, Vol. 9, November 1972, pp. 70–76.
5. *Proceedings of the IEEE*, Special issue on "The History of Electrical Engineering," Vol. 64, September 1976.
6. *IEEE Transactions on Electron Devices*. Special issue on "Historical Notes on Important Tubes and Semiconductor Devices," Vol. ED-23, July 1976.

ELECTRIC AND MAGNETIC FIELDS

7. Hayt, W. H. Jr., *Engineering Electromagnetics*, 3rd edition, McGraw-Hill, 1974.

CONDUCTION IN SOLIDS AND A VACUUM

8. Sproull, R. L. *Modern Physics*, 2nd edition, Wiley, 1963.

ELECTRICAL MEASUREMENTS

9. Wolf, S. *Guide to Electronic Measurements and Laboratory Practices*, Prentice-Hall, 1973.
10. Spitzer, F. L. and B. Howarth. *Principles of Modern Instrumentation*, Holt, Rinehart and Winston, 1972.

SOLID-STATE DEVICES

11. Grove, A.S. *Physics and Technology of Semiconductor Devices*, Wiley, 1967.
12. Streetman, B. G. *Solid State Electronic Devices*, Prentice-Hall, 1974.

INTEGRATED CIRCUITS

13. Camenzind, H. R. *Electronic Integrated System Design*, Van Nostrand Reinhold, 1972.

14. Grebene, A. B. *Analog Integrated Circuit Design*, Van Nostrand Reinhold, 1972.
15. Hamilton, D. J. and W. G. Howard. *Basic Integrated Circuit Engineering*, McGraw-Hill, 1975.

HANDBOOKS

16. Westman, H. P. (Ed.). *Reference Data for Radio Engineers*, 5th edition, Howard W. Sams & Co., Inc., 1968.
17. Fink, D. G. (Ed.). *Standard Handbook for Electrical Engineers*, 10th edition, McGraw-Hill, 1968.

PROBLEMS

1. Given a pair of parallel oppositely charged metal plates with a charge of $10^{-8}\,C/cm^2$, calculate the electric field between the plates assuming that the dielectric material between the plates is air. Also calculate the voltage between the two plates if they are separated by (a) 1 mm and (b) 0.1 mm.
2. Repeat Problem 1 only assume that the dielectric between the two plates is silicon dioxide, which has a relative dielectric constant $\epsilon_r = 4$ where $\epsilon = \epsilon_r \epsilon_o$.
3. In Figure 1.4 assume that the voltage between the two plates is 100 V and that the distance $x_d = 1$ cm.
   (a) Calculate the electric field $\mathscr{E}_x$.
   (b) Assuming that the electron has zero velocity at $x = 0$, find its velocity when it strikes the positively charged plate.
   (c) Calculate the time required for the electron to travel between the two plates.
4. Repeat Problem 3 with $x_d = 10$ cm.
5. Repeat Problem 3 with a voltage of 1000 V.
6. In Problem 3 show that when the electron reaches the positively charged plate its kinetic energy has been increased by ev where $ev = \frac{1}{2} m u_{x_d}^2$.
7. Sketch the following sinusoidal waveforms.

   (a) $\sin(2\pi \, 60t + 30°)$ (b) $5 \sin(2\pi \, 10^6 t - 45°)$
   (c) $2 \sin(2\pi \, 10^9 t + 90°)$ (c) $0.1 \cos(2\pi \, 10^4 t - 90°)$

   Scale the axes appropriately.
8. In the rotating generator given that $N = 50$, $B = 1$ Wb/m^2, the coil is 10 cm $\times$ 10 cm, and the coil is rotating at 3600 rpm (revolutions/minute). Calculate ω and the peak voltage V generated.
9. Given a solid cylinder that has 10^{15} positively charged particles per cm^3 with a mobility of 100 cm^2/V-s and 10^{14} negatively charged particles per cm^3 with a mobility of 300 cm^2/V-s, find the current flow in the cylinder if the electric field is 10 V/cm and uniform across the cylinder, the charge on each particle is 1.6×10^{-19} C, and the cross-sectional area of the cylinder is 0.1 cm^2.
10. Calculate the resistance per meter for each gauge of copper wire listed below. The resistivity of copper is 1.724×10^{-6} Ω-cm.

American Wire Gauge (Brown & Sharpe)	Diameter (cm)
10	0.2588
12	0.2053
14	0.1628
18	0.1024
24	0.05105

11. Repeat Problem 10, only now use aluminum wire that has a resistivity 2.824×10^{-6} Ω-cm.
12. Repeat Problem 10, only now use steel wire that has a resistivity 10^{-5} Ω-cm.
13. In metal the number of free electrons/cm^3

$$n = \frac{N_o dv}{A_t}$$

where $N_o = 6.025 \times 10^{23}$ molecules/gram-mole (Avogadro's number)
A_t = atomic weight (grams/mole)
d = density (grams/cm^3)
v = number of valence electrons/atom (free electrons)

Given that aluminum has a resistivity 2.824×10^{-6} Ω-cm, an atomic weight of 26.98 grams/mole, a density of 2.70 grams/cm^3 and three valence electrons/atom, calculate the mobility of the free electrons in aluminum.

14. Shown in Figure P1.14 is a plot of majority carrier mobility as a function of impurity concentration at 300 K. Calculate the resistivity of both n-type and p-type semiconductors for impurity concentrations of 10^{15}, 10^{16}, 10^{18}, and 10^{20} atoms/cm^3. For each impurity concentration find an approximate value of the mobility from Figure P1.14.

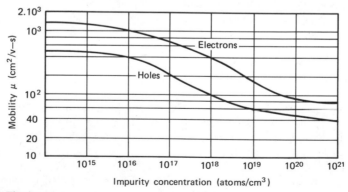

Fig. P1.14

15. Suppose that $A = 10\,\text{cm}^2$ and $d = 0.3\,\text{cm}$ in a vacuum tube. Sketch the i-v characteristic for $0 \leqslant v_b \leqslant 25$ V in 5-V increments.

16. In Problem 15 find the voltage across the vacuum tube when (a) $i = 1\,\text{mA}$ and (b) $i = 10\,\text{mA}$.

17. If the maximum deflection in a cathode-ray tube is to be $Y = \pm 5\,\text{cm}$ and $v_z = 1\,\text{kV}$, $v_y = \pm 100$ V, $l = 2\,\text{cm}$, and $d = 0.5\,\text{cm}$, find the length L required.

18. In an integrated circuit the sheet resistance is $1\,\text{k}\Omega/\square$. How many squares are needed for a 12-kΩ resistor? If the minimum width of the resistor is $10^{-2}\,\text{cm}$, what is the minimum area required by this resistor?

19. Repeat Problem 18 for a sheet resistance of $300\,\Omega/\square$.

20. A diffused resistor has a depth of $2\,\mu\text{m}$. Using the resistivities in Problem 14 calculate the sheet resistance for an n-type semiconductor with an impurity concentration of $10^{16}\,\text{atoms/cm}^3$. How many squares are required to make a 5-kΩ resistor? How much area is required if the minimum width of the resistor is $10\,\mu\text{m}$?

21. Repeat Problem 20 for an impurity concentration of $10^{18}\,\text{atoms/cm}^3$.

22. Repeat Problem 20 for a p-type material.

23. Repeat Problem 20 for a p-type material with an impurity concentration of $10^{18}\,\text{atoms/cm}^3$.

Chapter Two
Signals

In order to analyze circuits it is not only necessary to develop mathematical models for the electrical components, but it is also necessary to characterize the voltage and current waveforms mathematically. In this chapter we use the term signal to denote a current or voltage waveform. Basically, signals can be grouped into two classifications: periodic or nonperiodic. In the next several sections some typical periodic and nonperiodic signals and their characteristics are discussed. The reader should review complex number theory in Appendix B before proceeding.

2.1 PERIODIC SIGNALS

A signal $s(t)$ is said to be periodic with period T if $s(t) = s(t + T)$ for some finite T and all t. The case when $s(t)$ is *constant* can be considered in this category. The periodic waveforms shown in Figure 2.1 are quite common. All of these waveforms can be produced by electronic generators that consist of an oscillator plus additional waveshaping circuitry. The sinusoidal waveform is also generated by the electromagnetic rotating generator.

The sinusoidal waveform is very useful in testing the frequency response of a system. One varies the frequency ω of the waveform and observes the steady-state output as a function of ω. The square and pulse waveforms are useful in applications where sudden switching is needed and when we want

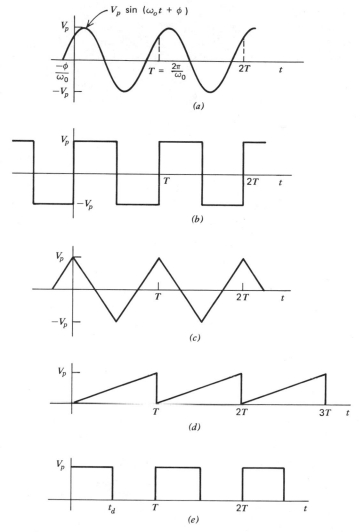

Figure 2.1 Typical periodic waveforms. (*a*) Sinusoidal waveform. (*b*) Square waveform. (*c*) Triangular waveform. (*d*) Sawtooth waveform. (*e*) Pulse waveform.

to observe the response of a system to a sudden change in the signal level. In the sawtooth waveform the voltage rises linearly and then switches instantly to its initial value. This waveform is useful for sweeping an electron beam across the face of a cathode-ray tube.

Typically the waveforms in Figure 2.1 can be characterized by their peak value V_p, period T, and phase (position of the wave with respect to the origin). One additional parameter is needed to characterize nonsymmetric waveforms. For example, we must specify the duration time t_d of the pulse in Figure 2.1e in addition to the repetition rate $1/T$ and peak value V_p. In addition to these parameters, the average and root-mean square (rms) value of a periodic waveform are useful parameters and are discussed next.

2.1.1 AVERAGE AND rms VALUES

The average value of a signal denotes the dc content of the signal, or, in other words, it indicates whether the weight of the signal is more positive than negative or vice versa. The average value of a periodic signal $s(t)$ with period T is defined as

$$S_{avg} = \frac{1}{T} \int_0^T s(t)dt \qquad (1)$$

For example, the average value of the sawtooth waveform is

$$S_{avg} = \frac{1}{T} \int_0^T \frac{V_p}{T} t \, dt = \frac{V_p}{2} \qquad (2)$$

This is the result one might expect since the area of the triangle in Figure 2.1d is $V_p T/2$. The average values of the sinusoid, square waveform, and the triangular waveform in Figure 2.1 are zero.

In Chapter Three we learn that the power dissipated in a conductor is proportional to the square of the current through the conductor or the square of the voltage across the conductor. If the waveforms are periodic, then the average power is proportional to the square of the *root-mean-squared (rms)* value or *effective value* S_{rms} of the voltage or current where

$$S_{rms} = \left[\frac{1}{T} \int_0^T s^2(t) \, dt \right]^{1/2} \qquad (3)$$

From this definition the rms value of a sinusoidal waveform

$$s(t) = S_p \cos(\omega_o t + \theta) \qquad (4)$$

is

$$S_{rms} = \left[\frac{\omega_o}{2\pi} \int_0^{2\pi/\omega_o} S_p^2 \cos^2(\omega_o t + \theta) \, dt \right]^{1/2} \qquad (5)$$

Let $\omega_o t + \theta = \tau$, then (5) becomes

$$S_{rms} = \left(\frac{S_p^2}{2\pi} \int_0^{\theta+2\pi} \cos^2 \tau \, d\tau \right)^{1/2}$$

and

$$S_{rms} = \left[\frac{S_p^2}{2\pi} \left(\frac{1}{2} \tau + \frac{1}{4} \sin 2\tau \right) \Big|_\theta^{\theta+2\pi} \right]^{1/2}$$

so that

$$S_{\text{rms}} = S_p/\sqrt{2} \qquad (6)$$

Thus, the rms value of a sinusoid is its peak value divided by the $\sqrt{2}$. This very important result is used frequently in subsequent chapters.

At this point it is appropriate to point out that voltmeters and ammeters measure either the dc or ac value of a periodic voltage or current waveform. Some meters are equipped with a range switch that allows you to select either the dc or ac value for measurement. The dc measurement indicates the average value of the voltage or current. However, the ac measurement is much more involved. For example, some modern electronic instruments use a thermal measurement technique in which the temperature in an element is measured. Since the temperature of the element is proportional to the average power dissipated in the element, the instrument is calibrated to indicate the rms value of the waveform. These instruments are said to give *true rms* values.

Typically in less expensive instruments the ac value indicated is found by first passing the signal through a full-wave rectifier circuit. This circuit is made of diodes and is discussed in Chapter Six. The output signal of this circuit is the absolute value of the input signal, that is, the negative part of the signal is inverted. For example, Figure 2.2 illustrates a full-wave rectified sinusoid. The instrument then measures the average value of this full-wave rectified signal. For example, a typical low-cost meter has a d'Arsonval movement which consists of a coil suspended between the poles of a permanent horseshoe magnet. When current flows through the coil it deflects at an angle proportional to the *average value* of the current in the coil. The instrument is then calibrated to read the *rms value of a sinusoid.*

For example, in Figure 2.2 the average value of the full-wave rectified sinusoid is

$$S_{\text{avg}} = \frac{\omega_o}{\pi} \int_0^{\pi/\omega_o} S_p \sin \omega_o t \, dt = \frac{2S_p}{\pi} \qquad (7)$$

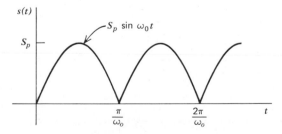

Figure 2.2   Full-wave rectified sinusoid.

However, the scale of the meter is calibrated to read

$$1.11 \times S_{\text{avg}}$$

where

$$1.11 = \frac{S_{\text{rms}}(\text{sinusoid})}{S_{\text{avg}}(\text{sinusoid})} = \frac{S_p/\sqrt{2}}{\dfrac{2 S_p}{\pi}} \tag{8}$$

Thus, in ac measurements these instruments indicate the true rms value of the signal only when the signal is a sinusoid with a dc value of zero. For example, consider the signal

$$s(t) = 3 + 3 \sin \omega_o t \tag{9}$$

which is plotted in Figure 2.3. Note that

$$|s(t)| = 3 + 3 \sin \omega_o t \tag{10}$$

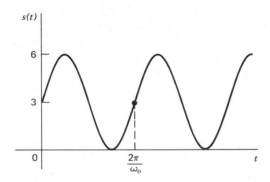

Figure 2.3   Sinusoid with a nonzero dc average.

$$|S|_{\text{avg}} = \frac{\omega_o}{2\pi} \int_0^{2\pi/\omega_o} (3 + 3 \sin \omega_o t)\, dt = 3 \tag{11}$$

and

$$S_{\text{rms}} = \left[\frac{\omega_o}{2\pi} \int_0^{2\pi/\omega_o} (3 + 3 \sin \omega_o t)^2 dt \right]$$

$$= \left[\frac{1}{2\pi} \int_0^{2\pi} (9 + 18 \sin \tau + 9 \sin^2 \tau)\, d\tau \right]^{1/2}$$

so that

$$S_{\text{rms}}(\text{true}) = (9 + 9/2)^{1/2} = 3.67 \tag{12}$$

Note that the true rms value is simply the square root of the sum of the squares of the rms values of each term in (9). Now an ac meter that senses the dc value of the full-wave rectified signal would indicate that

$$S_{rms}(\text{meter}) = S_{avg} \times 1.11 = 3.33 \tag{13}$$

Thus, in order to avoid ac measurement errors one must be knowledgeable of the waveform characteristics and meter circuitry when making ac measurements with an instrument that does not indicate true rms values.

One final result of note is that given a signal

$$s(t) = c_o + \sum_{n=1}^{N} c_n \sin(n\omega_o t + \theta_n) \tag{14}$$

which consists of a dc term c_o plus sinusoids of frequency ω_o and phase θ_1 and its harmonics with frequency $n\omega_o$ and phase θ_n, the rms value of the signal is

$$S_{rms} = \left[c_o^2 + \sum_{n=1}^{N} \left(\frac{c_n}{\sqrt{2}} \right)^2 \right]^{1/2} \tag{15}$$

that is, it is the square root of the sum of the squares of the rms values of each term. This result agrees with (12), but the general proof is left as an exercise. Equation 15 is also valid for a sum of sinusoids that are not harmonically related.

2.1.2 ADDITION OF SINUSOIDS OF THE SAME FREQUENCY

In many applications it is necessary to obtain the sum of several sinusoidal signals of the *same frequency*. The sum of sinusoids of the same frequency yields a single sinusoid with that frequency. The purpose of this section is to show how to determine the amplitude and phase of this composite sinusoidal waveform. For example, let us show that

$$s_1(t) = a \cos \omega_o t + b \sin \omega_o t \tag{16}$$

is equivalent to

$$s_2(t) = c \cos(\omega_o t + \theta) \tag{17}$$

where $c = \sqrt{a^2 + b^2}$ and $\tan \theta = -b/a$. To prove this let us use Euler's identity in Appendix B to represent the sinusoids in (16).

$$s_1(t) = a \left(\frac{e^{j\omega_o t} + e^{-j\omega_o t}}{2} \right) + b \left(\frac{e^{j\omega_o t} - e^{-j\omega_o t}}{2j} \right) \tag{18}$$

which can be grouped as

$$s_1(t) = \tfrac{1}{2}(a - jb)e^{j\omega_o t} + \tfrac{1}{2}(a + jb)e^{-j\omega_o t} \tag{19}$$

Now the second term is simply the complex conjugate of the first term, thus

$$s_1(t) = 2\text{Re}[\tfrac{1}{2}(a - jb)e^{j\omega_o t}] \tag{20}$$

The quantity $a - jb$ can be expressed in polar form by means of vector addition as shown in Figure 2.4 so that

$$s_1(t) = \text{Re}[\sqrt{a^2 + b^2}\ e^{j(\omega_o t + \theta)}] = c \cos(\omega_o t + \theta) \tag{21}$$

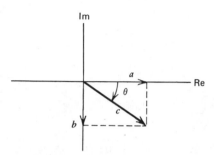

Figure 2.4   Vector addition.

where c and θ are as indicated in (17) so that

$$s_1(t) \equiv s_2(t) \tag{22}$$

We can think of the complex number $c\, e^{j(\omega_o t + \theta)}$ as a vector with magnitude c and angle θ at $t = 0$, which is rotating with angular velocity ω_o as shown in Figure 2.5. In (21) the real part operation simply yields the projection of this vector on the real axis.

We could arrive at the above result using trigonometric identities, but vector addition is much simpler as we will now show. Write (16) in the form

$$s_1(t) = a \cos \omega_o t + b \cos(\omega_o t - 90°) \tag{23}$$

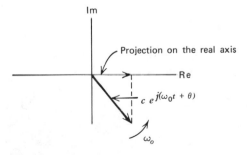

Figure 2.5   Sinusoid represented as a rotating vector.

Now we express each term of this sum by its *phasor*. A *phasor* is a vector that denotes the amplitude and phase of the sinusoid. For example, the first term in (23) can be expressed as $a\underline{/0°}$ or $ae^{j0°}$ and the second term as $b\underline{/-90°}$ or $be^{-j90°}$. Given the sum of sinusoids *with identical frequencies* the amplitude and phase of the composite signal is simply found by summing the phasors of each term.

As a final example, suppose that we are given a signal

$$s(t) + a\,\sin(\omega_o t + \phi_1) + b\,\sin(\omega_o t + \phi_2) \tag{24}$$

From Euler's identity

$$s(t) = a\left[\frac{e^{j(\omega_o t+\phi_1)} - e^{-j(\omega_o t+\phi_1)}}{2j}\right] + b\left[\frac{e^{j(\omega_o t+\phi_2)} - e^{-j(\omega_o t+\phi_2)}}{2j}\right]$$

or

$$s(t) = \frac{1}{2j}[(a\,e^{j\phi_1} + b\,e^{j\phi_2})e^{j\omega_o t} - (a\,e^{-j\phi_1} + b\,e^{-j\phi_2})e^{-j\omega_o t}]$$

Since the difference between a number and its conjugate is j times twice its imaginary part we obtain

$$s(t) = \text{Im}[(a\,e^{j\phi_1} + b\,e^{j\phi_2})e^{j\omega_o t}] \tag{25}$$

The addition of the phasors $a\underline{/\phi_1}$ and $b\underline{/\phi_2}$ is shown in Figure 2.6 and yields the phasor $c\underline{/\phi_3}$ so that

$$s(t) = c\,\sin(\omega_o t + \phi_3) \tag{26}$$

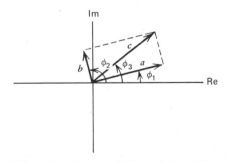

Figure 2.6   Phasor addition of sinusoids of the same frequency.

In conclusion, when adding sinusoids of the same frequency simply convert all of the terms to the cosine function using the identity

$$\sin(\omega_o t + \theta) \equiv \cos(\omega_o t + \theta - 90°) \tag{27}$$

or convert all of the terms to the sine function using the identity

$$\cos(\omega_o t + \theta) \equiv \sin(\omega_o t + \theta + 90°) \tag{28}$$

and add the phasors in the complex plane. It is not necessary to use Euler's identity. Euler's identity was only used to verify the above statement.

As an example consider the signal

$$s(t) = 3 \cos \omega_o t + 4 \sin \omega_o t \tag{29}$$

which we can write as

$$s(t) = 3 \cos \omega_o t + 4 \cos (\omega_o t - 90°)$$

To find the amplitude and phase of $s(t)$ we add the phasors $3\underline{/0°}$ and $4\underline{/-90°}$ as shown in Figure 2.7 and obtain the phasor

$$5\underline{/-53.1°}$$

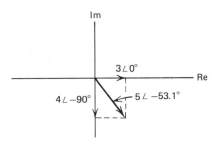

Figure 2.7   An example of phasor addition.

so that

$$s(t) = 5 \cos(\omega_o t - 53.1°) \tag{30}$$

Alternatively we could have written (29) as

$$s(t) = 3 \sin(\omega_o t + 90°) + 4 \sin \omega_o t \tag{31}$$

and we add the phasors $3\underline{/90°}$ and $4\underline{/0°}$ and obtain the phasor $5\underline{/36.9°}$ so that

$$s(t) = 5 \sin(\omega_o t + 36.9°) \tag{32}$$

Note that identities (27) and (28) verify that (32) and (30) are equivalent.

2.1.3 FREQUENCY SPECTRUM AND THE FOURIER SERIES

The concept of the frequency spectrum of a signal is a very useful one. For example, consider the amplitude modulated (AM) signal

$$s(t) = (V_c + V_s \cos \omega_s t) \cos \omega_c t \tag{33}$$

The amplitude of $\cos \omega_c t$ is modulated by $V_s \cos \omega_s t$. The resultant waveform is illustrated in Figure 2.8 for $V_c > V_s$ and $\omega_c \gg \omega_s$. The dashed line indicates the envelope within which the signal is contained, that is, since $\cos \omega_c t$ varies between ± 1, the envelope is $\pm(V_c + V_s \cos \omega_s t)$ and has period $2\pi/\omega_s$. Using trigonometric identities we can write (33) as

$$s(t) = V_c \cos \omega_c t + \frac{V_s}{2}[\cos(\omega_c + \omega_s)t + \cos(\omega_c - \omega_s)t] \tag{34}$$

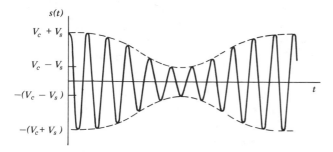

Figure 2.8 AM signal.

From this expression we see that the frequency spectrum of the signal has components at ω_c, $\omega_c + \omega_s$, and $\omega_c - \omega_s$. The component at ω_c is called the carrier frequency and $\omega_c - \omega_s$ and $\omega_c + \omega_s$ are called the sidebands. The frequency spectrum is illustrated pictorially in Figure 2.9. In standard broadcast stations the carrier frequency $f_c = \omega_c/2\pi$ lies between 550 and 1600 kHz. The frequency spectrum of speech lies between 300 and 4000 Hz, hence each station is allotted a frequency band of $f_c \pm 5$ kHz.

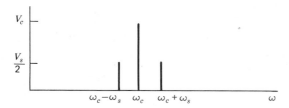

Figure 2.9 Frequency spectrum of the AM signal.

Allotting different carrier frequencies to each station in a given area not only allows one to separate messages at the receiver by means of filters, but also allows for more efficient transmission. For example, to efficiently radiate a 1-kHz tone over a wide area a one-quarter wavelength $(\frac{1}{4}\lambda)$ dipole antenna is required where $\lambda = f/v$, f is the frequency of the wave, and v is its velocity. Assuming that the wave travels at the speed of light ($v = 3 \times$

10^8 m/s), then the antenna would have to be 75 km in length, but only a 75-m antenna is required to efficiently radiate a 1000-kHz tone.

In order to find the frequency content of a periodic waveform it is often necessary to find the Fourier series representation of the signal, that is, the signal is represented by a sum of sinusoids consisting of the dc content (average value) of the signal, a fundamental frequency component at $\omega_o = 2\pi/T$, and harmonics at $n\omega_o$. For all practical purposes any peridoc signal $f(t)$ can be represented by the Fourier series, which is defined as

$$f(t) = a_o + \sum_{n=1}^{\infty} a_n \cos n\omega_o t + \sum_{n=1}^{\infty} b_n \sin n\omega_o t \qquad (35)$$

where $f(t) = f(t + T)$, $T = 2\pi/\omega_o$. The term a_o represents the dc component of the signal, and the terms a_n and b_n determine the magnitude and phase of the $n\omega_o$ frequency component of the signal. These coefficients are computed as follows.

The cosine and sine function are orthogonal, that is,

$$\int_0^T \sin k\omega_o t \sin l\omega_o t \, dt = \int_0^T \cos k\omega_o t \cos l\omega_o t \, dt = 0 \qquad (36)$$

where k and l are nonnegative integers and $k \neq l$ and

$$\int_0^T \sin k\omega_o t \cos l\omega_o t \, dt = 0 \qquad (37)$$

for all k, l. Applying properties (36) and (37) we can solve for a_n and b_n in (35). If we integrate (35) over the period T, then the dc component

$$a_o = \frac{1}{T} \int_0^T f(t) \, dt \qquad (38)$$

Note that the dc component a_o is the average value of the signal. If we multiply (35) by $\cos n\omega_o T$ and integrate over the period T, then

$$a_n = \frac{2}{T} \int_0^T f(t) \cos n\omega_o t \, dt \qquad (39)$$

Similarly, if we multiply (35) by $\sin n\omega_o T$ and integrate over the period,

$$b_n = \frac{2}{T} \int_0^T f(t) \sin n\omega_o t \, dt \qquad (40)$$

where a_n and b_n are the amplitudes of the cosine and sine wave at the frequency $n\omega_o$.

Let us apply (38 to 40) to the signal $v_2(t)$ in Figure 2.10 obtained by passing a sine wave through a half-wave rectifier circuit. This circuit clips the

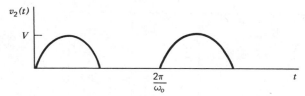

Figure 2.10   Half-wave rectifier output.

negative part of the signal. First, we obtain the dc component

$$a_o = \frac{\omega_o}{2\pi} \int_0^{\pi/\omega_o} V \sin \omega_o t \, dt = \frac{V}{2\pi} \Big|_0^{\pi/\omega_o}$$

$$\boxed{a_o = V/\pi}$$

Next we calculate the fundamental frequency component.

$$a_1 = \frac{\omega_o}{\pi} \int_0^{\pi/\omega_o} V \sin \omega_o t \cos \omega_o t \, dt = \boxed{0}$$

$$b_1 = \frac{\omega_o}{\pi} \int_0^{\pi/\omega_o} V \sin^2 \omega_o t \, dt = \boxed{\frac{V}{2}}$$

Next we compute the harmonic content. For $n \geq 2$ let $\tau = \omega_o t$, then

$$a_n = \frac{1}{\pi} \int_0^{\pi} V \sin \tau \cos n\tau \, d\tau$$

$$= \frac{V}{\pi} \left(+\frac{\cos(n-1)\tau}{2(n-1)} - \frac{\cos(n+1)\tau}{2(n+1)} \right) \Big|_0^{\pi} = \begin{cases} 0 & n \text{ odd} \\[2ex] \dfrac{-2V}{\pi(n^2-1)} & n \text{ even} \end{cases}$$

$$b_n = \frac{1}{\pi} \int_0^{\pi} V \sin \tau \sin n\tau \, d\tau = 0$$

so that

$$\boxed{v_2(t) = \frac{V}{\pi} + \frac{V}{2} \sin \omega_o t - \frac{2V}{\pi} \sum_{n=2,4,6\ldots} \frac{1}{n^2-1} \cos n\omega_o t} \tag{41}$$

Note that the output frequency spectrum contains a dc component V/π plus components at the frequencies ω_o, $2\omega_o$, $4\omega_o$, and so on. The summation of the first four terms of (41) is shown in Figure 2.11. Note that the waveform already closely approximates the half wave in Figure 2.10.

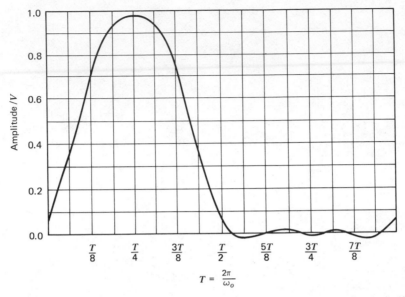

$$T = \frac{2\pi}{\omega_o}$$

Figure 2.11 Summation of the first four terms of the Fourier series for the half wave.

As a final example let us use the Fourier series to obtain the frequency spectrum of the pulse in Figure 2.1e. Assume that $t_d = T/2$, then the dc component is

$$a_o = \frac{1}{T} \int_0^{T/2} V_p \, dt = \boxed{\frac{V_p}{2}}$$

For $n \geqslant 1$ we obtain

$$a_n = \frac{2}{T} \int_0^{T/2} V_p \cos \frac{n2\pi t}{T} \, dt$$

$$= \frac{2}{T} V_p \frac{\sin \dfrac{n2\pi}{T} t}{\dfrac{n2\pi}{T}} \Bigg|_0^{T/2} = \boxed{0}$$

and

$$b_n = \frac{2}{T} \int_0^{T/2} V_p \sin n \frac{2\pi t}{T} \, dt = \frac{-2V_p}{T} \frac{\cos \dfrac{n2\pi t}{T}}{\dfrac{n2\pi}{T}} \Bigg|_0^{T/2}$$

so that

$$b_n = \begin{cases} 0, & n \text{ even} \\ \dfrac{2V_p}{n\pi}, & n \text{ odd} \end{cases}$$

Therefore, the Fourier series of this pulse is

$$v(t) = \left(\frac{1}{2} + \sum \frac{2}{n\pi} \sin \frac{n2\pi t}{T}\right) V_p \tag{42}$$
$$n = 1, 3, 5, \ldots$$

Both equations [(41) and (42)] are used in the analysis of diode circuits in Chapter Six.

2.2 NONPERIODIC SIGNALS

Typical nonperiodic signals are the step function and the exponential function. The step function, denoted by $u(t-a)$, is illustrated in Figure 2.12 and is defined as

$$u(t-a) = \begin{cases} 0, & t < a \\ \\ 1, & t \geq a \end{cases} \tag{43}$$

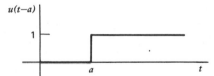

Figure 2.12   Step function.

It is primarily used to represent a sudden change in an independent voltage or current source. The step function is also frequently used to test how fast a circuit responds to a sudden change in the input excitation. For example the 5 volt, 3 second pulse in Figure 2.13 is expressed mathematically as

$$s(t) = 5u(t) - 5u(t-3) \tag{44}$$

The exponential function e^{st}, where $s = \alpha + j\omega$, is also a very important function because the transient solution of circuits containing linear constant energy storage elements (inductors and capacitors) has this

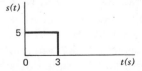

Figure 2.13   Pulse.

waveshape. Also, in the previous section we have used Euler's identity to represent the sinusoid in terms of the exponential function $e^{j\omega t}$. The exponential function is plotted in Figure 2.14 for real exponents ($s = \alpha$). Often when $\alpha < 0$ the exponential function is expressed in the form $e^{-t/\tau}$ where $\tau = 1/|\alpha|$ and is called the time constant. The time constants specifies how fast the function decays to zero. For example, in one time constant the function is $e^{-1} = 0.37$, and in four time constants $e^{-4} = 0.02$. In some circuits the transient

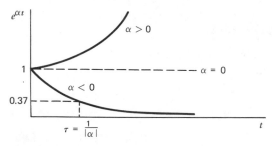

Figure 2.14 Exponential function.

response is a damped oscillation of the form $e^{\alpha t} \cos(\omega_o t + \phi) = \text{Re}[e^{j(\omega_o t + \phi)}] \cdot e^{\alpha t}$ where $\alpha < 0$. The $\cos(\omega_o t + \phi)$ term oscillates between ± 1 with period $T = 2\pi/\omega_o$ and the amplitude decays at the rate $e^{-t/\tau}$ where $\tau = 1/|\alpha|$. See Figure 2.15.

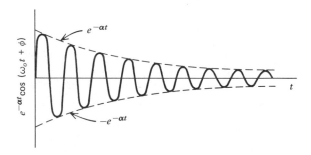

Figure 2.15 Damped sinusoid ($\alpha > 0$).

Finally, sometimes a signal is very erratic and cannot be described by a simple mathematical function (Figure 2.16). In such cases the signal is considered to be a random variable and it is described by its statistical properties. Fluctuations in voltage and current caused by the random motion of electrons in electrical devices is called noise and is treated as a random variable. Also, speech waveforms can frequently be treated as

Figure 2.16   Random signal.

random. The discussion of the characterization of these signals requires a background in probability theory and random processes.

2.3 SAMPLED SIGNALS

The technological realization of large-scale integrated circuits has resulted in a significant decrease in digital hardware cost. This has made the processing of signals digitally feasible in many instances. In order to process a signal digitally, the signal must first be characterized by a sequence of binary words. Each word approximates the value of the signal at a given time instant. These words are called quantized samples of the signal and they are uniformly spaced in time. The samples must be quantized (rounded or truncated) because only a finite number of bits are available for characterizing the level of the signal. In Figure 2.17 a continuous signal is shown, which is sampled every T seconds. The

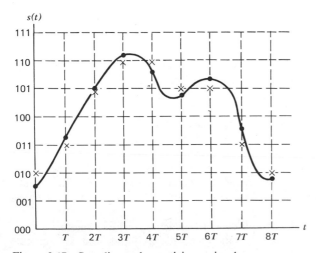

Figure 2.17   Sampling and quantizing a signal.

samples are rounded to the nearest quantization level. The quantized sample is indicated with an x at each time instant. In this example a three-bit word is used, therefore there are eight (2^3) quantization levels. The sampling rate is $1/T$.

In sampling a signal we must sample at a rate equal to or greater than the Nyquist rate, which is twice the highest frequency component of the signal. Sampling at a lower rate causes high frequency components to appear as low frequency components as shown in Figure 2.18. This is called aliasing.

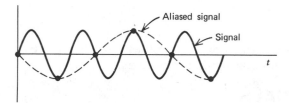

Figure 2.18   Aliasing of a signal.

Often a signal is passed through a filter to band limit its frequency spectrum so that an unnecessarily high sampling rate is not required and aliasing is avoided.

REFERENCES

1.   J. B. Cruz, Jr. and M. E. Van Valkenburg, *Signals in Linear Circuits*, Houghton Mifflin, 1974, Chapters 1–3.
2.   M. E. Van Valkenburg, *Network Analysis*, 3rd edition, Prentice-Hall, 1974, Chapter 15.

PROBLEMS

1.   In Figure 2.1e let the duration of the pulse $t_d = aT$ where $o < a < 1$. Express the average and rms value of this waveform in terms of a.
2.   The pulse waveform in Figure 2.1e has an average value of 12 V and an rms value of 15.5 V. Find V_p and a where $a = t_d/T$.
3.   Find the average and rms values of the signal $f(t) = 5 + \sin 2\pi t$.
4.   Find the rms value of the square wave in Figure 2.1b.
5.   Find the rms value of the sawtooth waveform in Figure 2.1d.
6.   Suppose that a sawtooth voltage waveform in Figure 2.1d is measured with a voltmeter with a d'Arsonval movement. The meter indicates 5 V on the ac scale. What is the average value of the voltage? Does your answer depend on the shape of the waveform?
7.   Suppose that the signal in Problem 3 is measured with a meter with a d'Arsonval movement, a full-wave rectifier on the ac input, and the ac scale

calibrated to read sinusoidal rms values. What would the meter indicate on the ac scale of this instrument?

8. Repeat Problem 7 for the pulse in Figure 2.1e with $V_p = 10$ V and $t_d = T/2$. What is the true rms value of this signal?

9. Use vector addition of phasors to find the amplitude c and phase θ of the signals below.

    (a) $4 \cos \omega_o t + 3 \sin \omega_o t$ (c) $10 \cos(\omega_o t + 60°) + 5 \cos(\omega_o t - 60°)$

    (b) $4 \cos \omega_o t - 3 \sin \omega_o t$ (d) $4 \cos \omega_o t + \cos \omega_o t$

10. Use Equations 36 and 37 to verify Equation 15.

11. Plot the frequency spectrum of the following signals.

    (a) $5 + 3 \cos \omega_o t + \cos 3\omega_o t$ (c) $10 + 4 \cos \omega_o t + 3 \sin \omega_o t$

    (b) $V(\cos \omega_s t)(\cos \omega_o t)$ (d) $5 \cos \omega_o t + 2 \sin 2\omega_o t$

12. Find the Fourier series for the square wave in Figure 2.1b.

13. Sketch the first three terms of (42) and compare with Figure 2.1e. Using the first five terms of (42) plot the frequency spectrum (see Figure 2.9) of the pulse waveform.

14. Find the Fourier series for the sawtooth waveform in Figure 2.1d.

15. Find the Fourier series for the full-wave rectified sine function in Figure 2.2. Note that the period of the original waveform was $2\pi/\omega_o$, but the period of the rectified waveform is π/ω_o.

16. For the functions below sketch the corresponding waveforms.

    (a) $f(t) = u(t) - u(t - 5 \mu s)$

    (b) $f(t) = u(t) + u(t - 1) + 2u(t - 2) - 5u(t - 3) + u(t - 5)$

    (c) $f(t) = u(t + 2) - u(t - 2)$

17. Express the waveform in Figure P2.17 in terms of a summation of step functions.

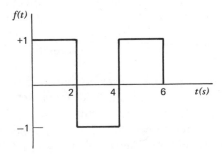

Figure P2.17

18. Repeat Problem 16 for the functions below.

    (a) $5e^{-10^3 t}$ (d) $\text{Re}[e^{-10^3 t}(e^{j2\pi \cdot 10^3 t})]$

    (b) $5e^{-10^6 t}$ (e) $\text{Im}[e^{-10^3 t}(e^{j2\pi \cdot 10^3 t})]$

    (c) $5e^{10^3 t}$ (f) $5e^{-10^3 t}(\sin 2\pi \cdot 10^2 t)$

Chapter Three
Circuit Analysis: Models and Kirchhoff's Laws

To model the electrical circuit for the purpose of analysis, it is common to model each of the components by a lumped element or set of lumped elements whose characteristics are described by mathematical relationships between the voltage or voltages across the terminals of the element and the current or currents entering or leaving the terminals of the element. These characteristics are frequently determined by electrical measurement techniques.

In addition to the mathematical relations that describe the electrical behavior of the device at its terminals, there is an additional set of equations that is determined by the topology of the circuit—that is, the manner in which the components are interconnected. These topological constraint equations are determined by Kirchhoff's laws and are *completely independent* of the individual electrical characteristics of the components.

In this chapter we introduce the lumped element model, power and energy in lumped elements, the ideal resistor and the ideal source lumped elements, Kirchhoff's laws, the independence of the Kirchhoff equations, and solution techniques for linear resistive circuits. The reader may want to review Appendix A on the properties and solution of algebraic equations before reading this chapter.

48

3.1   THE LUMPED CIRCUIT

The lumped electrical circuit is an interconnection of lumped electrical elements, and the lumped electrical element is one whose electrical behavior is modeled by the voltage and current relationships at its terminals. (When spatial effects are included in the model the element is said to be *distributed*.) A multiterminal lumped element is shown in Figure 3.1. In the lumped element it is always assumed that the rate of flow of charge entering the element is equal to the rate of flow of charge leaving the element—that is,

$$i_1 + i_2 + i_3 + i_4 + i_5 + i_6 = 0 \qquad (1)$$

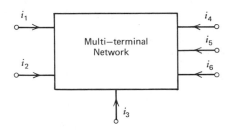

Figure 3.1   Multiterminal lumped element.

This assumption is referred to as *space-charge neutrality*, and experimental evidence confirms this assumption that the element is electrically neutral.

The multiterminal element is sometimes called an *N-terminal network* where N is the number of terminals. Sometimes the terminals of the multiterminal element can be paired so that the current entering one terminal is equal to the current leaving the other terminal. In this case the lumped element is called a *multiport* or *N-port* where N is the number of ports. Figure 3.2 illustrates a 3-port. Note that an N-port has $2N$ terminals.

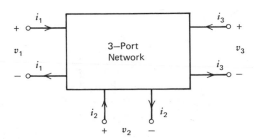

Figure 3.2   Multiport example.

3.1.1 THE BRANCH

The *branch* is a two terminal lumped element (1-port) and is the basic building block of most circuits. Even multiterminal elements are usually modeled by a network of branches, and practically all the circuits we analyze in this book contain only branches. Given the branch in Figure 3.3 there are four different combinations for assigning the $v - i$ directions. In Figure 3.3 we show the *standard reference system* in which—after the current direction is assigned—the *voltage polarity is always assigned positive at the terminal into which the current is flowing.* The assignment of the current direction is arbitrary. If you assign a current direction in your circuit model that is opposite to the flow of current in the actual circuit,

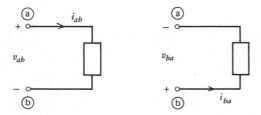

Figure 3.3 The branch and *i-v* assignment.

your analysis will simply yield a negative value for that current. *The standard reference system in Figure 3.3 is very important since all of our mathematical relationships between the branch voltage and the branch current for an electrical element will be defined with respect to this reference.* Note that our space-charge-neutrality assumption allows us to assign only one current to the two-terminal branch. The current entering one terminal must leave at the other terminal. In Figure 3.3, $v_{ab} = - v_{ba}$ and $i_{ab} = - i_{ba}$.

The branch can be connected to a circuit N as shown in Figure 3.4a or the terminals can be reversed as shown in Figure 3.4b. The branch is *bilateral* if both connections shown in Figure 3.4 yield identical electrical responses. If

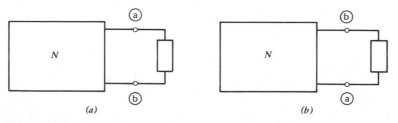

Figure 3.4 Two possible connections of a branch to a circuit.

the branch is nonbilateral than the connections in Figure 3.4 do not give equivalent results. Nonbilateral components are usually marked at their terminals with a code so that the user can differentiate between the two terminals. A good example of a nonbilateral element is a battery. One terminal is marked as the ($+$) pole and the other as the ($-$) pole. Connecting a battery improperly between a pair of terminals in a circuit can have serious consequences. The characteristics of a bilateral element are symmetric about the origin as shown in Figure 3.5. The variables x and y are (1) current and voltage for resistive elements and sources, (2) charge and voltage for capacitors, or (3) flux linkages and current for inductors.

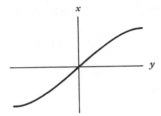

Figure 3.5   Bilateral branch.

3.2 ENERGY AND POWER IN LUMPED ELEMENTS

In Chapter One we defined the voltage between two points as the work done per unit charge moved between these two points. Therefore, the total energy required or supplied by the lumped element in the time interval $[-\infty, t]$ is

$$w(t) = \int_0^q v \, dq \tag{2}$$

where q is the total charge moved through the lumped element in that time interval. But $dq = i\,dt$, hence the total energy either supplied or absorbed by the element is

$$w(t) = \int_{-\infty}^t vi \, dt = w(0) + \int_0^t vi \, dt \tag{3}$$

We define power as the work done per unit of time.

$$p(t) = \frac{dw}{dt} = vi \tag{4}$$

The MKS unit of power is joules/second or the watt in honor of the Scottish inventor James Watt.

If the voltage and current polarities are assigned according to the *standard reference system,* it follows from our definition of voltage and current in Chapter One that if

$$p(t) = v(t)i(t) > 0 \tag{5}$$

in a given time interval, then the element is *absorbing energy* in that time interval. For example, in our discussion of current flow in a wire in Chapter One the vi product is always positive in our standard reference system so the wire always absorbs energy. However, if

$$p(t) = v(t)i(t) < 0 \tag{6}$$

in a given time interval, then the element is *supplying energy* to the circuit in that interval.

If we do not use the standard reference system, that is, reverse the assigned polarity of v or assigned direction of flow of i in Figure 3.3, for example as shown in Figure 3.6, then $vi > 0$ in Figure 3.6 implies that the element is supplying energy to the circuit, and $vi < 0$ implies that the element is absorbing energy in a given time interval.

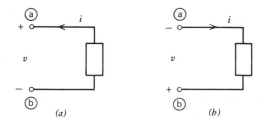

(a)   (b)

Figure 3.6   Nonstandard assignment of branch current and voltage.

In the case when the time-varying waveforms are periodic with period T, then it is convenient to speak of the average power

$$P_{avg} = \frac{1}{T} \int_0^T p(t) \, dt \tag{7}$$

For example, suppose that $v(t) = V_m \sin(\omega t + \beta)$ and $i(t) = I_m \sin(\omega t + \alpha)$, then

$$P_{avg} = \frac{\omega}{2\pi} \int_0^{2\pi/\omega} V_m I_m \sin(\omega t + \beta) \sin(\omega t + \alpha) \, dt$$

$$= \frac{1}{2\pi} V_m I_m \int_0^{2\pi} \sin(\tau + \beta) \sin(\tau + \alpha) \, d\tau \tag{8}$$

where $\tau = \omega t$. Equation 8 can be written as

$$P_{avg} = \frac{1}{2\pi} V_m I_m \int_0^{2\pi} \frac{1}{2} [\cos(\beta - \alpha) - \cos(2\tau + \beta + \alpha)] \, d\tau$$

$$= \frac{1}{2} V_m I_m \cos(\beta - \alpha) \tag{9}$$

Equation 9 states that the average power is equal to half the product of the peak amplitude of the *sinusoidal waveforms* times the cosine of their phase differences. For sinusoidal waveforms it is common to write (9) as

$$P_{avg} = V_{rms} I_{rms} \cos(\beta - \alpha) \tag{10}$$

where $V_{rms} = V_m/\sqrt{2}$ and $I_{rms} = I_m/\sqrt{2}$ are the *effective* or *root-mean-square* (rms) value of the *sinusoidal waveform* defined in Chapter Two.

For example, if the voltage supplied to your residence is 120 V ac, the 120 V means rms volts. The peak value of the waveform is 170 V. The rms value of the waveforms is used to simplify power calculations. Suppose you have an electrical appliance that requires 1200 W, then assuming no phase difference between the voltage and current ($\alpha = \beta$) the *rms current* drawn by the appliance is

$$I_{rms} = \frac{1200 \text{ W}}{120 \text{ V}} = 10 \text{ A} \tag{11}$$

Typically the power company charges by the kilowatt-hour. So that if this appliance is operated for three hours, the energy consumed in that time interval is

$$w(t) - w(t - 3h) = (1200 W)(3h)$$

$$= 3600 Wh \tag{12}$$

If the power company charges \$0.05 per kilowatt-hour then the cost of operating this appliance for three hours is

$$(3.6 kWh)(\$0.05/kWh) = \$0.18 \tag{13}$$

In the next section we introduce some of the ideal lumped elements that are used to model the characteristics of electrical components.

3.3 SOME IDEAL LUMPED ELEMENTS

Electrical circuits are composed of physical devices whose electrical properties can be quite complex. However, for the purposes of mathematical analysis, experience has shown that the electrical properties of these devices can usually be modeled with five basic *ideal* lumped elements: the resistor, the capacitor, the inductor, the voltage source, and the current source. The resistor dissipates energy (electromagnetic radiation in the form of heat, light, radio waves, etc.). The capacitor stores energy in an electric field. The inductor stores energy in a magnetic field. The voltage source and current source denote a source of energy, such as a battery or generator.

Now bear in mind the difference between models for mathematical analysis and actual devices. When you purchase a resistor, or capacitor, for example, these components are not ideal, that is, at high frequency an actual resistor can be inductive, an inductor can appear to be capacitive at high frequencies, and both the actual capacitor and inductor do dissipate some energy. Thus, typically an actual component may have to be modeled by two or more ideal elements. The accurate modeling of physical components is frequently more difficult than the subsequent analysis. This modeling is usually accomplished by measurement techniques. Also, the more complex the model, the more difficult and costly the analysis. Thus, it is important to include only those elements in the model that have an effect on the desired operation of the circuit.

Below the resistor and source branches are discussed. The capacitor and inductor branches are discussed in Chapter Eight.

3.3.1 THE IDEAL RESISTOR

An ideal resistor is any two-terminal element whose characteristics can be described completely in the $i - v$ plane independent of the waveform of the voltage or current. A resistor may be linear or nonlinear, time invariant or time varying as illustrated in Figure 3.7. For consistency these characteristics are always measured in the *standard reference* system, that is, *if the assumed current direction is from terminal a to terminal b then the voltage at terminal a is assumed to be positive with respect to terminal b.*

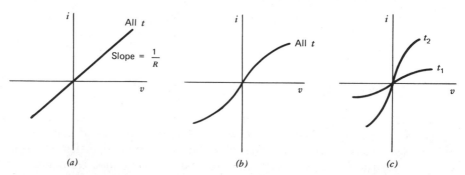

Figure 3.7 Resistor characteristics. (*a*) Linear and time invariant. (*b*) Nonlinear and time invariant. (*c*) Nonlinear and time varying.

For the linear resistor we see that

$$v = Ri \qquad (14)$$

where R is the resistance in ohms $(\Omega) =$ volts/amp. If the slope is positive in Figure 3.7a then R is positive, and if the slope is negative then R is negative. With this convention

$$p(t) = vi = i^2R = v^2/R \qquad (15)$$

Hence, if $R > 0$ then $p(t) > 0$ for all t. The positive resistor dissipates energy.

The reciprocal of this resistance is called conductance

$$G = \frac{1}{R} \text{ siemens (mho)}[1] \qquad (16)$$

The symbol for the ideal linear resistor is given in Figure 3.8a. The number 10 above the element indicates that its resistance is $+10$ ohms. Note that the linear resistor is a bilateral element. In analysis a current direction and voltage polarity are assigned to the resistor. Figure 3.8b illustrates all the possible assignments. Note that $i_1 = -i_2$ and $v_1 = -v_2$. Since the resistance is defined in the standard reference system we must write its branch constraints as

$$v_1 = 10i_1$$

or

$$v_2 = 10i_2$$

or $\qquad (17)$

$$v_2 = -10i_1$$

or

$$v_1 = -10i_2$$

If the resistance R varies with time we may say that the resistor is linear and time varying.

It is important to note that a short circuit between two terminals is said

(a)   (b)

Figure 3.8   The linear, time-invariant resistor. (a) Symbol. (b) Possible polarity assignments.

[1] Although the international standard unit is siemens, it is more common to find mho in U.S. literature.

to exist when the resistance between the terminals is zero. From (14) if $R = 0$ then the voltage between the terminals is zero. At the other extreme an open circuit means that the resistance is infinite (zero conductance) so that the current through the element is zero.

For the nonlinear resistance we define the incremental resistance measured at the point (i_0, v_0) in Figure 3.9 as

$$r = \frac{dv}{di}\bigg|_{i_0,\, v_0}. \tag{18}$$

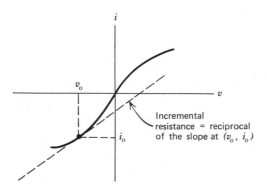

Figure 3.9   Incremental resistance.

Note that the incremental resistance is the reciprocal of the slope of the i-v characteristic and depends on (i_o, v_o). The reciprocal of r is again called conductance g. the characteristic of nonlinear resistors must be expressed in the form $f(v, i) = 0$. In particular, if v is a single-valued function of i (for every value of i there is a unique v), then $v = f_1(i)$, or if i is a single-valued function of v then $i = f_2(v)$, for example,

$$i = 5v^3 \tag{19}$$

Because of the difficulty of analyzing circuits with nonlinear resistors, approximation techniques or computer programs are frequently used.

3.3.2 IDEAL SOURCES

There are two types of ideal sources of electrical energy that we use in modeling circuits—the voltage source and the current source. The voltage source is a branch whose potential difference across its terminals at any given time is independent of the current through the element. Its v-i characteristic is illustrated in Figure 3.10a. Note that this element has *zero incremental resistance*—the change in voltage for any change in current is

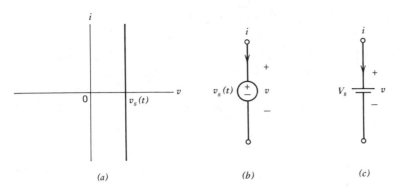

Figure 3.10 Voltage source. (*a*) *i-v* characteristic. (*b*) Time varying or constant voltage source symbol. (*c*) Constant voltage source symbol.

zero. *If* $v_s = 0$ *then the voltage source is a short circuit.* Also, note that v_s may be a function of time, for example, $v_s(t) = V \sin \omega t$. Thus, the characteristic is always parallel to the current axis, but its intersection with the voltage axis may change with time. Commonly used circuit symbols are given in Figures 3.10*b* and 3.10*c*. The symbol in Figure 3.10*c* is often used when the voltage source is constant with respect to time. We call this a dc (direct current) source.

Another type of source is the current source. Its *v-i* characteristic is given in Figure 3.11*a*, and its circuit symbol is shown in Figure 3.11*b*. This element has *zero incremental conductance*—for any change in v the change in i is zero. Thus, the characteristic is a line parallel to the voltage axis, and the i axis intercept i_s may vary with time or be time independent (dc). *If* $i_s = 0$ *then the current source is an open circuit.*

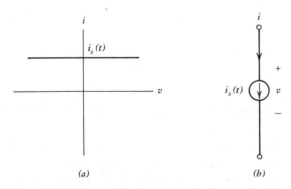

Figure 3.11 Current source. (*a*) *i-v* characteristic. (*b*) Symbol.

3.4 THE CIRCUIT GRAPH

In the next section we introduce Kirchhoff's laws, the basic laws of circuit theory, which yield circuit equations that depend only on the circuit topology, that is, the manner in which the lumped elements are interconnected, but not on the specific characteristics of the elements themselves. When a circuit consists only of branches, there is a shorthand notation for the topology called the circuit *graph*. A circuit and its graph are shown in Figure 3.12. Each branch in the circuit is represented by a line segment in its graph. The circuit illustrated has four branches so its graph has four line segments. *The junction of two or more branches is called a node.* Our example has three nodes. Note that in the circuit model the wires connecting terminals are ideal, that is, they offer no resistance to current flow and the voltage drop across them is zero.

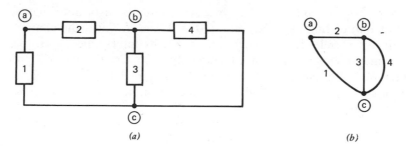

Figure 3.12   A circuit and its graph. (*a*) Circuit. (*b*) Graph.

In a circuit the branches form paths to the nodes. *A loop in a circuit is any closed path in which no node in this path is crossed more than once.* There are three different loops in Figure 3.12: loop 123, loop 124, and loop 34 (the numbers indicate the branches in the loop).

A *planar* circuit is one whose graph can be drawn such that no branches intersect each other. Figure 3.13 is an illustration of a graph of a circuit that is nonplanar. In integrated circuit design it is desirous to have planarity, since crossovers are costly and troublesome in integrated circuit fabrication.

In planar circuits we can define a special type of loop called the mesh. *A mesh is a loop that does not contain any other loops.* The graph in Figure 3.14 has three meshes that are denoted as *a*, *b*, and *c*. For obvious reasons sometimes the meshes are called *windows*.

Finally, in our discussion of the independence of the Kirchhoff equations we consider only circuits whose graphs are connected and not hinged or disconnected (see Figure 3.15). From our discussion of space-charge neutrality the hinged network can be split into two separate graphs at the

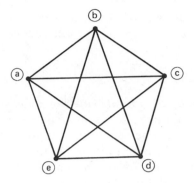

Figure 3.13   Nonplanar graph.

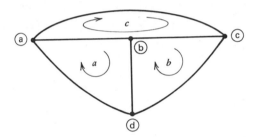

Figure 3.14   Meshes of a graph.

(a)   (b)

Figure 3.15   Disconnected and hinged graphs. (a) Disconnected. (b) Hinged.

hinged node. Hence both of the above cases can be treated as two separate circuits and need not be discussed any further in this book.

3.5 KIRCHHOFF'S LAWS

In the analysis of electrical networks two sets of equations must be satisfied. The first set consists of the topological constraints imposed by the given network configuration, that is, the particular interconnection of elements in the network. These constraints are independent of the element characteristics and depend only on the particular network configuration. The topological constraints are dictated by Kirchhoff's current law and voltage law and are discussed next. The second set of constraints are the element constraints, such as $v = Ri$ for a resistor or $v = 15$ V for a voltage source.

3.5.1 KIRCHHOFF'S CURRENT LAW

Kirchhoff's current law states that *the sum of all currents entering a node is equal to zero.* This law states that charge does not pile up on a node and is a direct result of the assumption of space-charge neutrality. Applying this law to the circuit in Figure 3.12, we first must orient the graph, that is, assign current directions. The current directions for each branch are assigned arbitrarily in Figure 3.16. The Kirchhoff's current law (KCL) equations are

$$\text{node } \textcircled{a} \quad : \quad -i_1 + i_2 \qquad\qquad = 0 \tag{20a}$$

$$\text{node } \textcircled{b} \quad : \quad \qquad -i_2 + i_3 + i_4 = 0 \tag{20b}$$

$$\text{node } \textcircled{c} \quad \quad i_1 \qquad -i_3 - i_4 = 0 \tag{20c}$$

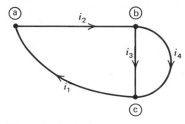

Figure 3.16   Assignment of current directions.

Note that we decided to refer to currents leaving the nodes as positive and currents entering the nodes as negative. Again, this is usually standard practice, but there is no reason why we could not multiply all or some (a mixed reference system) of the equations by -1.

3.5.2 KIRCHHOFF'S VOLTAGE LAW

Kirchhoff's voltage law states that *the sum of all the voltages around any loop is equal to zero.* This law states that a voltage cannot be created by simply traversing a loop and is a result of the fact that the electric field is conservative. Applying Kirchhoff's voltage law (KVL) to our example in Figure 3.16 we obtain

$$\text{loop } 123 \quad : \quad v_1 + v_2 + v_3 \qquad = 0 \tag{21a}$$

$$\text{loop } 34 \quad : \quad \qquad -v_3 + v_4 = 0 \tag{21b}$$

$$\text{loop } 124 \quad : \quad v_1 + v_2 \qquad + v_4 = 0 \tag{21c}$$

We have traversed each loop in the clockwise direction and used the standard reference convention so that the tail of the arrow is the positive terminal and the head of the arrow points to the negative terminal. Again, we could have used the clockwise direction for some loops and the counterclockwise direction for other loops (multiplication of the loop equation by -1).

3.6 INDEPENDENCE OF THE KIRCHHOFF EQUATIONS

To obtain a unique solution for the circuit, we need as many equations as we have unknowns and the equations must be independent. In our example we have four branches and therefore eight unknowns—the four branch currents and the four branch voltages. Now four equations are obtained from the branch constraints, that is, the voltage-current relation for each branch. These equations are independent of how the branch is connected to the circuit. If the circuit model has a unique solution, we must have eight independent equations. We need four more equations. These equations must come from the topological constraints, that is, the Kirchhoff equations. However, we have three KCL equations, (20), and three KVL equations, (21). We have too many equations! This means that (20) and (21) are not independent sets of equations or that our circuit does not have a unique solution. The former is more likely. Let us write (20) in the matrix form $\mathbf{A}_n \mathbf{i} = \mathbf{0}$ as shown below

$$
\begin{array}{c}
 \\
\text{ⓐ} \\
\text{ⓑ} \\
\text{ⓒ}
\end{array}
\begin{array}{cccc}
① & ② & ③ & ④
\end{array}
\begin{bmatrix}
-1 & 1 & 0 & 0 \\
0 & -1 & 1 & 1 \\
1 & 0 & -1 & -1
\end{bmatrix}
\begin{bmatrix}
i_1 \\ i_2 \\ i_3 \\ i_4
\end{bmatrix} = \mathbf{0}
\tag{20}
$$

The rows of $\mathbf{A}_n$ correspond to the nodes of the circuit and the columns of $\mathbf{A}_n$ correspond to the branches. The sum of any two rows in (20) is equal to the third row, hence one of the equations is dependent on the other two and should be eliminated.

The KVL equations have the form $\mathbf{M}\mathbf{v} = \mathbf{0}$. For example, (21) is written as

$$
\begin{array}{c}
 \\
\text{ⓐ} \\
\text{ⓑ} \\
\text{ⓒ}
\end{array}
\begin{array}{cccc}
① & ② & ③ & ④
\end{array}
\begin{bmatrix}
1 & 1 & 1 & 0 \\
0 & 0 & -1 & 1 \\
1 & 1 & 0 & 1
\end{bmatrix}
\begin{bmatrix}
v_1 \\ v_2 \\ v_3 \\ v_4
\end{bmatrix} = \mathbf{0}
\tag{21}
$$

where the rows correspond to the loops and the columns correspond to the branches. Note that row 3 is equal to the sum of row 1 and row 2.

Therefore (21c) is dependent on (21a) and (21b). Thus, we have two independent KCL equations in (20) and two independent KVL equations in (21).

Obviously we should not have to list all of the KCL and KVL equations and then seek out the maximum number of independent equations from this list. There must be a more systematic approach. There is, and it is presented in the next section.

3.6.1 INDEPENDENCE OF THE KCL EQUATIONS

Given an n-node-oriented graph that is connected and not hinged, then

the n KCL node equations of the graph are dependent, and any set of $n-1$ KCL node equations is independent.

Thus, to obtain the maximum number of independent KCL equations, we simply apply KCL to any $n-1$ of the n nodes.

The proof of the above statement is rather simple. First, a branch is connected between two nodes with its branch current leaving one node $(+i)$ and entering the other $(-i)$. Thus, when we sum the KCL equations over all of the nodes we obtain

$$\sum_{j=1}^{b} [(+i_j) + (-i_j)] \equiv 0$$

where b is the number of branches. This result is clearly seen in the matrix $\mathbf{A}_n$ in (20). Since every branch is connected between two nodes and the current in the branch leaves one node and enters the other, then each column contains a $+1$ and a -1. For example, branch 1 is connected between nodes ⓐ and ⓒ and the current leaves node ⓒ and enters node ⓐ, so that a $+1$ appears in position $(3, 1)$ and a -1 in position $(1, 1)$ of the matrix. Thus the sum of the rows of $\mathbf{A}_n$ are always zero so that all n-node equations form a dependent set. However, if we eliminate any one of the rows of $\mathbf{A}_n$, then the remaining rows will be independent as long as the node equation that was eliminated is connected to the rest of the circuit. For example, if we eliminate row ⓐ, then column 1 has only a $+1$ in position $(3, 1)$ and zeros elsewhere, and column 2 has only a -1 in position $(2, 2)$ and zeros elsewhere, because branches 1 and 2 are both connected to node ⓐ. There is no combination of rows ⓑ and ⓒ that will eliminate the nonzero elements in columns 1 and 2. Thus, rows ⓑ and ⓒ are independent. Similarly, we can show that rows ⓒ and ⓑ, and rows ⓐ and ⓒ are independent. This argument can be repeated for any subset of nodes. *Thus, in an n-node-connected circuit any n − 1 KCL node equations form an independent set.*

3.6.2 INDEPENDENCE OF THE KVL EQUATIONS

Given a planar circuit

(a) **There are $b - n + 1$ meshes.**
(b) **The $b - n + 1$ KVL mesh equations are independent.**
(c) **All other KVL loop equations are dependent on the mesh equations.**

The above result ties everything into a neat package because now we have a total of b independent topological equations and two simple procedures for writing down these equations.

To prove (a) we use induction. A single branch has two nodes and forms no meshes.[2] It has $b - n + 1$ meshes where $b - n + 1 = 0$. Each time we add a branch to the circuit without adding another node we form another mesh. If we add a branch and a node we form no new meshes. Hence, there are $b - n + 1$ meshes in a connected circuit. This result is illustrated in Figure 3.17.

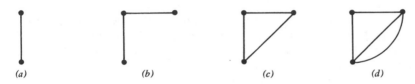

(a)   (b)   (c)   (d)

Figure 3.17   (a) $b = 1$, $n = 2$, $b - n + 1 = 0$ meshes.
                (b) $b = 2$, $n = 3$, $b - n + 1 = 0$ meshes.
                (c) $b = 3$, $n = 3$, $b - n + 1 = 1$ mesh.
                (d) $b = 4$, $n = 3$, $b - n + 1 = 2$ meshes.

To prove (b) we note that two meshes can share branches. If we sum the KVL mesh equations the voltage drops associated with branches common to two meshes will cancel. However, given any collection of meshes it is impossible for all branches to be common to two meshes. There must be some exterior branches belonging to only one mesh. These mesh voltages cannot cancel when the KVL equations for the meshes are summed. For example, in Figure 3.16 the mesh equations, (21a) and (21b), in the matrix form $Mv = 0$ are shown below.

$$
\begin{array}{c}
 \\
\text{ⓐ} \\
\text{ⓑ}
\end{array}
\begin{array}{cccc}
① & ② & ③ & ④ \\
\end{array}
\begin{bmatrix}
1 & 1 & 1 & 0 \\
0 & 0 & -1 & 1
\end{bmatrix}
\begin{bmatrix}
v_1 \\ v_2 \\ v_3 \\ v_4
\end{bmatrix} = 0
$$

---

[2] We exclude the possibility of the two terminals of a branch being shorted together since this situation is of no interest in circuit analysis.

Note that columns 1, 2, and 4 have only one nonzero elements because they correspond to exterior branches that are contained in one and only one mesh. There is no combination of rows of M that will result in all zeros in a row, since there is no way that the variables v_1 and v_2 (exterior branch voltages) can be eliminated from the first equation. Similarly, the variable v_4 cannot be eliminated from the second equation since it is an exterior branch and has zeros elsewhere in column 4 that corresponds to branch 4.

Finally, (c) is rather trivial since any loop in a planar network must contain one or more meshes. In Figure 3.16 loop 124 is the sum of meshes 123 and 34. Thus (21c) is equal to (21a) plus (21b).

3.7 CIRCUIT–CIRCUIT MODEL—CIRCUIT ANALYSIS

Let us put together the concepts in this chapter to analyze the following simple circuit. A variable intensity light circuit is shown in Figure 3.18. A 3-V battery is connected to a 1-kΩ potentiometer, which is a 1-kΩ resistor with a wiper arm (terminal c). The light is connected between the negative terminal of the battery and the wiper arm c. To find our circuit model let us assume that the light bulb has the characteristic shown in Figure 3.19. The smoke ball is to remind us that all physical devices have a power rating that must not be exceeded. One of the purposes of analysis is to check this condition. Let us proceed to model this bulb with a 50-Ω linear resistor as shown by the dashed line in Figure 3.20. If the wiper on the potentiometer is three quarters of the distance to terminal b in Figure 3.18, then, assuming a linear relation between wiper distance and resistance, we arrive at the circuit model in Figure 3.21. In Figure 3.22 we assign branch numbers, node numbers, and polarities to the branches. Note that there are

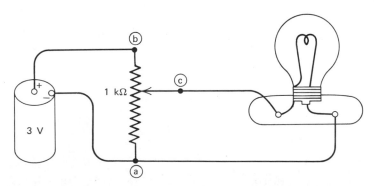

Figure 3.18   Variable intensity light.

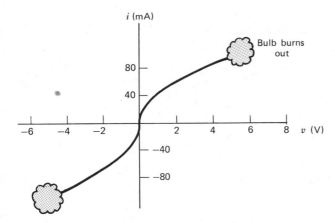

Figure 3.19 $i\text{-}v$ characteristic of a 2-V 60-mA light bulb.

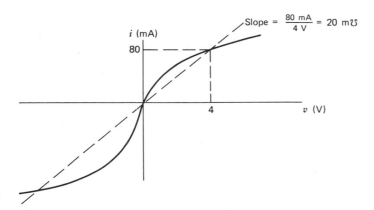

Figure 3.20   Linear modeling of the light bulb.

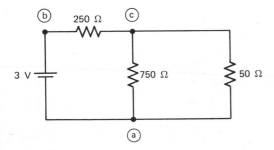

Figure 3.21   Circuit model.

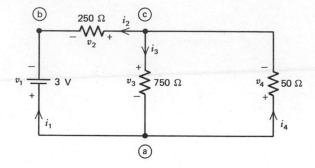

Figure 3.22   Assignment of polarities.

three nodes and four branches. The two independent KCL equations are

$$\text{node } \textcircled{b} \; : \; -i_1 - i_2 \qquad\qquad = 0 \qquad\qquad (22a)$$
$$\text{node } \textcircled{c} \; : \qquad\quad i_2 + i_3 - i_4 = 0 \qquad\qquad (22b)$$

The independent KVL equations for the meshes are

$$v_1 - v_2 + v_3 = 0 \qquad\qquad (23a)$$
$$- v_3 - v_4 = 0 \qquad\qquad (23b)$$

Finally the four branch constraints are

$$v_1 = -3 \text{ V}, \qquad\qquad (24)$$
$$v_2 = 250 \, i_2 \qquad\qquad (25)$$
$$v_3 = 750 \, i_3 \qquad\qquad (26)$$

and

$$v_4 = 50 \, i_4 \qquad\qquad (27)$$

Note that $v_1 = -3$ V since it was assigned opposite to the actual battery polarity. Thus, we have four branches, eight unknowns, and eight equations to solve. Substituting (24) to (27) into (23) we obtain

$$- 250 \, i_2 + 750 \, i_3 \qquad = 3 \text{ V} \qquad\qquad (28)$$
$$- 750 \, i_3 - 50 \, i_4 = 0 \qquad\qquad (29)$$

Use (22), (28) and (29) to eliminate i_1 and i_3, then

$$- 1000 \, i_2 + 750 \, i_4 = 3 \text{ V} \qquad\qquad (30)$$
$$+ 750 \, i_2 - 800 \, i_4 = 0 \qquad\qquad (31)$$

Eliminating i_2 we obtain

$$750 \, i_4 - 1000 \left(\frac{800}{750}\right) i_4 = 3 \text{ V}$$

or

$$-\frac{950}{3} i_4 = 3 \text{ V} \tag{32}$$

Therefore,

$$i_4 = -9.47 \text{ mA} \tag{33}$$

Back substitution yields

$$
\begin{array}{ll}
i_1 = \;\;\;\; 10.1 \text{ mA} & v_2 = -2.53 \text{ V} \\
i_2 = -10.1 \text{ mA} & v_3 = \;\;\; 0.47 \text{ V} \\
i_3 = \;\;\; 0.63 \text{ mA} & v_4 = -0.47 \text{ V}
\end{array} \tag{34}
$$

The power in the light

$$p_4 = i_4 v_4 = 4.45 \text{ mW} \tag{35}$$

If the bulb is lit for one hour, then the energy dissipated by the light is

$$w_4 = (4.45 \text{ mW})(1 \text{ h}) = 0.00445 \text{ watt-hour} = 16 \text{ J} \tag{36}$$

where the joule has units of watt-seconds. The power in the battery is

$$p_1 = v_1 i_1 = -30.3 \text{ mW} \tag{37}$$

which, in our standard reference system, means that the battery is supplying energy to the circuit. The power in the potentiometer is

$$p_2 + p_3 = v_2 i_2 + v_3 i_3 = 25.85 \text{ mW} \tag{38}$$

Note that $\Sigma_{i=1}^4 p_i = 0$, as we might expect from the principles on which Kirchhoff's laws were formulated. A rigorous proof of this result is given in Chapter Fifteen.

3.8 REDUCTION OF THE NUMBER OF KIRCHHOFF EQUATIONS

In the above example we assigned a different current and voltage to each branch; however, we could assign the currents and voltages to the circuit in a manner that reduces the number of unknowns, and hence the number of KCL and KVL equations. For example, in Figure 3.23a the branches are in series (the same current flows through the branches). Thus, in Figure 3.23a we assign a current i to each series branch in the same direction. There is no need to write the KCL equations for nodes ⓐ and ⓑ.

Similarly, in Figure 3.23b the elements are in parallel (the same voltage is across each branch). With the voltage assignment shown we do not have to write a KVL equation for meshes 12 and 23.

Finally, in Figure 3.23c it is clear that by the current assignment KCL is

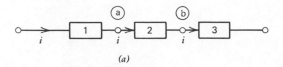

(a)

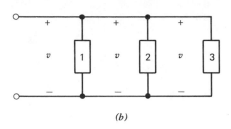

(b)

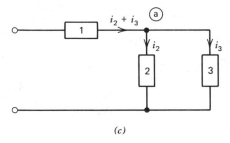

(c)

Figure 3.23   Assignment of current and voltage in order to reduce the number of Kirchhoff equations. (a) Series elements. (b) Parallel elements. (c) Series–parallel elements.

already satisfied at node ⓐ, therefore there is no need to write a KCL equation for this node.

For example, consider the circuit in Figure 3.24a. The circuit has five branches, three nodes, and three meshes. Thus, the 10 equations required would be, 2 KCL node equations, 3 KVL equations for the meshes and the five branch constraints. However, this number can be reduced considerably by a judicious assignment of variables. For example, in Figure 3.24b the voltage source current is assigned by mentally summing the current at node ⓐ. Also, the voltage v_1 has been assigned to both the parallel 5-Ω and 10-Ω resistors, and the voltage v_3 has been assigned to both the parallel current source and the 8-Ω resistor by mentally writing the KVL equations. Thus, only 1 KCL node equation and 1 KVL mesh equation remain.

$$-i_1 - i_2 + i_3 = 0.5 \text{ A} \qquad (39)$$

and

$$v_1 + v_3 = 10 \text{ V} \qquad (40)$$

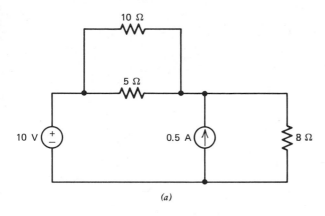

(a)

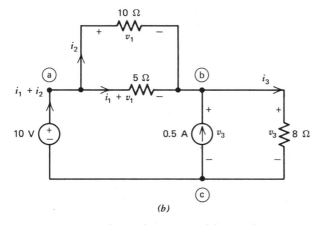

(b)

Figure 3.24   Reducing variables by judicious assignment.

Note that the current source branch constraint and the voltage source branch constraint have already been used in the above equations, hence there remains only the three resistor branch constraints.

$$v_1 = 5i_1 \tag{41a}$$

$$v_1 = 10i_2 \tag{41b}$$

and

$$v_3 = 8i_3 \tag{41c}$$

Thus we have only 5 equations in five unknowns, rather than 10 equations.

Upon the substitution of (41) into (39) we obtain two equations in two unknowns.

$$-\frac{v_1}{5\,\Omega} - \frac{v_1}{10\,\Omega} + \frac{v_3}{8\,\Omega} = \frac{1}{2}\,\text{A}$$

$$v_1 + v_3 = 10\text{ V}$$

or

$$\begin{bmatrix} -\dfrac{3}{10} & \dfrac{1}{8} \\[2mm] 1 & 1 \end{bmatrix} \begin{bmatrix} v_1 \\[1mm] v_3 \end{bmatrix} = \begin{bmatrix} 0.5 \\[1mm] 10 \end{bmatrix} \tag{42}$$

If we multiply the first equation by 10/3 and add it to the second equation we obtain

$$\begin{bmatrix} -\dfrac{3}{10} & \dfrac{1}{8} \\[2mm] 0 & \dfrac{34}{24} \end{bmatrix} \begin{bmatrix} v_1 \\[1mm] v_3 \end{bmatrix} = \begin{bmatrix} 0.5 \\[2mm] \dfrac{35}{3} \end{bmatrix} \tag{42b}$$

so that

$$v_3 = \frac{140}{17}\text{ V} \tag{43}$$

the substitution of (43) into the first equation of (42) yields

$$v_1 = \frac{30}{17}\text{ V} \tag{44}$$

and the substitution of (43) and (44) into (41) yields

$$i_1 = \frac{6}{17}\text{ A} \tag{45a}$$

$$i_2 = \frac{3}{17}\text{ A} \tag{45b}$$

$$i_3 = \frac{35}{34}\text{ A} \tag{45c}$$

To check the above answers, substitute (45) into (39), that is, our solutions must always satisfy Kirchhoff's laws. If equality is not achieved, then an error has been made.

The power delivered by the voltage source is

$$(10\text{ V})(i_1 + i_2) = \frac{90}{17}\text{ W} \tag{46a}$$

and the power delivered by the current source is

$$\left(\frac{1}{2}A\right)(v_3) = \frac{70}{17}W \tag{46b}$$

Note that both of these sources *are delivering energy* to the circuit since $i_1 + i_2$ and v_3 are defined opposite to the standard reference system.

3.9 SCALING LINEAR RESISTIVE CIRCUITS

The Kirchhoff equations are always linear and algebraic. Therefore, if our branch constraints are linear and algebraic, then our circuit is described by a system of linear-algebraic equations. In such a case the homogeneity property defined in Appendix A can be put to good use. For example, there is a simple scaling operation that can reduce all of the currents in the circuit by the factor a and leave the voltages unchanged. For any given circuit we have a set of KCL equations

$$\mathbf{A}_n \mathbf{i} = 0 \tag{47}$$

a set of loop equations

$$\mathbf{Mv} = 0, \tag{48}$$

and a set of branch constraints, for example,

$$v_i = R_i i_i \tag{49}$$

Now if we divide all currents by $1/a$ but leave the voltages unchanged, then the above equations become

$$\mathbf{A}_n\left(\frac{1}{a}\mathbf{i}\right) \tag{47'}$$

$$\mathbf{Mv} = 0 \tag{48'}$$

and

$$v_i = (aR_i)(i_i/a) \tag{49'}$$

This result is applied to the circuit in Figure 3.24 with $a = 1000$. The scaled circuit is illustrated in Figure 3.25. Note that the current source is now 0.5 mA and each resistor has been increased by 1000. The voltages in this circuit are unchanged, that is, $v_1 = 30/17$ V and $v_3 = 140/17$ V. However, $i_1 = 6/17$ mA, $i_2 = 3/17$ mA and $i_3 = 35/34$ mA. Also, note that the power for every element is reduced by $1/a$, since v is unchanged but i is reduced by i/a.

Another useful scaling operation is demonstrated in the next example. Suppose that we have a 120-W light and a 1200-W appliance plugged into an electrical wall outlet with a source voltage of 120 V rms, 60 Hz. Let the

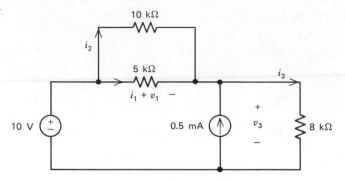

Figure 3.25   Scaled circuit.

resistance of the source and wires connecting the source to the wall outlet be 0.5 Ω. This circuit model is shown in Figure 3.26a. Let us assume that the bulb and appliances can be modeled by linear resistors R_1 and R_2, respectively, in Figure 3.26b. The circuit equations are

$$v_1 + v_3 = 120\sqrt{2}\sin 377t$$
$$v_3 - 0.5(i_1 + i_2) = 0$$
$$v_1 - R_1 i_1 = 0 \qquad\qquad (50)$$
$$v_1 - R_2 i_2 = 0$$

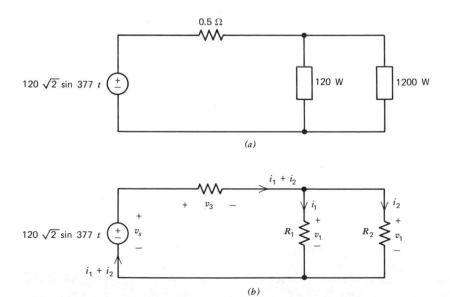

Figure 3.26   Analysis example.

This time let us multiply the circuit equations by a factor b so that (50) becomes

$$(bv_1) + (bv_3) = b\ 120\sqrt{2}\sin 377t$$
$$(bv_3) - 0.5(bi_1 + bi_2) = 0$$
$$(bv_1) - R_1(bi_1) = 0 \tag{51}$$
$$(bv_1) - R_2(bi_2) = 0$$

The resistor values are unchanged, but the homogeneity property yields the result that *if we scale every independent source by b, all of the branch currents and branch voltages will also be scaled by b.* The power is increased by b^2.

An additional useful result is the following. *In a linear resistive network if all the independent source waveforms are identical and have the same phase, then the time dependence can be eliminated from the equations.* For example, in the above problem the responses must all have a time dependence of the form $\sin 377t$; therefore, write each branch current and voltage in the form

$$i_j(t) = I_{jrms}\sqrt{2}\sin 377t \tag{52}$$

and

$$v_j(t) = V_{jrms}\sqrt{2}\sin 377t \tag{53}$$

respectively where j ranges over the branches. On the substitution of (52) and (53) into (51) and on the elimination of $\sqrt{2}\sin 377t$ from each term we obtain

$$V_{1rms} + V_{3rms} = 120$$
$$V_{3rms} - 0.5(I_{1rms} + I_{2rms}) = 0$$
$$V_{1rms} - R_1 I_{1rms} = 0 \tag{54}$$
$$V_{1rms} - R_2 I_{2rms} = 0$$

The equivalent circuit *model* for (54) is illustrated in Figure 3.27. This circuit is more convenient to analyze and yields

$$I_{1rms} = \frac{120\ W}{120\ V} = 1\ A \tag{55}$$

so that

$$R_1 = V_{1rms}/1\ A = 120\ \Omega \tag{56}$$

Similarly,

$$I_{2rms} = \frac{1200\ W}{120\ V} = 10\ A \tag{57}$$

so that

$$R_2 = 12\ \Omega \tag{58}$$

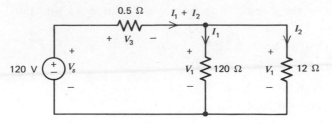

Figure 3.27   Model for rms variables.

Substitute (56) and (58) into (54) and solve for the rms voltages and currents in the circuit. Note that even though the response waveforms are sinusoidal we have reduced the analysis to a set of equations containing only the rms values by using the homogeneity property of linear algebraic equations.

In the next section, scaling is used as a very interesting aid in analyzing a very special and important type of circuit structure.

3.10 LADDER NETWORKS

A ladder network has the structure shown in Figure 3.28. If the ladder network consists of linear elements then the homogeneity property can be used to develop a simple analysis procedure for the ladder circuit. The

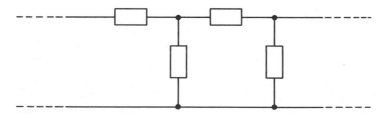

Figure 3.28   Ladder structure.

procedure is so simple that a computer program can easily be written to carry out the operations. To develop the algorithm consider the circuit in Figure 3.29. The procedure is as follows:

1. Assume 1 A flows in the last branch of the ladder, that is,

$$i_5 = 1 \text{ A}$$

2. Compute the voltage across the branch,

$$v_5 = (4 \text{ } \Omega)(1 \text{ A}) = 4 \text{ V}$$

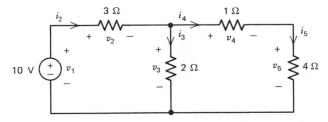

Figure 3.29 Ladder example.

3. Sum currents at the node

$$i_4 = i_5 = 1 \text{ A}$$

4. Compute the voltage drop in the series arm,

$$v_4 = (1 \, \Omega)(1 \text{ A}) = 1 \text{ V}$$

5. Compute the voltage in the next shunt arm by KVL,

$$v_3 = v_4 + v_5 = 5 \text{ V}$$

6. Compute the current in the shunt arm,

$$i_3 = 5 \text{ V}/2 \, \Omega = 2.5 \text{ A}$$

7. Use KCL to compute the current in the next series arm

$$i_2 = i_3 + i_4 = 3.5 \text{ A}$$

8. Compute the voltage drop in the series arm

$$v_2 = (3 \, \Omega)(3.5 \text{ A}) = 10.5 \text{ V}$$

9. Use KVL to compute the voltage in the next shunt arm,

$$v_1 = v_2 + v_3 = 15.5 \text{ V}$$

This is the input voltage, therefore the algorithm stops.

10. Since the input voltage is not 15.5 V but 10 V we need to scale the input source

$$\text{scale factor} = \frac{\text{actual voltage of source}}{\text{computed voltage of source}} = \frac{10}{15.5}$$

11. Since the circuit is linear and if the input is scaled by the above scale factor, then the responses must also be scaled by this factor, that is

$$v_2 = \frac{10}{15.5}(10.5 \text{ V}) \qquad v_3 = \left(\frac{10}{15.5}\right)(5 \text{ V})$$

$$v_4 = \left(\frac{10}{15.5}\right)(1 \text{ V}) \qquad v_5 = \left(\frac{10}{15.5}\right)(4 \text{ V})$$

and

$$i_2 = \left(\frac{10}{15.5}\right)(3.5 \text{ A}) \qquad i_3 = \left(\frac{10}{15.5}\right)(2.5 \text{ A})$$

$$i_4 = \left(\frac{10}{15.5}\right)(1 \text{ A}) \qquad i_5 = \left(\frac{10}{15.5}\right)(1 \text{ A})$$

Note that in the above example steps 3 to 6 can be done iteratively on a computer until the source branch is reached, then the scaling can be done.

3.11 PROPAGATION TIME AND RADIATION

In formulating the laws for lumped circuit analysis we have neglected spatial dimensions. It is only fitting that we conclude this chapter with a discussion of spatial effects. For example, consider the circuit in Figure 3.30 in which two circuits are connected by a long pair of wires. Suppose that circuit N_A initiates a signal v_A, which is to be received by circuit N_B.

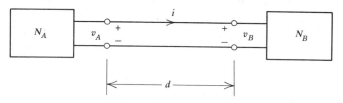

Figure 3.30   Circuit elements connected by a pair of wires.

You may want to assume that the wires are merely short circuits, hence $v_A(t) = v_B(t)$. This assumes that $v_A(t)$ travels down the line with infinite velocity. Such is not the case. For example, the speed of light travels 3×10^{10} cm/s. If we suppose that $v_A(t)$ propagates with this velocity and that $d = 100$ cm, then it takes 3.33 ns for the signal to reach N_B. If all of the voltages, currents, and charges in the circuit remain essentially constant during this time [e.g., $v_A(t_1) \approx v_A(t_1 + 3.33 \times 10^{-9})$], then we can neglect the propagation time of the signals and state that $v_A(t) = v_B(t)$, assuming that the wires have zero resistance to the current flow.

In wave theory the equation that relates the velocity of a wave front to its frequency and wavelength is

$$v = f\lambda$$

where v (cm/s) is the velocity of the wave, f is its frequency (cycles/s), and λ is its wavelength (cm). Thus, if $v = 3 \times 10^{10}$ cm/s, and if

(a) $f = 10^3$, then $\lambda = 3 \times 10^7$ cm.
(b) $f = 10^6$, then $\lambda = 3 \times 10^4$ cm.
(c) $f = 10^9$, then $\lambda = 30$ cm.

In circuit theory, as a rule of thumb, we conclude that if the dimensions of our circuit are much less than $\lambda/4$, where the value of λ is determined as above and f is the frequency of the signal in the circuit, then the propagation time may be neglected.

As the dimensions of the circuit approach one quarter of the wavelength of the signals, then we must not only include the propagation time of signals, but we must also consider the fact that wires can become good radiators. For example, Figure 3.31 illustrates a dipole antenna whose purpose is to radiate electromagnetic waves. For example, if we want to transmit a 1-MHz radio signal efficiently then the dipole should be one quarter the length of the electric wave. Thus, the length of one arm of our dipole antenna should be 75 m. This radiation is not always desirable, and to prevent it high frequency signals are often guided by coaxial cable or waveguides.

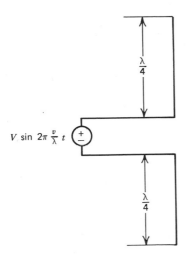

Figure 3.31   Half-wave dipole antenna.

When we include spatial effects in the determination of the characteristics of the electrical device model, we say that the element is distributed. The transmission line and the antenna are examples of distributed elements. Maxwell's field theory equations can be used to arrive at circuit models for these elements. For example, the power lost due to radiation can be accounted for by a resistor in the circuit that models the antenna, and the transmission line can be represented by a box with rather complicated trigonometric terminal characteristics. Thus, all of these phenomena can be put neatly into our lumped circuit theory package. But due to the introductory nature of this book we will not consider the modeling of these distributed elements. Suffice to say that the circuit designer must be aware of these effects so that they may be included in his or her mathematical model when necessary, or else the analysis of the circuit model will not be in agreement with experimental observations. For example, when the circuit contains high frequency waveforms it is important to keep the interconnecting wires between components as short as possible in order to neglect these effects. Microwave integrated circuits have been built up to 4 GHz in which these distributed effects have been neglected.[3]

REFERENCES

1. Van Valkenburg, M. E. *Network Analysis*, 3rd edition, Prentice-Hall, 1974, Chapters 1–3.
2. Friedland, B., O. Wing, and R. Ash. *Principles of Linear Networks*, McGraw-Hill, 1961, *Chapters 1, 2, 4.*
3. *Desoer, C. A. and E. S. Kuh. Basic Circuit Theory*, McGraw-Hill, 1969, Chapters 1, 2.
4. Balabanian, N. *Fundamentals of Circuit Theory*, Allyn and Bacon, 1961, Chapters 1, 2.
5. Cruz, J. B. Jr. and M. E. Van Valkenburg. *Signals in Linear Circuits*, Houghton Mifflin, 1973, Chapters 1–6.
6. Skilling, H. H. *Electrical Engineering Circuits*, Wiley, 1965.
7. Skilling, H. H. *Electrical Networks*, Wiley, 1974.

PROBLEMS

1. Calculate the power in each of the elements in Figure P3.1 and state whether energy is being absorbed by the element or the element is supplying energy to the circuit (not shown) connected to its terminals.
2. Calculate the energy absorbed by the elements in Figure P3.2 in the time interval $(0, 2\pi/\omega)$. What does a negative sign indicate?

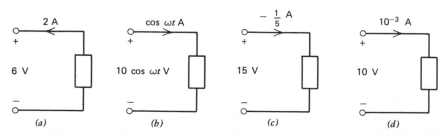

Figure P3.1

---

[3] M. Caulton and H. Sobol, "Microwave-Integrated-Circuit Technology—A Survey," *IEEE Journal of Solid-State Circuits*, Vol. SC-5, December 1970, pp. 292–302.

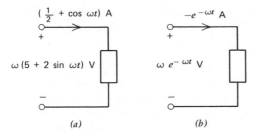

Figure P3.2

3. The President has ordered that 100 lights rated at 100 W each be turned off one hour earlier each night in the White House. If the electrical rate is $0.03/kWh, how much money has the President saved the taxpayers in 30 days?

4. For the assigned direction of v and i in Figure P3.4, write the branch constraint equation for each branch.

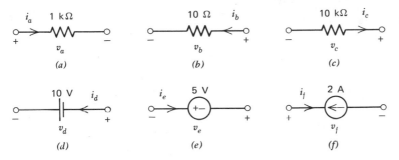

Figure P3.4

5. State if the following resistors are linear (L) or nonlinear (NL), time invariant (TI) or time varying (TV).

(a) $v = 10i + i^3$ (c) $i = -0.1v$
(b) $v = (2 + \cos 2\pi t)i$ (d) $v = (\sin \pi t)i + 10i^3$

6. In Problem 5 let $i = 1$ A and calculate the power dissipated in the resistors. The voltage and current have been assigned according to the standard reference system. Find the energy delivered or absorbed by these elements in the time interval $[0, 1s]$. State whether energy is absorbed by the element or delivered by the element.

7. Determine the incremental resistance of the following resistors for $i = 1$ A, and 3 A.

(a) $v = 2i + \dfrac{1}{3}i^3$ (b) $i = v^5$

8. Find the current and the incremental resistance of the following nonlinear resistor for $v = -1, 0, 0.2, 0.5, 0.6, 0.7, 0.8, 0.9$, and 1.0 V.

$$i = 10^{-12}(e^{40v} - 1)$$

9. Repeat Problem 8 for $i = 10^{-9}(e^{40v} - 1)$.

10. For the circuit shown in Figure P3.10: (a) determine the number of nodes, (b) determine the number of loops, (c) write KCL for each node, (d) write KVL for each loop, (e) form a collection of independent KCL equations (include the maximum possible number of equations) and show that these equations are independent, and (f) explain why the mesh equations are independent and show that the other loop equations depend on the mesh equations.

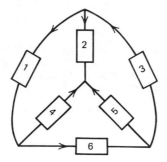

Figure P3.10

11. Repeat Problem 10 for the circuit in Figure P3.11.

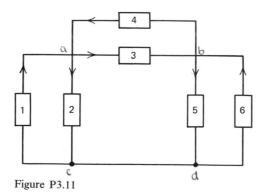

Figure P3.11

12. In the nonplanar graph in Figure 3.13 (not Figure P3.13) suppose that the branch connected between nodes ⓑ and ⓒ is removed. Is the remaining graph planar? (Try to redraw the graph). What if the branch between nodes ⓐ and ⓒ is removed instead of the branch between nodes ⓑ and ⓒ?

13. Are the KVL mesh equations and the KCL node equations below for the circuit in Figure P3.13 correct? Correct the incorrect equations.

$$v_1 + v_2 + v_4 = 0 \qquad i_1 + i_2 - i_6 = 0$$
$$v_4 + v_3 + v_5 = 0 \qquad i_2 + i_3 - i_4 = 0$$
$$v_6 - v_3 - v_2 = 0 \qquad i_3 + i_5 + i_6 = 0$$

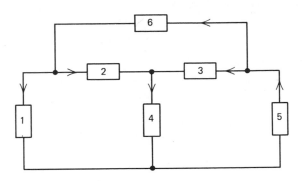

Figure P3.13

14. Find the voltage v_1 and the current i_1 in the circuits in Figure P3.14.

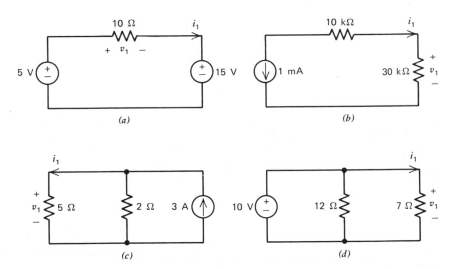

Figure P3.14

15. The voltages and currents in Figure P3.15 are defined such that there are four unknowns, v_1, v_2, i_1, and i_2. Write the four equations needed to obtain a solution.

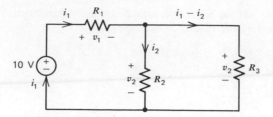

Figure P3.15

16. The voltages and currents in Figure P3.16 are defined such that we have six unknowns, v_2, v_3, v_4, v_x, i_1 and i_2. Four of the circuit equations needed to obtain a solution are listed below. What two equations are missing?

$$v_2 = R_2 i_1 \qquad v_4 = -R_4 i_2$$
$$v_3 = R_3 i_2 \qquad -10 + v_2 + v_x = 0$$

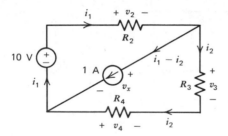

Figure P3.16

17. Define the voltage and current variables in the circuits in Figure P3.17 in order to minimize the number of KCL and KVL equations that must be written in order to solve the circuit equations. Find the branch voltages and currents in each circuit.

18. Typically, analog volt-ohmmeters have a d'Arsonval meter movement that consists of a coil suspended between the poles of a permanent horseshoe

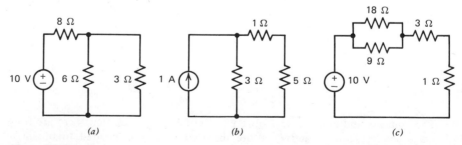

(a)   (b)   (c)

Figure P3.17

magnet. When a current is passed through the coil a magnetic field is set up around the coil that interacts with magnetic field from the permanent magnet causing the coil to rotate in a direction such that the opposite poles of the two fields are aligned. A spring is attached to the coil to oppose this rotation and to cause the deflection angle of the coil to be linearly proportional to the current in the coil. A pointer to indicate this deflection and a calibrated scale are added to complete the meter movement. If $50 \mu A$ of current are required to give a full-scale deflection, we say that the meter movement is 20,000 ohms/volt, that is, we need a total of $20,000 \Omega$ (including the resistance of the meter coil) in series with a 1-V source to yield a full-scale deflection. In Figure P3.18 we indicate the circuit of a voltmeter that is to be designed to operate over three ranges (0 to 1 V, 0 to 10 V, and 0 to 100 V). If R_m, the resistance of the coil, is 500Ω, compute the resistances R_1, R_2, and R_3 such that full-scale deflection is achieved on each of the ranges. For example, when 100 V is applied to the terminals a to b of the voltmeter, the current i_m should be $50 \mu A$.

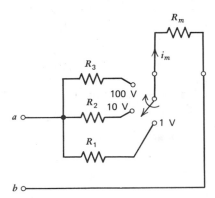

Figure P3.18

19. The voltmeter in Problem 18 is used to the measure the voltage v_x in the circuit below. What value of v_x does the meter indicate on the 1-V scale? What is the value of v_x with the meter disconnected?
20. In Problem 19 what value of v_x does the meter indicate on the 10-V scale?
21. Repeat Problem 19 with the 10-kΩ resistor replaced with a 1-kΩ resistor.

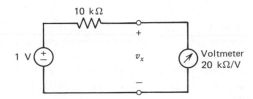

Figure P3.19

22. In Figure P3.22 the circuit of an ammeter is indicated. If the meter movement is 20 kΩ/V ($i_m = 50\ \mu$A yields full-scale deflection), calculate the shunt resistors R_1, R_2, and R_3 such the instrument can be used to measure currents in the 0 to 1 mA, 0 to 10 mA, and 0 to 100 mA range. What is the voltage drop across the meter at full-scale deflection? Assume $R_m = 500\ \Omega$.

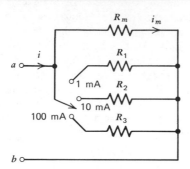

Figure P3.22

23. A typical ohmmeter circuit is illustrated in Figure P3.23. Express R_x as a function of R_1, R_2, R_m, and the current i_m in this ohmmeter circuit. If the meter resistance $R_m = 4000\ \Omega$ and $i_m = 50\ \mu$A for full-scale deflection, calculate R_1 and R_2 such that $i_m = 50\ \mu$A when $R_x = 0$ and $i_m = 25\ \mu$A (half-scale deflection) when $R_x = 100\ \Omega$.

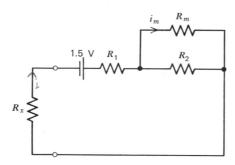

Figure P3.23

24. Repeat Problem 23 for $R_x = 1$ kΩ instead of 100 Ω.

25. In Figure 3.24 let the voltage of the voltage source be 3 V instead of 10 V. Solve for the branch currents and voltages and find the power delivered by each source.

26. Redraw the circuit in Figure P3.26 with the elements scaled so that the currents in all the branches of the circuit are reduced by a factor of 1000, but the voltages in the circuit are unchanged.

27. Repeat Problem 26 for the circuit in Figure P3.16.

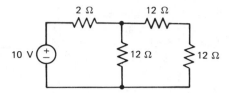

Figure P3.26

28. Write the equations for the circuit in Figure P3.28 in terms of the rms voltages and currents. Do not solve.

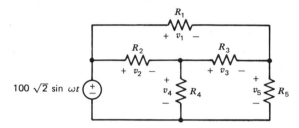

Figure P3.28

29. Figure P3.29 illustrates the model of an electrical toaster (T) and frying pan (FP) plugged into the same line that is protected by a 20-A fuse (rms current). The toaster requires 1200 W of power and the frying pan 1500 W. Will the fuse blow? Assume that the currents are in phase with the voltage.

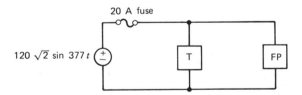

Figure P3.29

30. In Figure P3.30 find the voltage v_s.

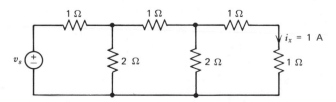

Figure P3.30

31. Use the ladder method to find the current i_x in Figure P3.30 if $v_s = 10\,\text{V}$.
32. Use the ladder method of analysis to find the branch currents i_1, i_2, i_o and voltages v_1, v_2, v_o of the circuit in Figure P3.32.

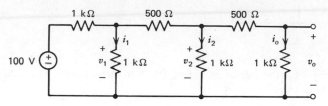

Figure P3.32

Chapter Four
Node and Mesh Methods for Analysis

In Chapter Three we saw that a circuit with b branches had $2b$ variables, and we required $2b$ independent equations in order to obtain a unique solution. Even though some shortcuts were suggested in writing the equations we still had a fairly large number of equations to solve. This chapter introduces two methods for writing circuit equations that substantially reduce the number of variables, and hence the number of equations that need to be solved. The methods are the *node method* and the *mesh method*. The node method is the more common of the two methods and is used in many computer programs that analyze electrical circuits. However, in cases in which a circuit has fewer meshes than nodes the mesh method may be desirable. These two methods are described in this chapter. Also, at the end of the chapter some additional topological relationships that exist among certain voltages and currents in the circuit graph are described. These topological properties are sometimes used to formulate the circuit equations in a computer program rather than the node method.

4.1 NODE VOLTAGES

In this section it is shown that any voltage can be expressed in terms of $n - 1$ node voltages. A *node voltage* is a voltage measured from a node to a reference node. For example, in Figure 4.1a let us choose node ⓓ as the reference node. Sometimes the reference node is called the datum (Latin

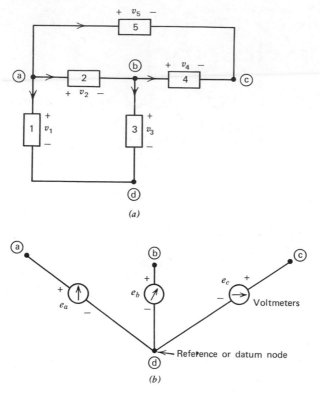

Figure 4.1 Node voltages. (*a*) Circuit. (*b*) Defining the node voltages of a circuit.

word meaning "what is given") node. With respect to this datum node there are $n - 1$ remaining nodes so that there are $n - 1$ node voltages in the circuit as shown in Figure 4.1*b*. In Figure 4.1*b* an ideal voltmeter (requires zero current) is shown connected between each node and the reference node. The branches in Figure 4.1*a* are not illustrated in Figure 4.1*b* for simplicity. The node voltages are labeled e_a, e_b, and e_c. The symbol e is used as opposed to v to point out that these voltages are node voltages and not necessarily branch voltages. For example, no branch is connected between node © and node ⓓ, but there is a node voltage between © and ⓓ. The node voltages are always assumed positive with respect to the reference node.

It is important to note that the voltmeters form a path from the reference node to every other node in the circuit, but form no loops. Since no loops are formed, no node voltage can be expressed as a function of the other node voltages, but every other voltage in the circuit can be expressed as a linear combination of the node voltages. For example, in Figure 4.1*a* we obtain, for the given reference directions,

$$v_1 = e_a$$
$$v_2 = e_a - e_b$$
$$v_3 = e_b \qquad (1)$$
$$v_4 = e_b - e_c$$
$$v_5 = e_a - e_c$$

Since a branch is connected between two nodes, *every branch voltage can be expressed in terms of the node voltages* by means of Kirchhoff's voltage law. This result is used later to develop a simplified approach for writing circuit equations. The method is called, appropriately, the node method.

Let us make one final observation. The $n - 1$ node voltage measurements yielded sufficient information to allow us to express every branch voltage in terms of these $n - 1$ measurements. However, *the $n - 1$ node voltages form a necessary set of voltages also*, because if we had only $n - 2$ node voltages or less then there is a branch connected to the node that is neither the reference node nor one of $n - 2$ nodes whose voltages are chosen as the variables. The voltage across this branch could not be expressed in terms of the $n - 2$ variables. For example, without the voltage e_c in Figure 4.1 we would not be able to compute v_4 and v_5 in equations (1).

4.2 NODE METHOD

The principle of the *node method* is as follows. For an n node circuit there are $n - 1$ independent KCL equations. Also, since every branch is connected between two nodes, we can use KVL to express each branch voltage as a difference of the two node voltages between which the branch is connected. Now suppose that every unknown branch current is a function of its branch voltage. Thus, write the $n - 1$ KCL equations; express each unknown branch current in terms of its branch voltage; express the branch voltages in terms of the node voltages. We now have $n - 1$ equations, which are independent if the circuit has a solution, and $n - 1$ unknowns.

For example consider the circuit in Figure 4.2 in which the nodes are labeled with small letters. Choose node ⓒ to be the reference or datum node and define the node voltages as shown in Figure 4.3. Next we write KCL for nodes ⓐ and ⓑ.

$$i_1 + i_2 + i_3 \quad = 0$$
$$\qquad\qquad -i_3 + i_4 = 0 \qquad (2)$$

Express each of the unknown branch currents in terms of the branch voltages.

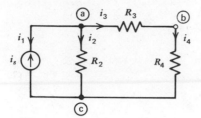

Figure 4.2   Node method example.

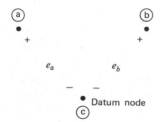

Datum node

Figure 4.3   Defining the node voltages (branches not shown).

$$-i_s + G_2v_2 + G_3v_3 \qquad = 0$$
$$\qquad\qquad -G_3v_3 + G_4v_4 = 0 \tag{3}$$

Note that the current i_s is a known quantity. Now each of the branch voltages is linearly related to the node voltages.

$$v_2 = e_a \qquad v_3 = e_a - e_b$$
$$v_4 = e_b \tag{4}$$

Thus, if we substitute (4) into (3) we obtain two equations in terms of the two node voltages.

$$-i_s + G_2e_a + G_3(e_a - e_b) = 0$$
$$\qquad -G_3(e_a - e_b) + G_4e_b = 0 \tag{5}$$

or, equivalently, we write (5) as

$$(G_2 + G_3)e_a - G_3e_b = i_s$$
$$-G_3e_a + (G_3 + G_4)e_b = 0 \tag{6}$$

The solution to Equation 6 is

$$e_a = \frac{\begin{vmatrix} i_s & -G_3 \\ 0 & G_3 + G_4 \end{vmatrix}}{\Delta} = \frac{i_s(G_3 + G_4)}{G_2(G_3 + G_4) + G_3G_4} \tag{7}$$

and

$$e_b = \frac{\begin{vmatrix} G_2 + G_3 & i_s \\ -G_3 & 0 \end{vmatrix}}{\Delta} = \frac{i_s G_3}{G_2(G_3 + G_4) + G_3 G_4} \tag{8}$$

Note that the circuit in Figure 4.2 has four branches and three nodes. If we wrote the four branch constraints, the two KCL equations and the two KVL equations, we would have eight equations in terms of the eight branch variables. The *node method* reduces the number of equations to be solved to $n - 1$, which is equal to 2 in the above example.

Now that the principle has been explained, let us consider several other examples in order to see the ease with which we can write the nodal equations. First, consider the circuit in Figure 4.4. We have labeled the nodes and let us designate node ⓓ as the reference node. Next we apply KCL to the remaining nodes using the reference that a current leaving a node is positive.

$$\begin{aligned}
\text{node } ⓐ \quad & : \quad G_3 e_a + G_2(e_a - e_b) - i_{s1} = 0 \\
\text{node } ⓑ \quad & : \quad G_2(e_b - e_a) + G_4 e_b + G_5(e_b - e_c) + i_{s1} = 0 \\
\text{node } ⓒ \quad & : \quad G_6 e_c + G_5(e_c - e_b) = 0.
\end{aligned} \tag{9}$$

Equivalently, we write

$$\begin{aligned}
(G_2 + G_3)e_a - G_2 e_b + 0 \cdot e_c &= i_{s1} \\
-G_2 e_a + (G_2 + G_4 + G_5)e_b - G_5 e_c &= -i_{s1} \\
0 \cdot e_a - G_5 e_b + (G_5 + G_6)e_c &= 0
\end{aligned} \tag{10}$$

The current i_{s1} represents the value of the independent current source and is assumed to be known.

Let us assume that $R_2 = 5\,\Omega$, $R_3 = 3\,\Omega$, $R_4 = 6\,\Omega$, $R_5 = 2\,\Omega$, $R_6 = 1\,\Omega$, and

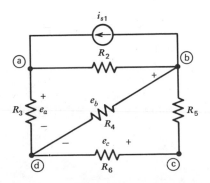

Figure 4.4   Second nodal example.

$i_{s1} = 2$ A, then (10) becomes

$$\left(\frac{1}{5} + \frac{1}{3}\right) e_a - \frac{1}{5} e_b = 2 \text{ A}$$

$$-\frac{1}{5} e_a + \left(\frac{1}{5} + \frac{1}{6} + \frac{1}{2}\right) e_b - \frac{1}{2} e_c = -2 \text{ A}$$

$$-\frac{1}{2} e_b + \left(\frac{1}{2} + 1\right) e_c = 0$$

The above set of equations can be solved by determinants or by Gaussian elimination. Let us use the latter. The augmented matrix for this set of equations is

$$\begin{bmatrix} \dfrac{8}{15} & -\dfrac{1}{5} & 0 & \vdots & 2 \\[2ex] -\dfrac{1}{5} & \dfrac{13}{15} & -\dfrac{1}{2} & \vdots & -2 \\[2ex] 0 & -\dfrac{1}{2} & \dfrac{3}{2} & \vdots & 0 \end{bmatrix}$$

which, by Gaussian elimination, reduces to

$$\begin{bmatrix} \dfrac{8}{15} & -\dfrac{1}{5} & 0 & \vdots & 2 \\[2ex] 0 & \dfrac{19}{24} & -\dfrac{1}{2} & \vdots & -\dfrac{5}{4} \\[2ex] 0 & 0 & \dfrac{45}{38} & \vdots & -\dfrac{15}{19} \end{bmatrix}$$

or

$$\frac{8}{15} e_a - \frac{1}{5} e_b = 2$$

$$\frac{19}{24} e_b - \frac{1}{2} e_c = -\frac{5}{4}$$

$$\frac{45}{38} e_c = -\frac{15}{19}$$

The solution to the above set of equations is found by back substitution.

$$e_c = -\frac{2}{3} \text{ V}$$

$$e_b = -2 \text{ V}$$

$$e_a = 3 \text{ V}$$

Now that we know the node voltages we can find any other current or voltage in the circuit. For example, the current flowing through the resistor R_2 from node ⓑ to node ⓐ is $(e_b - e_a)/R_2$ or $(-2 V - 3 V)/5 \Omega = -1$ A. The power delivered by the current source is $2 A(e_a - e_b) = 2 A[3 V - (-2 V)] = 10$ W.

Finally, we should point out that the node method is relatively easy to program on a computer. To see this, express (10) in the form

$$Ge = i_s \qquad (11)$$

where G is a $n - 1 \times n - 1$ array, e and i_s are vectors of dimension $n - 1$. For example, in the previous circuit

$$G = \begin{array}{c} \text{ⓐ} \\ \text{ⓑ} \\ \text{ⓒ} \end{array} \begin{matrix} \overset{\text{ⓔ}_a}{} & \overset{\text{ⓔ}_b}{} & \overset{\text{ⓔ}_c}{} \\ \begin{bmatrix} G_2 + G_3 & -G_2 & 0 \\ -G_2 & G_2 + G_4 + G_5 & -G_5 \\ 0 & -G_5 & G_5 + G_6 \end{bmatrix} \end{matrix} \qquad (12)$$

$$e = \begin{bmatrix} e_a \\ e_b \\ e_c \end{bmatrix} \quad (13) \qquad \text{and} \qquad i_s = \begin{array}{c} \text{ⓐ} \\ \text{ⓑ} \\ \text{ⓒ} \end{array} \begin{bmatrix} i_{s1} \\ -i_{s1} \\ 0 \end{bmatrix} \qquad (14)$$

The vector e represents the unknown node voltages. Note that a given row of G and i_s corresponds to a node in the circuit about which the currents are summed to zero. The columns of G correspond to the node voltages. Typically the nodes are numbered 1 through $n - 1$ with "0" corresponding to the reference or ground node. The arrays G and i_s are filled in as follows. Consider the connection shown in Figure 4.5. The resistor R_k is connected between nodes ⓘ and ⓙ. The current leaving node ⓘ through R_k is

$$G_k(e_i - e_j) \qquad (15)$$

and the current leaving node ⓙ through R_k is

$$G_k(e_j - e_i) \qquad (16)$$

Thus, in the array G in row i we write G_k in column i and $-G_k$ in column j. Similarly, in row j we write G_k in column j and $-G_k$ in column i as

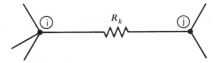

Figure 4.5 Resistor connection.

illustrated below

$$
G =
\begin{array}{c}
 \\
i \\
\\
j \\
\\
\end{array}
\begin{bmatrix}
& \overset{i}{\vdots} & & \overset{j}{\vdots} & \\
\cdots & G_k & \cdots & -G_k & \cdots \\
& \vdots & & \vdots & \\
\cdots & -G_k & \cdots & G_k & \cdots \\
& \vdots & & \vdots & \\
\end{bmatrix}
\tag{17}
$$

Note that if node ⓘ or ⓙ is the reference node then we delte that row and column since we do not write KCL for the reference node and the reference node voltage is 0, since it is measured with respect to itself.

To fill in the vector $\mathbf{i}_s$ consider the connection in Figure 4.6. The current source i_m is connected between nodes ⓘ and ⓙ. Thus, we have i_m amperes flowing into node ⓘ and leaving node ⓙ. Therefore,

$$
\mathbf{i}_s =
\begin{array}{c}
ⓘ \\
\\
\\
ⓙ \\
\end{array}
\begin{bmatrix}
i_m \\
\vdots \\
\vdots \\
-i_m \\
\end{bmatrix}
\tag{18}
$$

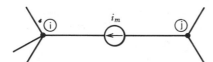

Figure 4.6 Current source connection.

Note that currents flowing into a node are positive since $\mathbf{i}_s$ is on the opposite side of the equality sign in (11).

This array technique is very simple and probably is a less-error-prone technique for writing down the nodal equations even if you do not have a computer. In this book our examples never exceed three nodal equations; however, computer programs using this approach have solved circuits with hundreds of nodes.

4.3 THE NODE METHOD WITH VOLTAGE SOURCES

In the previous section we deliberately avoided the analysis of circuits with voltage sources, because the current in a voltage source is an unknown variable that cannot be directly expressed in terms of the node voltages. However, this does not render the node method useless for circuits with voltage sources, but calls for some modifications.

Consider the circuit in Figure 4.7. Again we will call ⓓ the datum node. Also, note that we have assigned a direction and name for the current in the voltage source. Thus,

$$
\begin{array}{lll}
\text{node ⓐ} & : & G_1 e_a \qquad\qquad\quad + i_v = 0 \\
\text{node ⓑ} & : & (G_2 + G_3) e_b - G_3 e_c - i_v = 0 \\
\text{node ⓒ} & : & -G_3 e_b + (G_3 + G_4) e_c \;\; = 0
\end{array} \tag{19}
$$

Note that we have three equations and four unknowns, the fourth unknown being the current in the voltage source. However, a voltage source places a

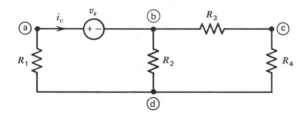

Figure 4.7   Nodal example with a voltage source.

constraint on the node voltages that is, in Figure 4.7

$$
e_a - e_b = v_s \tag{20}
$$

We now have four equations and four unknowns. Finally the current i_v can be easily eliminated by summing the equations for node ⓐ and node ⓑ, the two nodes between which the voltage source is connected. Also, e_a or e_b can be eliminated by substituting (20) into (19), yielding

$$
\begin{aligned}
(G_1 + G_2 + G_3) e_b - G_3 e_c &= -G_1 v_s \\
-G_3 e_b + (G_3 + G_4) e_c &= 0
\end{aligned} \tag{21}
$$

Thus, if some of the obvious variables are eliminated, such as the current in the voltage source, and one of the node voltages of a node to which the voltage source is connected, a voltage source actually decreases the number of nodal equations by one.

Circuits that contain voltage sources with one terminal connected to the reference node are even easier to handle. Consider the example in Figure 4.8 in which the negative terminal of the voltage source is connected to the circuit ground terminal. If we choose the reference terminal to be the

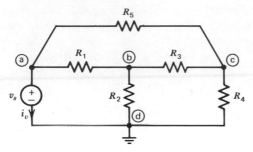

Figure 4.8 Grounded voltage source.

ground terminal of this circuit, then the node equations are

$$
\begin{array}{lll}
\text{node } \textcircled{a} & : & i_v + (G_1 + G_5)e_a - G_1 e_b - G_5 e_c \quad = 0 \\
\text{node } \textcircled{b} & : & -G_1 e_a + (G_1 + G_2 + G_3)e_b - G_3 e_c = 0 \\
\text{node } \textcircled{c} & : & -G_5 e_a - G_3 e_b + (G_3 + G_4 + G_5)e_c = 0
\end{array}
\tag{22}
$$

plus the voltage source constraint

$$
e_a = v_s \tag{23}
$$

Since one terminal of the voltage source v_s is connected to the reference node $\textcircled{d}$, we know that the voltage $e_a = v_s$, thus we make the substitution directly. Furthermore, there is no need to write the node $\textcircled{a}$ equation unless we need to obtain i_v. Thus, (22) and (23) reduce to

$$
\begin{aligned}
(G_1 + G_2 + G_3)e_b - G_3 e_c &= G_1 v_s \\
-G_3 e_b + (G_3 + G_4 + G_5)e_c &= G_5 v_s
\end{aligned}
\tag{24}
$$

Thus, whenever a circuit contains elements whose constraints are such that the current cannot be expressed as a function of the node voltages, the current appears as an unknown in the nodal equations and the branch constraints of those elements are appended to the nodal equations.

In the above example let $R_1 = 2\,\Omega$, $R_2 = 6\,\Omega$, $R_3 = 2\,\Omega$, $R_4 = 2\,\Omega$, $R_5 = 6\,\Omega$, and $v_s = 5$ V, then (24) becomes

$$
\frac{7}{6} e_b - \frac{1}{2} e_c = \frac{5}{2}
$$

$$
-\frac{1}{2} e_b + \frac{7}{6} e_c = \frac{5}{6}
$$

Solving by determinants

$$e_b = \frac{\begin{vmatrix} \dfrac{5}{2} & -\dfrac{1}{2} \\ \dfrac{5}{6} & \dfrac{7}{6} \end{vmatrix}}{\begin{vmatrix} -\dfrac{7}{6} & -\dfrac{1}{6} \\ -\dfrac{1}{2} & \dfrac{7}{6} \end{vmatrix}} = \frac{\dfrac{40}{12}}{\dfrac{40}{36}} = 3 \ V$$

and

$$e_c = \frac{\begin{vmatrix} \dfrac{7}{6} & \dfrac{5}{2} \\ -\dfrac{1}{2} & \dfrac{5}{6} \end{vmatrix}}{\dfrac{40}{36}} = \frac{\dfrac{80}{36}}{\dfrac{40}{36}} = 2 \ V$$

If we are interested in the power supplied by the voltage source then we can solve the nodal equation for node ⓐ in (22) to obtain the current delivered by the voltage source.

$$i_v = -\left(\frac{1}{2}+\frac{1}{6}\right)(5 \ V) + \frac{1}{2}(3 \ V) + \frac{1}{6}(2 \ V) = -1.5 \ A$$

The power delivered by the voltage source is $(v_s)(-i_v) = 7.5 \ W$.

4.4 MESH CURRENTS

In this section we define, for planar circuits, a current called a mesh current and show that every branch current can be uniquely defined in terms of the mesh currents. Since a planar circuit has $b - n + 1$ meshes, every planar circuit will have $b - n + 1$ mesh currents. A *mesh current* is simply defined as the current circulating in a mesh. For example, consider the three mesh circuit in Figure 4.9a. In Figure 4.9b we have drawn its graph and oriented the branches by assigning a current direction to each branch. In addition, we have defined three circulating mesh currents i_a, i_b, and i_c. Now every branch current can be expressed in terms of these mesh

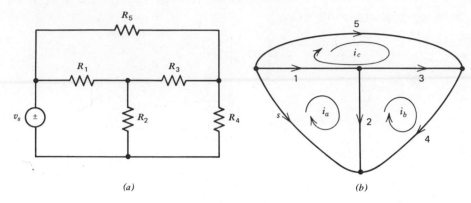

Figure 4.9   Mesh current example. (a) Circuit. (b) Oriented circuit graph.

currents by means of Kirchhoff's current law, that is,

$$
\begin{aligned}
i_s &= -i_a \\
i_1 &= i_a - i_c \\
i_2 &= i_a - i_b \\
i_3 &= i_b - i_c \\
i_4 &= i_b \\
i_5 &= i_c
\end{aligned}
\tag{25}
$$

Let us now show that these mesh currents are independent of one another. Let us write the KVL equation for each of the meshes. We obtain

$$
\begin{matrix}
\text{mesh } a: \\
\text{mesh } b: \\
\text{mesh } c:
\end{matrix}
\begin{bmatrix}
-1 & 1 & 1 & 0 & 0 & 0 \\
0 & 0 & -1 & 1 & 1 & 0 \\
0 & -1 & 0 & -1 & 0 & 1
\end{bmatrix}
\begin{bmatrix}
v_s \\ v_1 \\ v_2 \\ v_3 \\ v_4 \\ v_5
\end{bmatrix}
= \mathbf{0}
\tag{26}
$$

which we write as

$$
M\mathbf{v} = \mathbf{0}
\tag{27}
$$

Note that the three rows of the matrix M are independent since the KVL mesh equations were shown to be independent in Chapter Three. Also, note that the columns of M correspond to the branches of the circuit. For example, branch 2 is contained in mesh a and mesh b and its current flow is in the direction of i_a, but opposite to i_b thus we have a $+1$ in column 3, row a and -1 in column 3, row b. Because of the structure of this matrix

M it follows that we can use the transpose of this matrix to express the branch currents in terms of the mesh currents.

$$M^T i_\mathbf{m} = i_\mathbf{b} \tag{28}$$

that is,

$$
\begin{bmatrix}
-1 & 0 & 0 \\
1 & 0 & -1 \\
1 & -1 & 0 \\
0 & 1 & -1 \\
0 & 1 & 0 \\
0 & 0 & 1
\end{bmatrix}
\begin{bmatrix}
i_a \\
i_b \\
i_c
\end{bmatrix}
=
\begin{bmatrix}
i_s \\
i_1 \\
i_2 \\
i_3 \\
i_4 \\
i_5
\end{bmatrix}
\tag{29}
$$

Note that (29) is identical to (25). Also, the columns of M^T are independent since they are the rows of M, which tells us that the mesh currents i_a, i_b, and i_c are independent. We conclude that in a planar circuit every branch current can be uniquely expressed in terms of the $b - n + 1$ mesh currents and no less. This conclusion is utilized in the next section to develop another efficient method for writing circuit equations.

4.5 MESH METHOD

Recall that the mesh KVL equations are linearly independent and any other loop equation is dependent on the $b - n + 1$ mesh equations. In the *mesh method* we define a circulating current for each mesh as shown in Figure 4.10. The standard reference is to choose the currents circulating in the clockwise direction, but this is not necessary. With the branch polarities referenced as indicated the KVL mesh equations are

$$
\begin{aligned}
-15\,\text{V} + v_2 + v_3 &= 0 \\
-v_3 + v_4 &= 0
\end{aligned}
\tag{30}
$$

but

$$
\begin{aligned}
v_2 &= 10 i_2 \\
v_3 &= 10 i_3 \\
v_4 &= 10 i_4
\end{aligned}
\tag{31}
$$

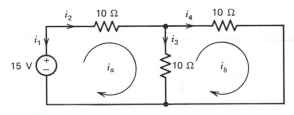

Figure 4.10   Mesh method example.

In the previous section we showed that every branch current can be expressed as a linear combination of mesh currents. In this example,

$$i_2 = i_a$$
$$i_3 = i_a - i_b \tag{32}$$
$$i_4 = i_b$$

Substitute (32) into (31) and (31) into (30), then we obtain

$$20i_a - 10i_b = 15$$
$$-10i_a + 20i_b = 0 \tag{33}$$

The solution to (33) is

$$i_a = 1 \text{ A}$$
$$i_b = 0.5 \text{ A} \tag{34}$$

Upon substitution of (34) into (32) and (31) we easily obtain the branch variables.

Now that the basic principle has been explained, let us illustrate how to write the mesh equations directly from the circuit without all of the intermediate steps. Consider the circuit in Figure 4.11:

$$
\begin{aligned}
\text{mesh } a \quad &: \quad -v_s + R_1(i_a - i_c) + R_2(i_a - i_b) = 0 \\
\text{mesh } b \quad &: \quad R_2(i_b - i_a) + R_3(i_b - i_c) + R_4 i_b = 0 \\
\text{mesh } c \quad &: \quad R_5 i_c + R_3(i_c - i_b) + R_1(i_c - i_a) = 0
\end{aligned} \tag{35}
$$

which reduces to

$$
\begin{aligned}
\text{mesh } a \quad &: \quad (R_1 + R_2)i_a - R_2 i_b - R_1 i_c = v_s \\
\text{mesh } b \quad &: \quad -R_2 i_a + (R_2 + R_3 + R_4)i_b - R_3 i_c = 0 \\
\text{mesh } c \quad &: \quad -R_1 i_a - R_3 i_b + (R_1 + R_3 + R_5)i_c = 0
\end{aligned} \tag{36}
$$

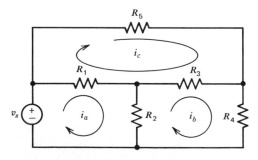

Figure 4.11   Second mesh method example.

Note that in mesh a the current i_a flows through R_1 and R_2, therefore R_1 and R_2 are coefficients of i_a, while there is coupling between mesh a and mesh b through R_2 and i_b flows in the opposite direction of i_a. Therefore, the coefficient of i_b is $-R_2$. Similarly, there is coupling between mesh a and mesh c through R_1 and the coefficient of i_c is $-R_1$ since i_c is in the opposite direction of i_a through R_1. Examine mesh equations b and c in (36) to see how easy it is to write these equations without any intermediate steps.

4.6 MESH METHOD WITH CURRENT SOURCES

The *mesh method* assumes that each unknown branch voltage can be expressed as a function of its branch current. Recall the definition of a current source and see that the voltage drop across the current source cannot be expressed in terms of mesh currents unless the other branch and topological constraints are also considered. Thus, the mesh analysis of circuits with current sources proceeds as follows. The circuit in Figure 4.12 will serve as our example. The mesh equations are

$$
\begin{aligned}
\text{mesh } a \quad &: \quad -v_1 + R_2(i_a - i_b) = 0 \\
\text{mesh } b \quad &: \quad R_2(i_b - i_a) + R_4(i_b - i_c) + R_3i_b = 0 \\
\text{mesh } c \quad &: \quad R_4(i_c - i_b) + R_5i_c + R_6i_c = 0
\end{aligned} \tag{37}
$$

Note that we have three equations and four unknowns because of the unknown voltage v_1 across the current source. The fourth constraint is the current source constraint which is

$$
i_a = -i_s \tag{38}
$$

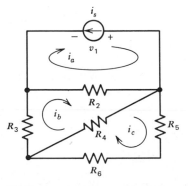

Figure 4.12   Mesh method with current sources.

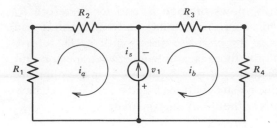

Figure 4.13 Mesh method with current sources between meshes.

Upon substituting (38) into (37) we obtain

$$\text{mesh } a \quad : \quad -R_2 i_b - v_1 = R_2 i_s$$
$$\text{mesh } b \quad : \quad (R_2 + R_3 + R_4)i_b - R_4 i_c = -R_2 i_s \qquad (39)$$
$$\text{mesh } c \quad : \quad -R_4 i_b + (R_4 + R_5 + R_6)i_c = 0$$

The last two equations can be solved for i_b and i_c, that is, a current source in a mesh determines that mesh current and reduces the number of mesh equations by one. If the voltage v_1 across the current source is desired, then i_b and i_c can be substituted into the equation for mesh a.

Let us consider one final example of mesh analysis with current sources, one that has a current source common to two meshes as shown in Figure 4.13. The two mesh equations are

$$\text{mesh } a \quad : \quad (R_1 + R_2)i_a \qquad\qquad - v_1 = 0$$
$$\text{mesh } b \quad : \qquad\qquad\qquad (R_3 + R_4)i_b + v_1 = 0 \qquad (41)$$

and the constraint on the mesh currents due to the current source is

$$i_b - i_a = i_s \qquad (42)$$

We can add (40) and (41) to eliminate v_1, and by means of (42) we can eliminate the current i_b to obtain

$$(R_1 + R_2 + R_3 + R_4)i_a = -(R_3 + R_4)i_s \qquad (43)$$

To conclude, the mesh method is essentially KVL with the voltages expressed as functions of a new set of variables called mesh currents. If a voltage is not a known source and cannot be expressed in terms of the mesh currents then it must be left in the equations as an unknown, and the element's constraints must be introduced as auxiliary equations.

4.7 ADDITIONAL METHODS FOR FORMULATING CIRCUIT EQUATIONS

Unfortunately, the mesh concept can only be applied to planar circuits. Furthermore, even in planar circuits it is difficult to write a computer algorithm to locate meshes. In this section we introduce some additional topological concepts that are very useful for the formulation of a set of $n-1$ independent KCL equations and $b-n+1$ independent KVL equations. The techniques described below are employed in several computer programs for circuit analysis.

4.7.1 TREE VOLTAGE AND FUNDAMENTAL LOOPS

We learned that every branch voltage could be expressed in terms of the $n-1$ node voltages and the measurement of *the node voltages formed a path to each node from the reference node, but formed no loops*. Now any $n-1$ branch voltages yield the same result provided that these $n-1$ branches connect all the nodes. If they connect all the nodes they cannot form any loops. *A collection of branches that connect all of the nodes but form no loops is called a tree.* If one more branch is added to the tree a loop will be formed. Thus we conclude that every tree has $n-1$ branches. For example, in Figure 4.14 all possible trees of the circuit in Figure 4.1a are indicated. The solid lines indicate the branches in the tree and the dashed lines indicate the remaining branches of the circuit graph and are called *links*. Each tree in Figure 4.14 has three branches since there are four nodes, and $(b-n+1)=2$ links. If we consider the tree in Figure 4.14a, the tree voltages are v_1, v_2, and v_4.

Now each link forms a unique loop with some of the branches in the tree. Therefore, the link voltage can be expressed as a linear combination of tree branch voltages. For example, in Figure 4.14a link branch 5 forms a loop with tree branches 2 and 4, and link branch 3 forms a loop with tree branches 1 and 2. Applying KVL to these loops we obtain

and

$$v_3 = -v_2 + v_1$$
$$v_5 = v_2 + v_4$$

(44)

The loop formed by a link with the tree branches is called a fundamental loop. Since there are $b-n+1$ links, we have $b-n+1$ fundamental loops. Furthermore, *the KVL equations for the fundamental loops are independent* since each KVL fundamental loop is formed by a unique link. Finally, *any other KVL loop equation in the circuit is dependent on the KVL fundamental loop equations.* To prove this last statement consider the tree chosen in Figure 4.14h. The fundamental loops are shown in Figure 4.15. Usually the direction of the loop is determined by the direction of the current in the link. In this circuit the link voltages are

$$v_2 = v_1 - v_3$$
$$v_4 = -v_1 + v_3 + v_5$$

(45)

Now let us show that the KVL equation for loop 542 is dependent on the

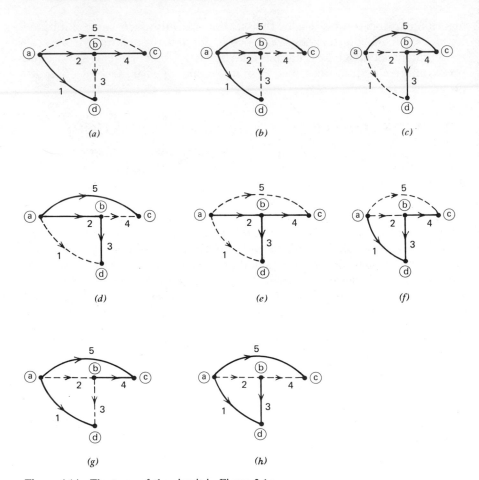

Figure 4.14   The trees of the circuit in Figure 3.1a.

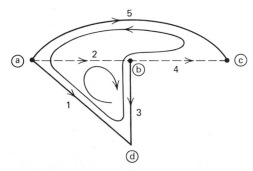

Figure 4.15   Fundamental loops.

fundamental loop KVL equations (45). The KVL loop 542 equation is

$$v_5 - v_4 - v_2 = 0 \tag{46}$$

Let us replace the link voltages in (46) by tree voltages by means of (45).

$$v_5 - (-v_1 + v_3 + v_5) - (v_1 - v_3) = 0 \tag{47}$$

Equation 47 contains *only tree branch voltages and tree branches form no loops*. Therefore, since (47) must be satisfied for all types of branch constraints, (46) must be dependent on (45). A closer look at (47) yields

$$0 \equiv 0 \tag{48}$$

that is, the left side of (47) is identically zero as it should be because since tree branches form no loops their sum need not satisfy KVL. Thus, the terms in (47) must cancel which establishes the dependency.

Let us summarize by noting that the circuit voltages can be expressed in terms of $n - 1$ voltages and no less than $n - 1$ voltages provided that these voltages when measured form no loops. Also, instead of writing the $b - n + 1$ independent KVL mesh equations we could select a tree and write the $b - n + 1$ KVL fundamental loop equations. The mesh approach seems simpler and thus preferable to the human mind, however it is much easier to write a computer algorithm to find a tree (a set of $n - 1$ branches that form a path to each node) than it is to locate meshes with a computer algorithm. Also, when a circuit graph is nonplanar the mesh concept no longer applies and the tree concept must be used to write a set of $b - n + 1$ independent KVL equations.

4.7.2 FUNDAMENTAL CUTSETS

Let us now show that every branch current can be expressed as a linear combination of link currents for a given tree. To do so we need to introduce the concept of a cutset. A *cutset* is a collection of branches that, when cut, separate the circuit into two pieces and if any branch in this set is not cut the circuit remains connected. Given a tree of a graph, *a fundamental cutset is a cutset that consists of only one tree branch and the remaining branches are links*. Since there are $n - 1$ tree branches, we have $n - 1$ fundamental cutsets for a given tree. For example, the three fundamental cutsets of the graph of Figure 4.14a are illustrated in Figure 4.16. Cutset 3 consists of branches 235. Note that only one tree branch is included in each fundamental cutset.

Since the circuit is cut into two pieces the sum of currents entering either piece must be zero. In Figure 4.16 this assertion of space-charge-neutrality

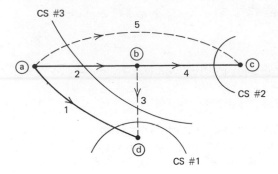

Figure 4.16   Fundamental cutsets.

yields

$$\begin{aligned} \text{CS 1} &: \quad i_1 = -\,i_3 \\ \text{CS 2} &: \quad i_4 = -\,i_5 \\ \text{CS 3} &: \quad i_2 = i_3 - i_5 \end{aligned} \tag{49}$$

Note that *every tree branch current can be expressed as a linear combination of link currents.* Thus, every branch current can be expressed as a linear combination of $b - n + 1$ currents if those currents are the links of a tree. Furthermore, we can replace the $n - 1$ independent KCL node equations by the $n - 1$ fundamental cutset equations. The fundamental cutset equations are independent since *a different tree branch current is contained in each equation.* Also, *any other KCL cutset or node equation can be expressed as a linear combination of the $n - 1$ KCL fundamental cutset equations, that is, it is dependent on the $n - 1$ KCL fundamental cutset equations.* To prove this statement let us return to our example in Figure 4.16. Let us show that the KCL equation at node ⓐ

$$i_1 + i_2 + i_5 = 0 \tag{50}$$

depends on (49). Replace each tree branch current in (50) by the corresponding relation in (49). We obtain

$$(-\,i_3) + (i_3 - i_5) + i_5 = 0$$

or

$$0 \equiv 0 \tag{51}$$

This identity must be the outcome because after elimination of the tree branch currents we have an equation [e.g. (51)] that consists *only of link currents.* The link currents do not form a cutset so that they do not have to sum to zero by Kirchhoff's current law. We conclude that the left side of (51) must be identically zero (dependent on the fundamental cutset equations) since (51) must be satisfied for all types of branch constraints. Finally, the

$b - n + 1$ link currents are not only sufficient but necessary to determine the tree branch currents, because with any fewer link currents we would not be able to determine all of the other branch currents from the KCL cutset equations.

In concluding this section note that we have established that every voltage in a circuit can be expressed in terms of the $n - 1$ voltages of a given tree in a circuit, and that any branch current can be expressed in terms of the $b - n + 1$ link branch currents with respect to that tree. Furthermore, it is relatively easy to write a computer program that will search through a list of branches and the nodes to which they are connected and separate them into a set of tree branches and a set of link branches and form the $b - n + 1$ KVL fundamental loop equations and the $n - 1$ KCL fundamental cutset equations. In addition we have the b branch constraints for a total of $2b$ equations. These $2b$ equations can be systematically reduced by eliminating variables. For example, if we can express all of the branch voltages in terms of the $b - n + 1$ link currents, then upon substitution of these equations into the $b - n + 1$ KVL fundamental loop equations we have $b - n + 1$ equations in terms of the $b - n + 1$ unknown link currents analogous to the mesh method. Similarly, if we can express all of the branch currents in terms of the tree voltages, then we can express the $n - 1$ KCL fundamental cutset or $n - 1$ KCL node equations in terms of the $n - 1$ tree voltages analogous to the node method. The tree and link concept is used in some computer programs to formulate the circuit equations. Also, the tree and link concept can be used to simplify some analysis problems as shown below.

Consider the circuit in Figure 4.17a which contains a current source common to both meshes. Using the tree-link concept we will show how to find all branch currents from one equation. To do this we must choose the current source branch as one of the link branches. Let us choose branch 4 as the remaining link branch. The tree and link branches are shown in Figure 4.17b. In Figure 4.17c the two fundamental loops are illustrated. Note that the fundamental loop current i_5 is the only current in the current source branch since this branch was chosen as one of the link branches. Thus $i_5 = 1$ A. Summing the voltage drops around the other fundamental loop to zero in Figure 4.17c we obtain

$$R_1(i_4 + 1 \text{ A}) + R_2(i_4 + 1 \text{ A}) + R_3 i_4 + R_4 i_4 = 0$$

so that

$$i_4 = \frac{-(R_1 + R_2)}{R_1 + R_2 + R_3 + R_4} \cdot 1 \text{ A}$$

Thus, when a current source is placed in a link branch, the fundamental loop current for that link is known. For similar reasons, we can show that

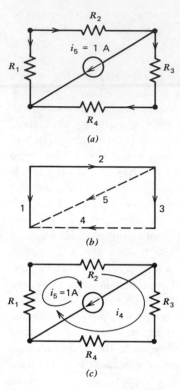

Figure 4.17 (*a*) Common mesh current source example. (*b*) Selection of the tree and link branches. (*c*) Fundamental loop analysis.

voltage sources should be placed in the tree branches when expressing the currents in the fundamental cutset equations as functions of the tree branch voltages.

REFERENCES

1. Cruz, J. B. Jr. and M. E. Van Valkenburg. *Signals in Linear Circuits*. Chapters 1, 13, Houghton Mifflin Co., 1974.
2. Desoer, C. A. and E. S. Kuh. *Basic Circuit Theory*. Chapters 6, 9, 10, 11, McGraw-Hill, 1969.

PROBLEMS

1. In Figure P4.1, a voltmeter is used to measure the voltage between nodes ⓐ and ⓒ and the voltage between nodes ⓑ and ⓒ. Thus, node ⓒ is the reference node. We obtain $e_a = 3$ V and $e_b = 2$ V. Compute the currents i_x and i_y, and the power delivered by the current source from the above measurements.

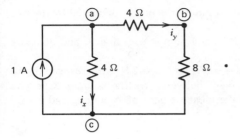

Figure P4.1

2. Use the node method to compute e_a and e_b in Figure P4.1. Your answer should agree with the measurements in Problem 1.
3. In Figure P4.3 measurements yield that the node voltages $e_b = 3$ V and $e_c = 5$ V. These voltages are measured with respect to the ground node ⓓ. Compute the currents i_{10} and i_{28} and the power delivered by the voltage source.

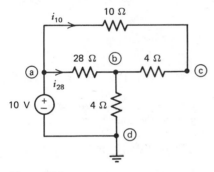

Figure P4.3

4. Use the node method to compute e_b and e_c in Figure P4.3. Your answer should agree with the measured values in Problem 3.
5. In Figure P4.5 measurements yield that the node voltages $e_a = 3$ V and $e_c =$

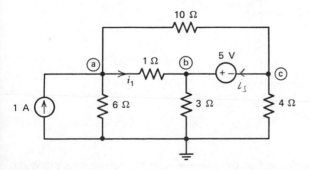

Figure P4.5

   -2 V with respect to the ground node. Compute the power delivered by the two
   sources. Find the current i_1.

6. Repeat Problem 4, only use Figure P4.5 and compute e_a and e_c and compare
   with Problem 5.

7. In Figure P4.7 the node voltages $e_a = -1$ V and $e_b = 2$ V with respect to the
   ground node. Compute i_x and the power delivered by the sources. Can you
   determine the node voltage e_a from the measurement of e_b and e_c and KVL?
   Explain your answer.

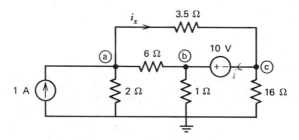

Figure P4.7

8. Repeat Problem 4, only use Figure P4.7 and compute e_a and e_b and compare
   with Problem 7.

9. In Problem 1 suppose that node ⓐ was chosen as the reference node.
   Determine e_b and e_c with respect to node ⓐ from the measurements given in
   Problem 1.

10. Choose node ⓓ as the reference node in Figure P4.10 and use the node
    method to solve for e_a, e_b, and e_c.

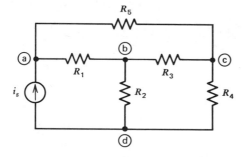

Figure P4.10

11. In Figure P4.11 use the node method to solve for e_b and e_c with respect to node
    ⓓ.

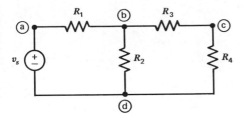

Figure P4.11

12. In Figure P4.12 use the node method to solve for the node voltage e_b with respect to ground. What is the power delivered by each of the sources?

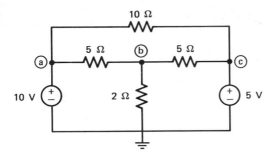

Figure P4.12

13. Use the node method to compute the voltage v_A and the current i_v in Figure P4.13.

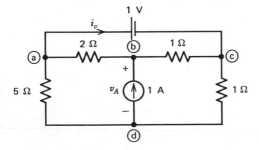

Figure P4.13

14. Use the node method to compute the power supplied by each of the sources in Figure P4.14.

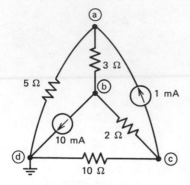

Figure P4.14

15. Repeat Problem 14 for the circuit in Figure P4.15.

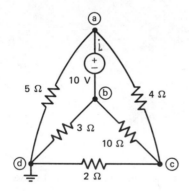

Figure P4.15

16. The polarity of both sources in Figure P4.13 are reversed, as shown in Figure P4.16. How is $\hat{v}_A$ and $\hat{i}_v$ in Figure P4.16 related to v_A and i_A in Figure P4.13?

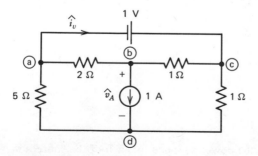

Figure P4.16

17. Set up the conductance matrix and the current vector i_s (the **G** array and i_s vector in Equation 11) for the circuit in Figure P4.10, using the guidelines expressed in Equations 17 and 18.
18. Repeat Problem 17 for the circuit in Figure P4.14.
19. In Figure P4.19 the mesh currents $i_a = 2$ A and $i_b = (2/3)A$, compute the branch voltages v_3 and v_2 from i_a and i_b.

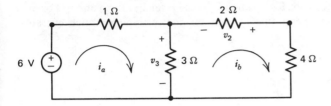

Figure P4.19

20. Use the mesh method to solve for the mesh currents i_a and i_b in Figure P4.19. Your answer should agree with the values given in Problem 19.
21. In Figure P4.21 the mesh current $i_b = 3$ mA, compute the node voltages e_a and e_b from the mesh currents.

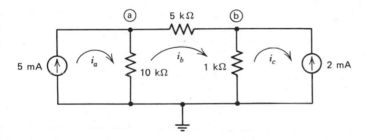

Figure P4.21

22. Use the mesh method to solve for the mesh current i_b in Figure P4.21. Your answer should agree with the current given in Problem 21.
23. Use the mesh method to determine the currents i_{28} and i_{10} and the power supplied by the 10 V source in Figure P4.3.
24. Use the mesh method to solve for the node voltage e_b and the power supplied by each of the voltage sources in Figure P4.12.
25. Use the mesh method to determine the branch currents and the power supplied by the 10 V source in Figure P4.15.
26. Use the mesh method to find i_x and i_y in Figure P4.1.
27. Use the mesh method to find i_x and the power supplied by each of the sources in Figure P4.7.
28. Use the mesh method to find i_v and v_A in Figure P4.13.

29. Use the mesh method to find the branch currents and the power supplied by each of the sources in Figure P4.14.

30. In Figure P4.30 the branches 3, 6, and 5 form a tree, and branches 1, 2, and 4 are the links of this tree. Given that $v_3 = 2$ V, $v_5 = -1$ V, $v_6 = -3$ V, $i_1 = 5$ A, $i_2 = -3$ A, and $i_4 = 2$ A, compute the remaining branch currents and voltages. Explain why you were able to compute the remaining branch currents and voltages with the above information. Would you have been able to compute all the remaining branch voltages if you were given one less branch voltage? Explain. Given one less branch current would you have been able to compute the remaining branch currents. Why?

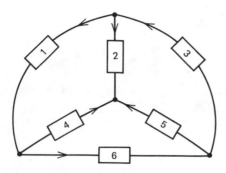

Figure P4.30

31. Given $v_2 = 10$ V, $v_5 = -5$ V, $i_1 = 1$ A, $i_3 = 3$ A, $i_4 = 2$ A, and $i_6 = -1$ A in Figure P4.31, repeat Problem 30.

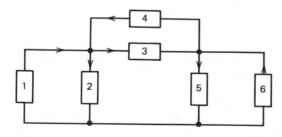

Figure P4.31

32. In Figure P4.32 the voltage sources form a tree. Express the remaining branch voltages in terms of the voltages of the tree formed by the voltage sources. Find the power delivered by each source.

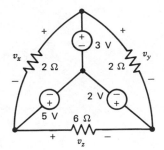

Figure P4.32

33. In Figure P4.33 select branches 1, 2, and 4 as the tree and express all other branch (link) voltages in terms of the tree voltages v_{s1}, v_{s2}, and v_4. Write KCL for node b and express the currents as a function of the tree voltages. You should have one equation and unknown, v_4. This node equation is one of the fundamental cutset equations. In order to solve for the currents in the voltage sources, write the fundamental cutset equations for each of the voltage source branches. Express the current in each of the voltage source branches as a function of v_{s1}, v_{s2}, and v_4.

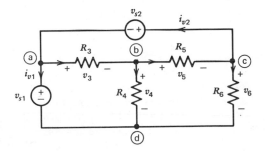

Figure P4.33

34. In Figure P4.33 select branches 1, 2, and 3 as the tree and write the fundamental cutset equations. Express the currents i_3, i_4, i_5, and i_6 in terms of v_{s1}, v_{s2}, v_3, i_{v1}, and i_{v2} so that you have three equations and three unknowns.

35. In Figure P4.35 choose branches 1, 2, and 3 to be the tree so that branches 4, 5, and 6 are the link branches. Since branches 4, 5, and 6 are independent current sources, we know the fundamental loop currents. Express currents i_1, i_2, and i_3 and voltages v_1, v_2, and v_3 in terms of the independent source currents i_4, i_5, and i_6.

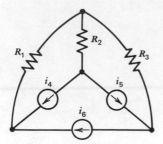

Figure P4.35

36. In Figure P4.36 choose branches 1, 2, and 4 to be the tree branches. Note that the current sources are in the set of link branches. Now write KVL for the fundamental loop formed by link branch 5 and the tree branches. Express these

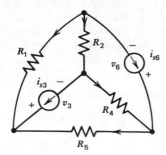

Figure P4.36

voltages in terms of i_{s3}, i_{s6}, and i_5. Thus, you have one equation and one unknown. In order to solve for the voltages v_3 and v_6 write the KVL equations for the remaining two fundamental loops. Express v_3 and v_6 in terms of the link currents.

Chapter Five
Equivalent Circuits

In the previous chapters we generated mathematical models for the electrical properties of a limited class of devices and with Kirchhoff's laws formulated the circuit equations. The equations have been algebraic and either linear or nonlinear depending on the device models. Frequently, certain properties of algebraic equations, linear algebraic equations in particular, can be used to replace part of the circuit by a simpler circuit that yields an equivalent response at a given set of terminals. This substitution not only simplifies the analysis, but frequently gives one more insight into the behavior of the circuit. In this chapter methods for the reduction of a circuit to an equivalent circuit or circuits at a defined set of terminals are discussed.

5.1 SUBSTITUTION

The substitution property simply stated is *given the voltage or current of a particular branch, that branch can be replaced by an independent voltage or current source whose value is equal to the given voltage or current, respectively, of that branch, and the other responses will be unaffected.* As an illustration consider the circuit in Figure 5.1. An analysis of this circuit yields $i_2 = -i_1 = 1$ A, $i_3 = i_4 = 0.5$ A, $v_2 = 10$ V, $v_3 = v_4 = 5$ V. If we replace branch 3 with an independent voltage source whose value is 5 V (equal to v_3) as shown in Figure 5.2a, then none of the responses are affected. For

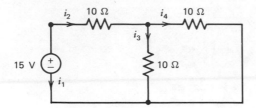

Figure 5.1   Substitution example.

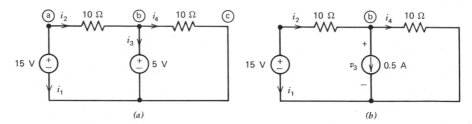

Figure 5.2   An illustration of substitution.

example, in Figure 5.2a we note that $i_2 = -i_1 = 1$ A, $i_4 = 0.5$ A, therefore, KCL applied to node ⓑ yields $i_3 = 0.5$ A. Similarly, we can replace branch 3 with an independent current source i_3 as shown in Figure 5.2b. Again, none of the responses are changed as seen from the node equation below.

$$\left(\frac{1}{10} + \frac{1}{10}\right)e_b = -0.5 + \frac{15\text{ V}}{10} \tag{1}$$

The solution to (1) is

$$e_b = v_3 = 5\text{ V} \tag{2}$$

so that $i_4 = 0.5$ A, and KCL yields $i_2 = -i_1 = 1$ A. All three of the above circuits have the same responses; hence they are *equivalent*.

The proof of the validity of the substitution property follows simply from that fact that *given a set of algebraic equations (linear or nonlinear) that have at least one solution (a set of values for the unknown variables which satisfy the equations), then replacing one of the unknown variables by its solution value certainly does not cause the solution values for the other unknown variables to change.*

5.2 SERIES-PARALLEL CIRCUITS

The procedure for reducing series-parallel combinations of elements to an equivalent network is very straightforward. The series connection is

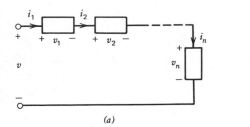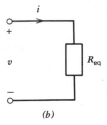

(a)   (b)

Figure 5.3   The series connection and its equivalent.

illustrated in Figure 5.3. The current through each element is the same and the voltage drop across the series combination is equal to the sum of the voltage drops across each element. Thus,

$$i = i_1 = i_2 = \cdots = i_n \tag{3}$$

and

$$v = v_1 + v_2 + \cdots + v_n \tag{4}$$

Equations 3 and 4 are valid for any elements in series. However, if in particular the elements are *linear resistors*, then

$$v = R_1 i + R_2 i + \cdots + R_n i \tag{5}$$

Hence, we can replace all *n* resistors by one resistor

$$v = R_{eq} i \tag{6}$$

where

$$\boxed{R_{eq} = R_1 + R_2 + \cdots + R_n} \tag{7}$$

In the parallel connection, Figure 5.4a, the voltage drop across all of the elements is equal to the voltage drop across each individual element, and the current into the parallel combination is equal to the sum of the currents

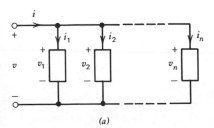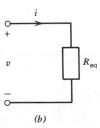

(a)   (b)

Figure 5.4   The parallel connection and its equivalent.

in each element. Thus,

$$v = v_1 = v_2 = \cdots = v_n \tag{8}$$

and

$$i = i_1 + i_2 + \cdots + i_n \tag{9}$$

For linear resistors we have that

$$i = \frac{v}{R_1} + \frac{v}{R_2} + \cdots + \frac{v}{R_n} \tag{10}$$

Thus, the equivalent resistance is

$$i = \frac{1}{R_{eq}} v \tag{11}$$

where

$$\boxed{\frac{1}{R_{eq}} = \frac{1}{R_1} + \frac{1}{R_2} + \cdots + \frac{1}{R_n}} \tag{12}$$

If there are only two linear resistive elements in parallel, then

$$R_{eq} = \frac{R_1 R_2}{R_1 + R_2} \tag{13}$$

Formula 13 is used frequently in circuit analysis, and one should take special note of it.

Finally, we say that elements are connected in series-parallel if the circuit can be reduced to an equivalent network by a sequence of series-parallel operations. The following example illustrates the above principles.

Example 1
Find the current delivered by the 10-V battery in Figure 5.5.

Solution
Since we are only interested in the current i delivered by the battery, let us reduce the series-parallel network to the equivalent resistance seen by

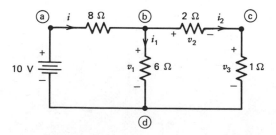

Figure 5.5 Series-parallel network.

the 10-V battery. The following sequence of figures illustrates the procedure. The 2-Ω and 1-Ω resistors are in series and can be combined into one 3-Ω resistor as shown in Figure 5.6a. Now this 3-Ω resistor is in parallel with the 6-Ω resistor; hence,

$$R_{eq} = \frac{3 \times 6}{3 + 6} = 2\ \Omega$$

This results in the equivalent circuit shown in Figure 5.6b. Note that now the 2-Ω resistor is in series with the 8-Ω resistor. Combining these series resistors gives us the final equivalent circuit shown in Figure 5.6c. Thus, the current delivered by the battery is

$$i = \frac{10\ V}{10\ \Omega} = 1\ A.$$

Returning to Figure 5.5 we see that the current through the 8-Ω resistor is 1 A; however, the currents through the 6-Ω, the 2-Ω, and the 1-Ω resistors are not known. We can find the current in these resistors by noting that the voltage drop across the 6-Ω resistor is determined by the 10-V source and the voltage drop across the 8-Ω resistor.

$$v_1 = -i8 + 10 = 2\ V$$

and

$$i_1 = \frac{2\ V}{6\ \Omega} = \frac{1}{3}\ A$$

Application of Kirchhoff's current law at node ⓑ gives

$$i_2 = i - i_1 = \frac{2}{3}\ A$$

In Figure 5.5 the 8-Ω and the 6-Ω resistors are not in series. Why?

We close by noting that the formulas (3), (4) and (8), (9) can also be applied to nonlinear components, but we do not usually obtain the simple results of (7) and (12).

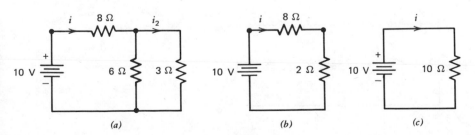

(a) *(b)* *(c)*

Figure 5.6   Series-parallel combinations.

5.3 DRIVING-POINT AND TRANSFER RELATIONS

The *driving-point (DP) relation* is defined as a relationship between the current and voltage of a one-port. For example, in Figure 5.7 by means of series-parallel combinations we can reduce the circuit to a single 10-Ω resistor (see Figure 5.5 and 5.6). Therefore, the DP relation can be expressed mathematically as

$$v = 10i \qquad (14)$$

Since v is a function of i, (14) is usually called the *DP function*. If (14) is expressed graphically, then the graphic representation of (14) is usually called the *DP characteristic*. The above relation is linear since the branch constraints of all the resistors in the circuit are linear. However, in Chapter Six we illustrate some nonlinear one-port DP relations.

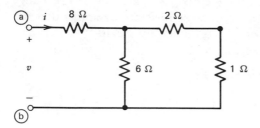

Figure 5.7   Driving-point example.

In the above example, the DP function was found by means of simple series-parallel reductions. However, the same result can be achieved by writing the circuit equations and eliminating all variables except v and i. In order to write the equations we can assume that a voltage source v or a current source i is connected to the one-port. For example, in Figure 5.8 a

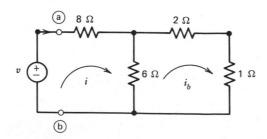

Figure 5.8   Calculation of the DP function.

voltage source has been connected to the port. The mesh equations are

$$14i - 6i_b = v$$
$$-6i + 9i_b = 0$$

(15)

The solution is

$$i = \frac{\begin{vmatrix} v & -6 \\ 0 & 9 \end{vmatrix}}{\Delta} = \frac{9v}{9 \times 14 - 36}$$

or

$$v = 10i$$

(16)

The equivalent circuit corresponding to this constraint is shown in Figure 5.9. We should realize that if the one-port contains independent sources, then i not only depends on v but also on the other independent sources in the circuit. The equivalent DP relation in this situation is discussed extensively in the next section.

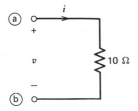

Figure 5.9   Equivalent circuit seen at terminals a-b.

The *transfer relation* is defined as a relation between a branch voltage or current at one pair of terminals to the branch voltage or current at a *different* pair of terminals. The four possible transfer relations are shown in Figure 5.10. Two transfer relationships which should be remembered because of their frequent usage are the *voltage divider function* and the *current divider function* (see Figure 5.11 and 5.12).

In determining the transfer characteristic for the voltage divider it is

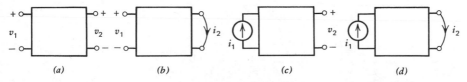

Figure 5.10   Transfer ratios. (*a*) Voltage-voltage. (*b*) Voltage-current. (*c*) Current-voltage. (*d*) Current-current.

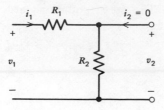

Figure 5.11 Voltage divider.

important to realize that we assume that $i_2 = 0$ in Figure 5.11. With this assumption,

$$i_1 = \frac{v_1}{R_1 + R_2} \tag{17}$$

Therefore, since $v_2 = i_1 R_2$, then

$$v_2 = \frac{R_2}{R_1 + R_2} v_1 \tag{18}$$

and (18) is the voltage divider function.

In Figure 5.12 we can show that

$$i_1 = \frac{R_2}{R_1 + R_2} i \tag{19}$$

or

$$i_2 = \frac{R_1}{R_1 + R_2} i \tag{20}$$

Both (19) and (20) are the current divider functions that are useful in computing the current in two parallel resistors given the total current i. Note that if $R_1 \gg R_2$ then most of the current flows through R_2. This observation is helpful in remembering (19) and (20).

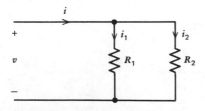

Figure 5.12 Current divider.

5.4 LINEAR CIRCUITS AND THEIR EQUIVALENTS

5.4.1 THÉVENIN AND NORTON EQUIVALENT CIRCUITS

In the previous sections we simplified a one-port without independent sources by an elimination of variables through series-parallel combinations or reduction of the circuit equations. We now discuss the case when the one-port contains independent sources and show that any linear one-port can be reduced to a simple one-port that contains a single resistor and single independent source. Consider the circuit in Figure 5.13. Two one-

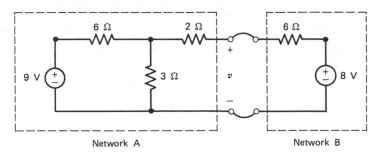

Network A   Network B

Figure 5.13   Interconnection of two 1-ports.

ports are connected together. In design, it is frequently useful to have a simple equivalent of a one-port that is connected to another one-port as in Figure 5.13. To find a simple equivalent of network A let us use the substitution property to replace B by a voltage source as shown in Figure 5.14. The mesh equations for this circuit are

$$9i_a - 3i_b = 9 \tag{21}$$
$$-3i_a + 5i_b = -v \tag{22}$$

where $i_b = -i$. Since we are only interested in the port voltage v and current i, let us eliminate the dependent variables i_a and i_b. To eliminate i_a

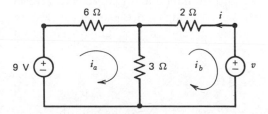

Figure 5.14   Reduction of 1-port A.

multiply (22) by 3 and add it to (21). The result is the equation

$$-4i = -v + 3 \qquad (23)$$

which is plotted in Figure 5.15. Network A constrains the solution to lie along the line in Figure 5.15. In Figure 5.16 two one-port circuits are illustrated that have the v-i characteristic shown in Figure 5.15; hence they are equivalent to the one-port A in Figure 5.13. In Figure 5.16a and b the

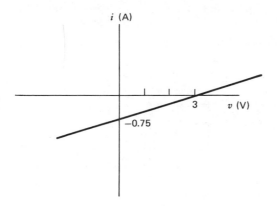

Figure 5.15 i-v characteristic of 1-port A.

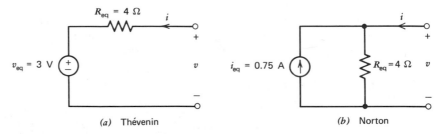

(a) Thévenin   (b) Norton

Figure 5.16   Equivalent circuits for 1-port A.

equivalent resistance is simply determined by the slope of the i-v characteristic and can be determined by setting all independent sources to zero in A (this causes the i-v characteristic to pass through the origin in Figure 5.15) and calculating the DP resistance as shown in Figure 5.17. From Figure 5.17 we find that

$$\frac{v}{i} = 4 \, \Omega \qquad (24)$$

which is the reciprocal of the slope in Figure 5.15. Next we note that v_{eq} in Figure 5.16a is simply the voltage at the v-axis intercept in Figure 5.15,

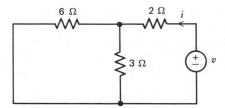

Figure 5.17   Determination of the equivalent resistance of 1-port A.

that is, it is the voltage across the terminals in network A when the current $i = 0$. This open-circuit voltage can be computed by opening the output terminals of A as shown in Figure 5.18. Finally, the current i_{eq} in Figure 5.16*b* is the current at the *i*-axis intercept in Figure 5.15, that is, it is the

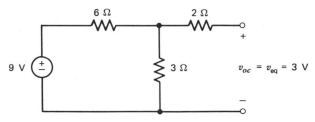

Figure 5.18   Determination of the equivalent voltage of 1-port A.

current between the terminals when $v = 0$. This short-circuit current can be computed by shorting the output terminals of A as shown in Figure 5.19. Note that the *i*-axis intercept is -0.75 in Figure 5.15 but $i_{eq} = +0.75$ since its reference direction is opposite that of the current *i*.

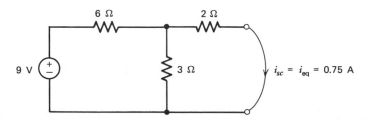

Figure 5.19   Determination of the equivalent current of 1-port A.

The equivalent circuit in Figure 5.16*a* is called the *Thévenin circuit*. The Thévenin equivalent voltage is the open circuit voltage. The circuit in Figure 5.16*b* is called the *Norton circuit*. The Norton equivalent current is the current obtained by shorting the output terminals of the one-port.

Now that we have found an equivalent circuit for A, the new circuit for

Figure 5.13 is shown in Figure 5.20. In Figure 5.21 we have plotted the i-v characteristic for network A and the i-v characteristic for network B. Their intersection determines the solution, that is, the constraints of B determine the operating point on the i-v characteristic of A in Figure 5.15. The solution is $i = 0.5$ A and $v = 5$ V. In conclusion, *any linear resistive one-port can be reduced to an equivalent one-port that contains one resistor and one independent source.*

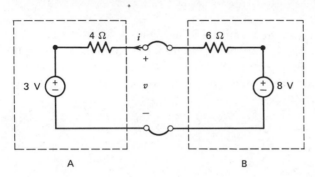

Figure 5.20   Equivalent port representation of the interconnection.

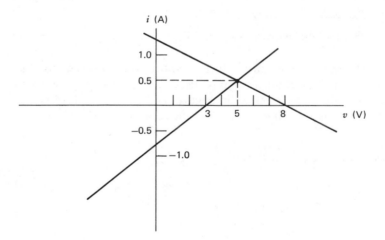

Figure 5.21   Graphical determination of the solution.

5.4.2 SOURCE TRANSFORMATIONS

The Thévenin and Norton equivalent circuits are very useful in network reduction. For example, consider the circuit in Figure 5.22a. The series combinations of the 6-Ω resistor and 9-V source are in shunt with the 3-Ω resistor. Therefore, by transforming the 6-Ω resistor and 9-V source branch

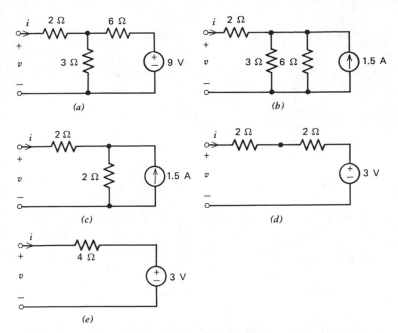

Figure 5.22 Reduction of a 1-port by source transformations and series/parallel combinations.

to a Norton equivalent, we obtain the circuit in Figure 5.22b. Next we combine the 6-Ω and 3-Ω resistors into an equivalent 2-Ω resistor as shown in Figure 5.22c. To complete our reduction the shunt combination of the 1.5-A source and the 2-Ω resistor are transformed to a Thévenin equivalent circuit as shown in Figure 5.22d so that the 2-Ω resistors can be combined into a single resistor. The final Thévenin equivalent of the one-port is shown in Figure 5.22e. A very practical example of the use of source transformations in network reduction follows.

The circuit in Figure 5.23a is a four-bit digital-to-analog (D/A) converter. The coefficients b_i are digitally controlled and can be set to "0" or "1." The voltage V_R is called the reference voltage. We wish to express the output voltage e_o (measured with respect to ground) as a function of the coefficients b_i. We could attempt the solution with the node or mesh method, but both of these methods require the solution of four equations in four unknowns. A eight-bit converter would result in eight equations. An easier approach, and an approach that also yields more insight into the behavior of the circuit, would be to use source transformations and series-parallel combinations as shown below.

First, transform all of the series $2R$-$b_i V_R$ branches to Norton equivalents as shown in Figure 5.23b. The reduction procedure is shown in Figure 5.24.

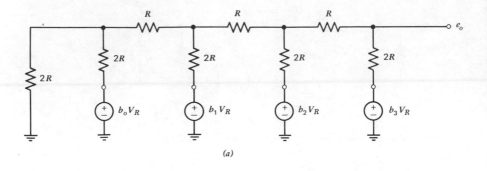

(a)

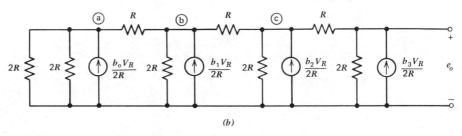

(b)

Figure 5.23   Digital-to-analog converter. (a) Voltage source model. (b) Current source equivalent.

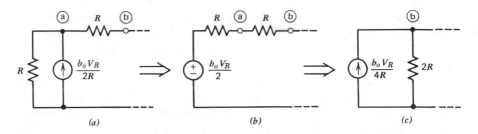

Figure 5.24   Reduction of the network.

The parallel $2R$ resistors at the left of the circuit in Figure 5.23 are combined to yield the left-most resistor R in Figure 5.24a. From Figure 5.24a to Figure 5.24b the current source is transformed back to a voltage source, then the series R resistors are combined, which eliminates node ⓐ and the Norton equivalent is found in Figure 5.24c. The reduced circuit is now drawn in Figure 5.25 in which the two parallel current sources have been combined. Note that we have a circuit equivalent to the circuit in Figure 5.23 at the output except that we have reduced the number of nodes and meshes by one. In the reduction process the source $b_0 V_R$ was reduced by half. The continuation of this reduction process yields the Thévenin equivalent circuit in Figure 5.26. Let the binary numbers b_3, b_2, b_1, b_0

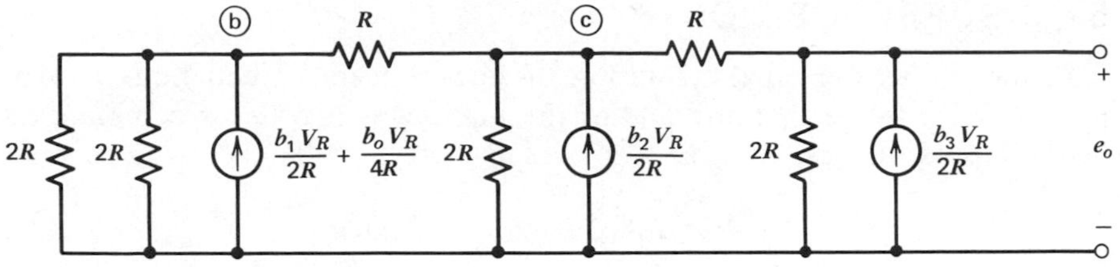

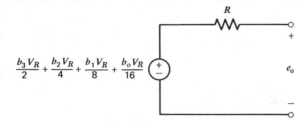

Figure 5.25   Combination of sources.

Figure 5.26   Thévenin equivalent of the digital/analog converter.

represent the digital input where b_0 is the least significant bit (LSB) and b_3 is the most significant bit (MSB). The open circuit analog output versus the binary input is listed in Table 5.1.

Table 5.1

Binary Input	Open Circuit Analog Output Voltage
0000	0
0001	$1/16\,V_R$
0010	$1/8\;\;V_R$
0011	$3/16\,V_R$
0100	$1/4\;\;V_R$
0101	$5/16\,V_R$
0110	$3/8\;\;V_R$
0111	$7/16\,V_R$
1000	$1/2\;\;V_R$
1001	$9/16\,V_R$
1010	$5/8\;\;V_R$
1011	$11/16\,V_R$
1100	$3/4\;\;V_R$
1101	$13/16\,V_R$
1110	$7/8\;\;V_R$
1111	$15/16\,V_R$

5.4.3 SUPERPOSITION

In the discussion on the solution of linear algebrcic equations in Appendix A we found that any one of the unknowns is a linear combination of the inputs, that is,

$$x_j = \frac{1}{\Delta}(\Delta_{1j}b_1 + \Delta_{2j}b_2 + \cdots \Delta_{nj}b_n) \tag{25}$$

where x_j denotes the response of interest, b_i represents the inputs, Δ is the system determinant, and Δ_{ij} is the ijth cofactor of the determinant. We can write (25) as

$$x_j^{(i)} = \frac{1}{\Delta}\Delta_{ij}b_i$$

so that

$$x_j = \sum_{i=1}^{n} x_j^{(i)} = \frac{1}{\Delta}\sum_{i=1}^{n} \Delta_{ij}b_i \tag{26}$$

that is, we can set all of the inputs to zero except for b_i and compute $x_j^{(i)}$. We can do this for $i = 1, 2, \ldots n$ and sum the responses to determine the complete response. In circuit analysis this is called the *superposition* principle.

As an example of the application of the superposition principle consider the circuit in Figure 5.27. Let us determine the open circuit voltage by means of superposition. First let us set the current source to zero. From

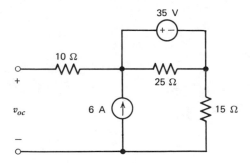

Figure 5.27   Superposition example.

Chapter Three recall that a zero ampere current source is an open circuit so that we obtain the circuit in Figure 5.28a. Since the terminals are open-circuited the current in the 10-Ω and 15-Ω resistors is zero. Thus the open-circuit voltage is

$$v_{oc}^{(1)} = 35 \text{ V} \tag{27}$$

where 35 V is the voltage drop across the 25-Ω resistor.

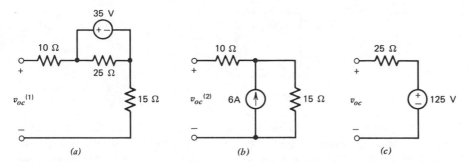

Figure 5.28 Application of superposition. (*a*) Current source zero. (*b*) Voltage source zero. (*c*) Thévenin equivalent.

Next let us set the voltage source to zero in Figure 5.27. Recall that a zero-volt source is a short circuit, thus we obtain the equivalent circuit in Figure 5.28*b*. In this figure the current in the 10-Ω resistor must be zero while the current in the 15-Ω resistor is 6 A so that

$$v_{oc}^{(2)} = (15\ \Omega)(6\ \text{A}) = 90\ \text{V} \tag{28}$$

By superposition the open-circuit voltage in Figure 5.27 is

$$v_{oc} = v_{oc}^{(1)} + v_{oc}^{(2)} = 125\ \text{V} \tag{29}$$

It is interesting to note that the analysis of the circuit in Figure 5.27 appeared somewhat formidable at first; however, since the circuit is linear, the superposition principle allowed us to reduce the analysis to two rather simple circuits.

In conclusion let us find the Thévenin equivalent circuit of Figure 5.27. We already know that the open circuit voltage is 125 V. The equivalent resistance can be found by setting the remaining source to zero in Figure 5.28*a* or *b*. We find that

$$R_{eq} = 10\ \Omega + 15\ \Omega = 25\ \Omega \tag{30}$$

This Thévenin equivalent circuit is shown in Figure 5.28*c*. Also note that the Thévenin equivalent of the 35-V voltage source in parallel with the 25-Ω resistor is simply a 35-V source, since the equivalent resistance of a voltage source is zero ohms and zero ohms in parallel with 25 ohms is zero ohms. *Thus, if we remove the 25-Ω resistor from Figure 5.27 we will not change any of the other voltages or currents in the circuit except that the current in the voltage source will now be*

$$i_s' = i_s - 35\ \text{V}/25\ \Omega \tag{31}$$

since the 25-Ω resistor has been removed.

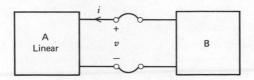

Figure 5.29   Connection of two 1-ports.

Superposition is frequently used to find the Thévenin equivalent circuit. Consider the connection of two one-ports in Figure 5.29 where A is linear. To establish the Thévenin equivalent circuit we use substitution and replace B by the current source i (Figure 5.30). Now $v = v^{(1)} + v^{(2)}$ where $v^{(1)}$

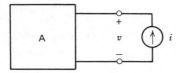

Figure 5.30   Substitution principle.

is the response due to the energy sources in A with $i = 0$ as shown in Figure 5.31a. Note that $v^{(1)} = v_{oc}$, the open circuit voltage of A. Now

$$\frac{v^{(2)}}{i} = R_{eq} \tag{32}$$

where R_{eq} is the equivalent resistance of A. Thus, by superposition,

$$v = v_{oc} + R_{eq}i \tag{33}$$

The circuit representation of (33) is shown in Figure 5.32.

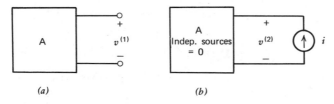

(a)   (b)

Figure 5.31   Superposition of the responses.

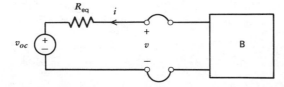

Figure 5.32   Thévenin equivalent of A.

5.4.4 SYMMETRICAL CIRCUITS

Sometimes linear circuits possess symmetry properties that can be used to reduce the amount of computation. For example, consider the circuit in Figure 5.33 in which we assume that A and A' are mirror images of one another. The employment of the superposition principle yields the two circuits in Figure 5.34, where

$$i_j = i_j^{(1)} + i_j^{(2)} \qquad j = 1, 2, 3 \tag{34}$$

Now, if $v_s = v'_s$ the symmetry conditions imply that

$$i_j^{(1)} = -i_j^{(2)} \qquad j = 1, 2, 3 \tag{35}$$

so that $i_j = 0$, $j = 1, 2, 3$. Since the connecting lines have zero current flow we can cut the circuit in half and analyze only the half-circuit in Figure 5.35.

It is also interesting to consider the symmetrical circuit with inputs of opposite polarities, Figure 5.36. Again we apply the superposition principle

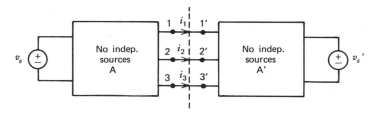

Figure 5.33   Symmetrical circuit.

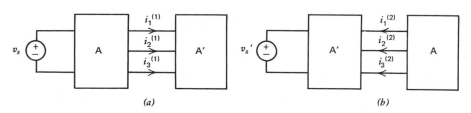

(a)   (b)

Figure 5.34   Application of superposition.

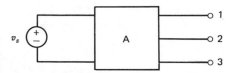

Figure 5.35   Half circuit.

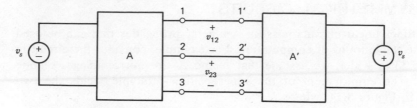

Figure 5.36 Antisymmetrical inputs.

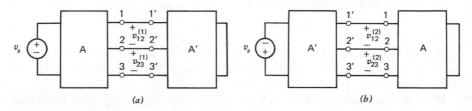

Figure 5.37 Application of superposition.

as shown in Figure 5.37. The voltages between any pair of terminals on the axis of symmetry is

$$v_{ij} = v_{ij}^{(1)} + v_{ij}^{(2)} \qquad i, j = 1, 2, 3 \tag{36}$$

However, symmetry conditions plus the homogeneity property (the input is multiplied by -1 in going from Figure 5.37a to Figure 5.37b) yields

$$v_{ij}^{(1)} = -v_{ij}^{(2)} \qquad i, j = 1, 2, 3 \tag{37}$$

so that $v_{ij} = 0$. This means that the connecting terminals can be shorted without effecting the circuit response. This generates the half circuit in Figure 5.38.

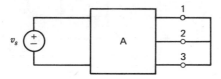

Figure 5.38 Half circuit, antisymmetrical input.

As an example of the application of this technique let us analyze the circuit in Figure 5.39. This circuit is redrawn in Figure 5.40. The independent current sources have been split in two with the requirement that

$$i_1 = i_1{}^a + i_1{}^b$$
$$i_2 = i_2{}^a + i_2{}^b \tag{38}$$

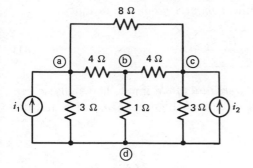

Figure 5.39   Symmetrical circuit example.

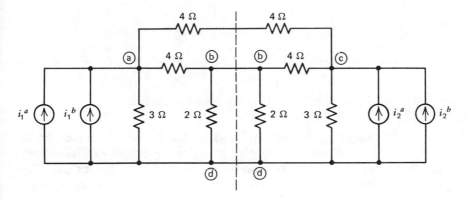

Figure 5.40.   Location of axis of symmetry.

We have introduced four unknowns, $i_1{}^a$, $i_1{}^b$, $i_2{}^a$, and $i_2{}^b$ and have only two constraints. In order to determine these unknowns uniquely, we need two additional constraints, and these constraints should be such to allow us to use our half circuits. Thus, choose

$$i_1{}^a = i_2{}^a$$

and

$$i_1{}^b = -i_2{}^b \tag{39}$$

The solution to (38) and (39) is

$$\boxed{i_1{}^a = \frac{i_1 + i_2}{2}} \tag{40}$$

which is referred to frequently as the *common mode* input and

$$i_1{}^b = \frac{i_1 - i_2}{2} \qquad (41)$$

where $2i_1{}^b$ is frequently called the *differential mode* input. The analysis can be carried out by means of the half circuits in Figure 5.41a and b and the responses can be superimposed.

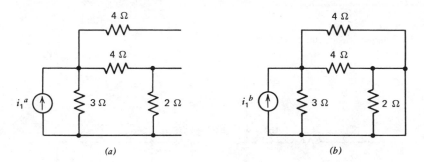

Figure 5.41   (a) Common-mode circuit. (b) Differential-mode circuit.

Finally, the symmetrical situation in which the wires are crossed needs to be discussed. To illustrate the point consider the circuit in Figure 5.42. (Note: the absence of a dot at the intersection of the wires means that they are not connected). First let us consider the common mode input ($v_s = v'_s$).

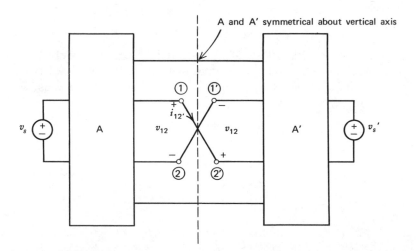

Figure 5.42   Symmetrical circuits with crossed connections.

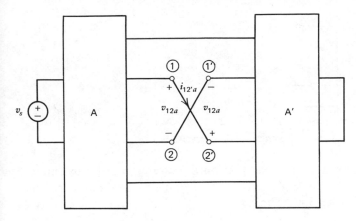

Figure 5.43   Response to one input.

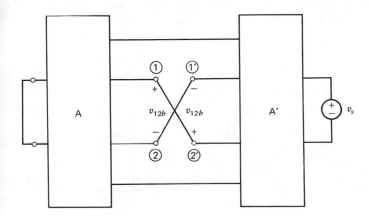

Figure 5.44   Response to second input, common mode.

Let us use superposition as shown in Figures 5.43 and 5.44. Thus,

$$v_{12} = v_{12a} + v_{12b} \tag{42}$$

However, because of the symmetry of the circuit

$$v_{12a} = -v_{12b} \tag{43}$$

so that

$$v_{12} = 0 \tag{44}$$

This result implies that cross-terminal connections can be shorted in the common-mode analysis as shown in Figure 5.45. Note that direct connections are opened as shown previously.

Next let us consider the differential mode case ($v_s = -v'_s$). Now we must

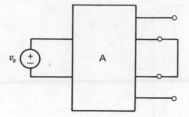

Figure 5.45 Equivalent circuit for common-mode analysis.

impose a horizontal symmetry condition on the circuit in order to obtain a simplification, that is, the circuit must be symmetrical about horizontal axis as shown in Figure 5.46. The current

$$i_{12'} = i_{12'a} - i_{2'1d} \tag{45}$$

where $i_{12'a}$ is determined from Figure 5.43 and $i_{2'1d}$ is determined from Figure 5.47. But our horizontal symmetry condition yields

$$i_{12'a} = i_{2'1d} \tag{46}$$

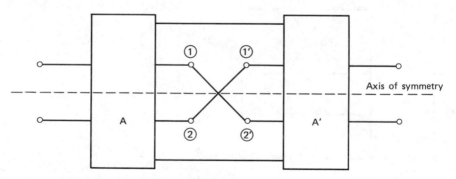

Figure 5.46 Horizontal symmetry constraint.

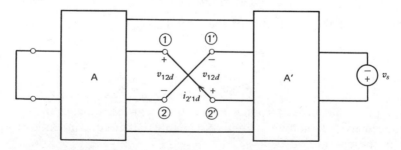

Figure 5.47 Response to second input, differential mode.

so that from (45) we obtain

$$i_{12'} = 0 \qquad (47)$$

A similar argument yields $i_{21'} = 0$ so that the cross wires can be opened in the differential mode *if the circuit is symmetrical about the horizontal axis.* As shown previously direct connections are shorted in the differential mode. See Figure 5.48. A circuit that has both horizontal and vertical symmetry is the balanced lattice circuit in Figure 5.49.

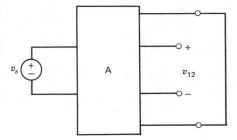

Figure 5.48   Equivalent differential-mode circuit.

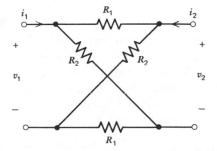

Figure 5.49   Balanced lattice circuit.

5.5 SOURCE SPLITTING

Recall that in the previous sections we showed that a circuit could be reduced to a simple equivalent circuit by series-parallel combinations and source transformations. However, source transformations can only be performed when the voltage source is in series with a single resistor or the current source is in parallel with a resistor. When this is not the case the analysis concept of source splitting is required. For example, the voltage source in Figure 5.50 is not in series with a single resistor. Therefore, it appears that the circuit cannot be reduced by simple source trans- formations and series-parallel combinations. However, let us split the

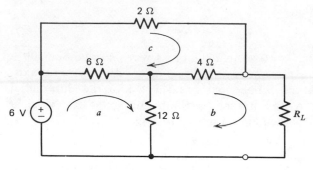

Figure 5.50   Voltage source-splitting example.

voltage source into two parallel sources as shown in Figure 5.51. The mesh equations for this circuit are

(a) $18i_a - 12i_b - 6i_c$ $= 6$ (48)

(b) $-12i_a + (16 + R_L)i_b - 4i_c = 0$ (49)

(c) $-6i_a - 4i_b + 12i_c$ $= 0$ (50)

(d) $+6 - 6$ $= 0$

or $0 \equiv 0$ (51)

Equations 48 to 50 are three equations in three unknowns and are identical to the mesh equations for the circuit in Figure 5.50. Therefore, if a unique solution exists for i_a, i_b, and i_c it is identical for both circuits. *However, (51) is a dependent equation which means that i_x is indeterminant, that is, when identical voltage sources are connected in parallel we have no way of calculating the current i_x flowing between them.* In Figure 5.51 we can only write that

$$i_2 + i_x + i_{s2} = 0 \qquad (52)$$

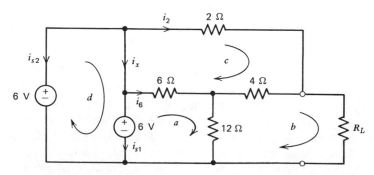

Figure 5.51   Voltage source splitting.

and

$$i_6 - i_x + i_{s1} = 0 \tag{53}$$

where i_x is the indeterminant current. The sum of (52) and (53) is

$$i_2 + i_6 + i_{s2} + i_{s1} = 0 \tag{54}$$

Since we can compute i_2 and i_6 from the mesh currents i_a and i_c, we conclude that *we cannot compute the currents i_{s1} and i_{s2}, but only their sum.*

Since the current i_x is indeterminant we could set it to zero and generate the equivalent circuit in Figure 5.52. The assumption $i_x = 0$ does not effect the other circuit responses. Note that the mesh equations for Figure 5.52

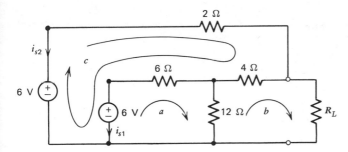

Figure 5.52   Equivalent circuit assuming $i_x = 0$.

are identical to (48) to (50). The circuit in Figure 5.52 can now be reduced by source transformations and series parallel operations as shown in Figure 5.53. Note that R_L is driven by a $3.5\,A$ source in parallel with a $8/5$ ohm resistor.

Figure 5.54 illustrates the general principle of voltage source splitting. A loop of voltage sources means that the loop current is indeterminant, that is, i_{x_1} and i_{x_2} in Figure 5.54b are indeterminant. If we assume that $i_{x_1} = i_{x_2} = 0$, we generate the circuit in Figure 5.54c. In comparing Figures 5.54a and 5.54c, essentially the voltage source is "pushed" into the branches connected to node ⓑ. Rather than "push" the voltage source into the branches connected to node ⓑ, we could "push" the voltage source in the opposite direction into the branches connected to node ⓐ.

Next let us consider the concept of current source splitting. Let us begin with Figure 5.55, and split the source as shown in Figure 5.56. It is assumed that the value of i_s is given. We introduce a new node ⓧ and the two current sources must be identical or else KCL is not satisfied at node ⓧ. The node equations for the circuit in Figure 5.56, assuming ⓓ is the datum

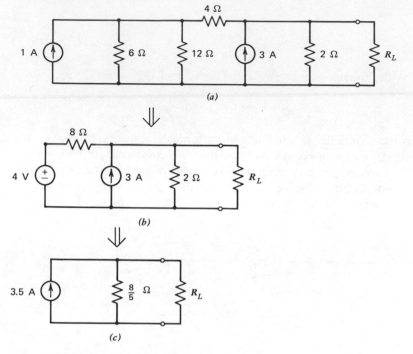

(a)

(b)

(c)

Figure 5.53   Reduction of the circuit.

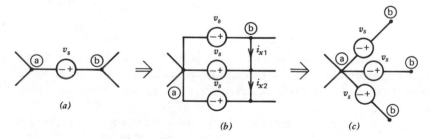

(a)

(b)

(c)

Figure 5.54   Voltage source splitting.

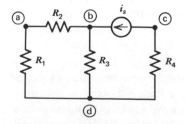

Figure 5.55   Current source-splitting example.

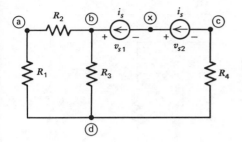

Figure 5.56   Splitting a current source.

node, are

$$\text{ⓐ} \quad : \quad \left(\frac{1}{R_1}+\frac{1}{R_2}\right)e_a - \frac{1}{R_2}e_b \quad = 0 \tag{55}$$

$$\text{ⓑ} \quad : \quad -\frac{1}{R_2}e_a + \left(\frac{1}{R_2}+\frac{1}{R_3}\right)e_b = i_s \tag{56}$$

$$\text{ⓒ} \quad : \quad +\frac{1}{R_4}e_c \qquad\qquad = -i_s \tag{57}$$

$$\text{ⓓ} \quad : \quad -i_s + i_s \qquad\qquad = 0$$

or

$$0 \equiv 0 \tag{58}$$

Equations 55 to 57 are identical to the node equations for the circuit in Figure 5.55 so that if a unique solution exists we have the same solution for both circuits. However (58) is a dependent equation (zero identity) so that the node voltage e_x is indeterminant and its value has no effect on the other responses. In Figure 5.56 we can only say that

$$e_b - e_c = v_{s1} + v_{s2} \tag{59}$$

Since the node voltage e_x is indeterminant and does not affect the other responses, we can assume that its voltage is identical to any other node voltage in the circuit allowing us to connect the two nodes with identical voltages together. For example, in Figure 5.57a we have connected node ⓧ to node ⓓ. In Figure 5.57b we have connected node ⓧ to node ⓐ. The node equations for both these circuits are identical to the Equations 55 to 58, so that the circuits are equivalent in the sense that all responses are identical, except that we can only say that the sum of the voltages across the split current sources is equal to the voltage across the current source in the original circuit.

In conclusion it should be noted that the topic of source splitting was not covered under Section 5.4, "*Linear Circuits and Their Equivalents*," because our proof only involves the KCL and KVL equations and source constraints. The other branch constraints can be *linear or nonlinear*. Below

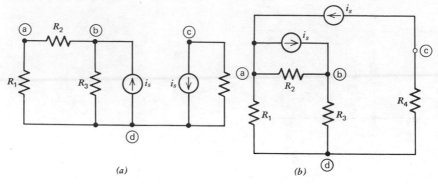

Figure 5.57 Equivalent circuits.

we present an example of how voltage source and current source splitting can be used in circuit analysis.

As an example, consider the circuit in Figure 5.58. Assume that only the current i_1 is of interest. This current will be found by reducing this circuit to a simplified model using source splitting and source transformations. The reduction technique is not unique. In Figure 5.59a we begin the reduction by splitting the voltage source (pushing it through node ⓒ; note that the voltage source could have been pushed through node ⓑ instead of ⓒ). In order to combine the source branches with the 3-Ω and 4-Ω resistors respectively, they must be converted to their Norton equivalents as shown in Figure 5.59b. Note that a series combination of a current source and a voltage source simply reduces to the current source. Now we have two sets of parallel branches in series. In order to reduce them we must convert the parallel branches to their Thévenin equivalents, which yields the simple circuit in Figure 5.59c. Thus, without the need for numerous circuit equations we see that

$$i_1 = -\frac{1}{6} \text{ A}$$

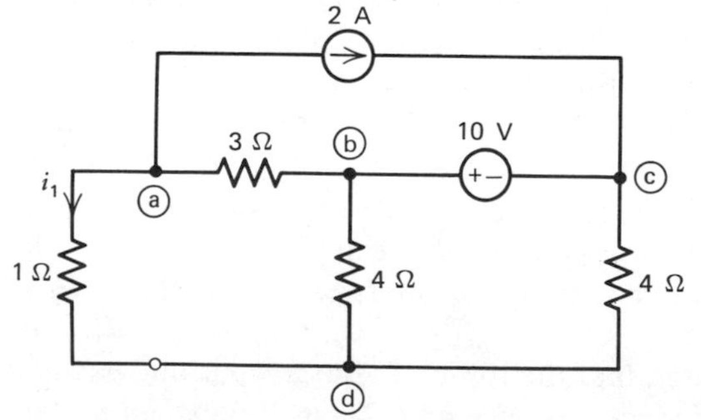

Figure 5.58 Source-splitting example.

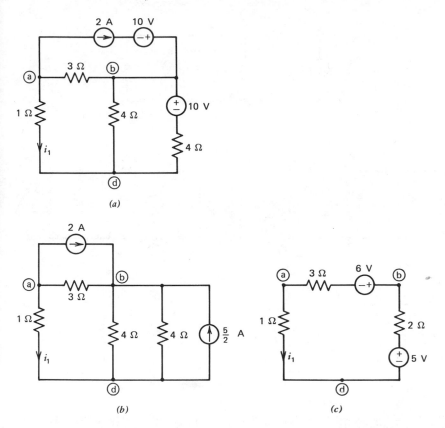

Figure 5.59   Circuit reduction with voltage splitting. (*a*) Voltage source pushed through node c. (*b*) Source branches replaced by Norton equivalents. (*c*) Source branches replaced by Thévenin equivalents.

Now the above reduction could begin by first splitting the current source instead of the voltage source as shown in Figure 5.60*a*. We can now use series/parallel reduction techniques. However, if we convert the 1-Ω resistor and 2-A source to a Thévenin equivalent we lose the branch current i_1. Therefore, we begin on the right side of Figure 5.60*a* and convert the 4-Ω resistor and 2-A source to a Thévenin equivalent, Figure 5.60*b*. We then combine the 10-V and 8-V source and convert the branch to its Norton equivalent as shown in Figure 5.60*c*. The parallel 4-Ω resistors reduce to a 2-Ω resistor and the Thévenin equivalent of this 2-Ω resistor and 9/2-A source is shown in Figure 5.60*d*. Finally, the 3-Ω and 2-Ω resistors are combined and the branch is reduced to its Norton equivalent, which yields the reduced circuit in Figure 5.60*e*. In this circuit, the total current leaving

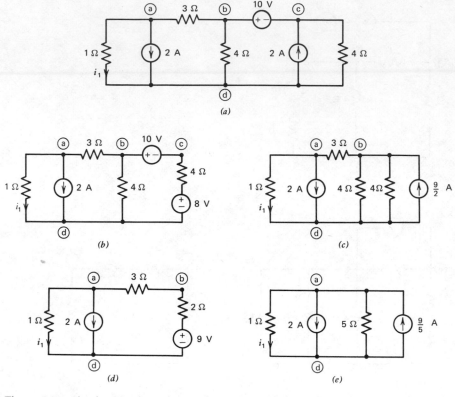

Figure 5.60 Circuit reduction with current source splitting. (a) Current source split and connected to node d. (b) Thévenin equivalent of right parallel branches. (c) Norton equivalent of series branches. (d) Thévenin equivalent of parallel branches. (e) Norton equivalent of series branches.

node ⓐ is 1/5 A. Thus, by the current divider relation,

$$i_1 = \left(\frac{5\,\Omega}{5\,\Omega + 1\,\Omega}\right)\left(-\frac{1}{5}\,A\right) = -\frac{1}{6}\,A$$

The circuit in Figure 5.58 could have been analyzed by means of the node method or mesh method. The superposition principle would also be very easy to apply to this problem. Being knowledgeable in all of these techniques gives more insight into electrical circuits and the ability to develop equivalent models that simplify analysis and design.

REFERENCES

1. Friedland, B., O. Wing, and R. Ash. *Principles of Linear Networks*, Chapter 3, McGraw-Hill, 1961.

PROBLEMS

1. Find the equivalent resistance of the one-ports in Figure P5.1.

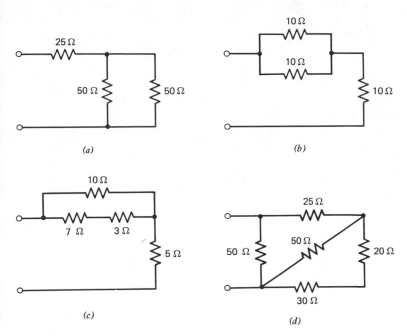

Figure P5.1

2. Use the current divider expression to find i_1 and i_2 in Figure P5.2a, and the voltage divider expression to find v_o in Figure P5.2b.

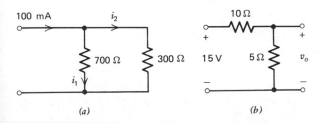

Figure P5.2

3. Use series/parallel reduction to find the power delivered by the source in Figure P5.3. Having found the voltage v_{ac}, compute v_{bc} by means of the voltage divider expression. Compute i_x by means of the current divider expression.

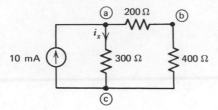

Figure P5.3

4. Use series/parallel reduction to determine the current i in Figure P5.4. Use the current divider expression to find i_1 and i_2.

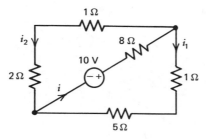

Figure P5.4

5. Find the Thévenin equivalent of the one-ports in Figure P5.5 by connecting a voltage source v to the port terminals and solving for the current i by means of the node or mesh method. What is the Norton equivalent?

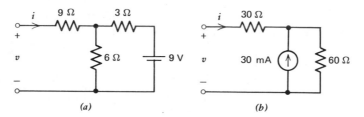

Figure P5.5

6. Repeat Problem 5, only connect a current source i to the port terminals and solve for v.
7. Repeat Problem 5 only find the equivalent resistance with the sources set to zero (series/parallel reduction may be used). Find the open circuit voltage and short circuit current in each case (the voltage divider and current divider results may be used when applicable).
8. Find the Thévenin and Norton equivalents of the one-ports in Figure P5.5 by source transformations and series/parallel reductions.

9. In Figure P5.9 what is the equivalent resistance of the one-port? What is the open-circuit voltage and the short-circuit current?

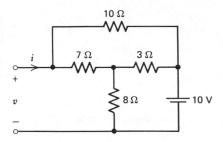

Figure P5.9

10. Find the Thévenin equivalent of the circuit in Figure P5.9 by connecting a voltage source v to the terminals of the one-port and solving for i.
11. Repeat Problem 10 only this time connect a current source i to the one-port and solve for v.
12. What are the Thévenin equivalent circuits for the one-ports in Figure P5.12? Does there exist a Norton equivalent circuit for either one-port? Explain.

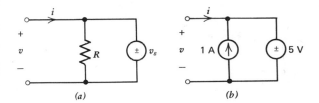

Figure P5.12

13. What are the Norton equivalent circuits for the one-ports in Figure 5.13? Does there exist a Thévenin equivalent circuit for either one-port? Explain.

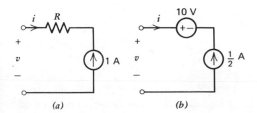

Figure P5.13

14. Find the Thévenin and Norton equivalents of the one-ports in Figure P5.14.

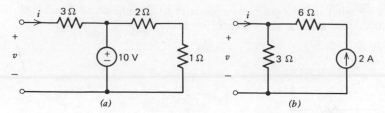

Figure P5.14

15. Use the principle of superposition to find the open-circuit voltage in Figure P5.15. What is the equivalent resistance of the one-port?

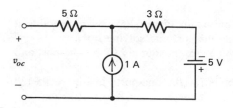

Figure P5.15

16. Repeat Problem 15 for the circuit in Figure P5.16.

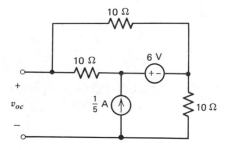

Figure P5.16

17. The Wheatstone bridge circuit is shown in Figure P5.17. It is used to measure the value of the resistor R_x. The resistor R_3 is adjusted until the current i_m through the galvanometer G is zero. (a) Remove the galvanometer and find the Thévenin equivalent circuit at terminals a and b. (b) Show that $i_m = 0$ with terminals a-b shorted when $R_1R_3 = R_2R_x$. Thus if R_1, R_2, and R_3 are precision resistors, then R_x can be determined very accurately using this null measurement method.

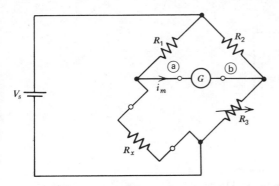

Figure P5.17

18. In the R-$2R$ ladder digital-to-analog converter suppose that $V_R = 10.24$ V. Furthermore, suppose that we want to be able to set the output voltage e_o anywhere between 0 and 10.24 V to within ± 20 mV. How many bits (coefficients b_i including b_o) are needed in the digital word? How many bits are needed if ± 5 mV accuracy is required?

19. Select the value of R in the R-$2R$ ladder digital-to-analog converter such that the output resistance (Thévenin equivalent resistance at the output port) is 2 kΩ. If $V_R = 10.24$ V what is the voltage e_o in Figure 5.23 if the binary input is 1011 and the output current is zero (open circuit). What is the voltage e_o if a 1-kΩ resistor is connected to the output terminals?

20. In Figure 5.39 and 5.40 assume that $i_1 = i_2$ so that $i_1{}^a = i_1$ and $i_1{}^b = 0$ and find the transfer relation $v_{ab}/i_1{}^a$ from the common-mode circuit. Next let $i_1 = -i_2$ so that $i_a = 0$ and $i_1{}^b = i$, and find the transfer relation $v_{ab}/i_1{}^b$. Now suppose that $i_1 = 5$ A and $i_2 = 3$ A, determine the common-mode current $i_1{}^a$ and one-half the differential mode current $i_1{}^b$. Find the response v_{ab} by means of superposition and the two transfer relations $v_{ab}/i_1{}^a$ and $v_{ab}/i_1{}^b$.

21. Use symmetrical circuit concepts to express the voltage v_o in the circuit in Figure P5.21 as a function of the common-mode input voltage and the differential-mode input voltage.

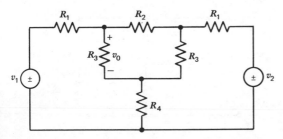

Figure P5.21

22. Use symmetrical circuit concepts to find the voltage v_o in the circuit in Figure P5.22.

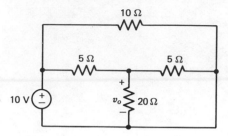

Figure P5.22

23. Repeat Problem 21 for the circuit in Figure P5.23.

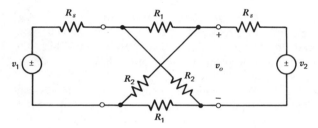

Figure P5.23

24. Use the concepts of voltage source splitting, source transformations, and series/parallel combinations to reduce the one-port in Figure P5.9 to its Thévenin equivalent.
25. Repeat Problem 24 for the circuit in Figure P5.16.
26. Use the concepts of current source splitting, source transformations, and series/parallel combinations to reduce the one-ports in Figure P5.26 to their Thévenin equivalents.

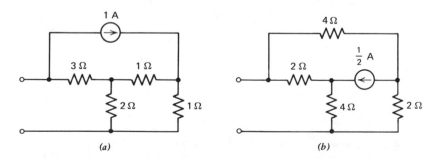

Figure P5.26

27. Repeat Problem 24 for the one-port in Figure P5.27.

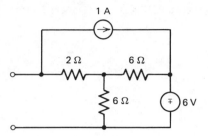

Figure P5.27

28. Repeat Problem 26 for the one-port in Figure P5.27.
29. Repeat Problem 24 for the one-port in Figure P5.29.

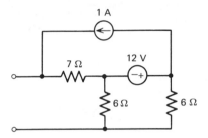

Figure P5.29

30. Repeat Problem 26 for the one-port in Figure P5.29.
31. Solve for the node voltage e_x in Figure P5.31 using network reduction principles.

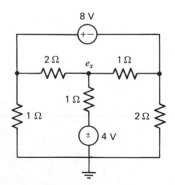

Figure P5.31

32. Solve for the voltage v_A and the current i_v in Figure P5.32 using network reduction principles.

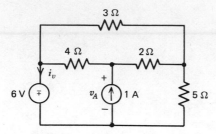

Figure P5.32

Chapter Six
The Diode

In the previous chapters our analysis was restricted to linear constant resistive circuits. Since the Kirchhoff equations are linear, linearity implies linear branch constraints. In such circuits any response can be expresssed as a summation of constants times the sources (superposition). The constants can be determined by evaluating determinants of the coefficient array of the linear equations. *In conclusion, in a linear constant circuit the response waveform is a linear sum of the input waveforms so that the frequency content of the responses is identical to the frequency content of the inputs.* If there is only one input, all responses have the same waveshape and frequency content as the input. When a circuit contains nonlinear branch constraints the response waveforms are no longer linearly related to the input waveforms and the frequency content is changed also. This ability to alter the waveshape and frequency content of a signal has many useful applications. For example, nonlinearities are deliberately introduced to rectify, limit, or modulate a signal. However, in some circuits, such as amplifiers, a faithful reproduction of the input signal is required and nonlinearities are undesirable since they cause distortion. To obtain nonlinear waveshaping the diode is the typical device used by the designer. We begin with the ideal diode.

The ideal diode is a very important nonlinear resistor that has the *i-v* characteristic shown in Figure 6.1. The ideal diode behaves as a switch, only it is controlled by the voltage across it and the current through it and

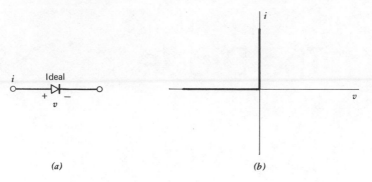

Figure 6.1   Ideal diode. (*a*) Symbol. (*b*) Characteristic.

not mechanically. For the assigned polarity of v and direction of i, the ideal diode has *zero resistance* (short circuit) in the forward direction ($i > 0$) and *zero conductance* (open circuit) in the reverse direction ($v < 0$). The diode has many useful applications, some of which are discussed below.

6.1 RECTIFYING, CLIPPING, AND LIMITING

The half-wave rectifier circuit in Figure 6.2 has the transfer characteristic shown in Figure 6.3. To verify this transfer characteristic assume that the diode is on (short circuit). Note that if $v_1 > 0$ then $i > 0$, and the diode is on so that $v_2 = v_1$. But $v_1 < 0$ implies $i < 0$, which is not allowed. This contradiction means that for $v_1 < 0$ the diode is off (open circuit) and $v_2 = 0$ for $v_1 \leqslant 0$. The output of the rectifier for an input $v_1(t) = V \sin \omega_o t$ is shown in Figure 6.3. Since the current through R can only flow in one direction note that the bottom half of the input voltage is clipped at the output. Also, note that the output is *not* sinusoidal, but it is periodic with period $2\pi/\omega_o$.

In order to find the frequency content of this periodic signal, the Fourier series representation is used. In Chapter Two it was shown that this half-wave signal has the Fourier series

$$v_2(t) = V\left(\frac{1}{\pi} + \frac{1}{2}\sin \omega_o t - \frac{2}{\pi}\sum_{n=2,4,6,\ldots}\frac{1}{n^2 - 1}\cos n\omega_o t\right) \tag{1}$$

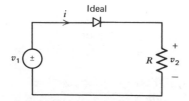

Figure 6.2   Half-wave rectifier.

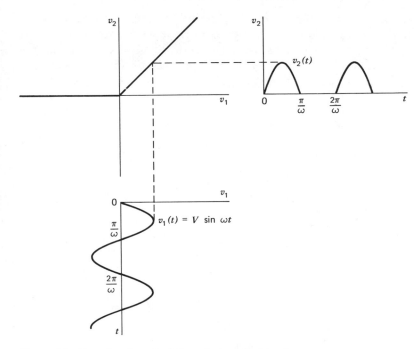

Figure 6.3 Transfer characteristic and sinusoidal input response.

Note that the frequency spectrum of the output voltage of the rectifier contains a dc component V/π plus components at the frequencies ω_o, $2\omega_o$, $4\omega_o$, and so on. In this application the rectifier circuit can be made into a dc power supply by passing the output signal through a filter that passes the dc component but shorts out all signals with frequencies higher than dc (zero), Figure 6.4. Note that it must be assumed that the input current to the filter is zero in order for the transfer characteristic in Figure 6.3 to be a valid description of the circuit response. A first-order approximation to the filter consists of a capacitor shunting the resistor. As will be seen later the capacitor is an open circuit to dc signals, and a short circuit to rapidly varying signals.

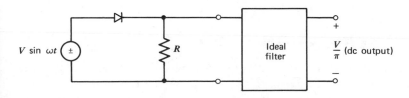

Figure 6.4 DC power supply circuit.

One final example in this section is the limiter or clipper circuit in Figure 6.5. When $v_1 > V$ the diode is on, and the diode is off for $v_1 \leqslant V$. The transfer characteristic for this device is shown in Figure 6.6. Note that the output signal is clipped at or limited to V volts. This circuit can be used to protect a system from excessive voltage peaks that could destroy it.

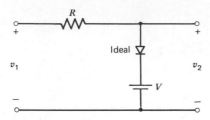

Figure 6.5   Limiter circuit.

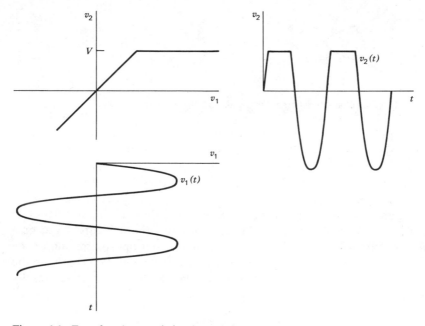

Figure 6.6   Transfer characteristic of the limiter.

6.2 COMMUNICATION CIRCUITS

In communication systems usually many signals are transmitted over the same channel. In order to separate the signals at the receiving end of the channel they must be transmitted in a time-multiplexed or frequency-multiplexed form, that is, each signal is allotted a certain time segment or a particular segment of the frequency spectrum. The frequency spectrum of a signal is changed by modulating a carrier signal with the information signal.

A simple amplitude modulator is shown in Figure 6.7. Let us assume that the information signal $g(t) = V_s \sin \omega_s t$ and $c(t)$ is the carrier signal shown in Figure 6.8 where $V_s < V$ and $\omega_s \ll \omega_c$. The purpose of the carrier signal is to switch the diode on and off. Thus the output voltage $v_o(t)$ is as shown in Figure 6.9. In order to find the frequency spectrum of this output signal note that this output can be represented by the expression

$$v_o(t) = (V + V_s \sin \omega_s t) \cdot s(t) \tag{2}$$

where $s(t)$ is the switching function in Figure 6.10. From Chapter Two the

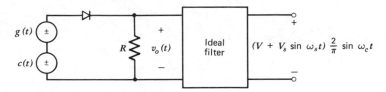

Figure 6.7   Amplitude modulator circuit.

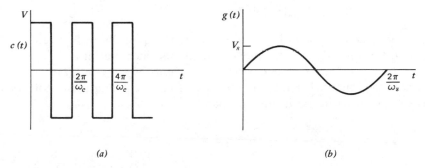

(a)   (b)

Figure 6.8   Input signals. (a) Carrier signal. (b) Information signal.

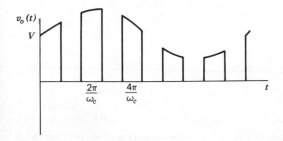

Figure 6.9   Output signal.

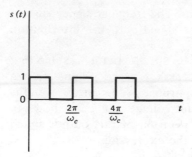

Figure 6.10   Switching signal.

Fourier series representation of this pulse is

$$s(t) = \frac{1}{2} + \frac{2}{\pi} \sum_{n=1,3,5,7,\ldots} \frac{1}{n} \sin n\omega_c t \qquad (3)$$

The substitution of (3) into (2) yields

$$v_o(t) = \frac{V}{2} + \frac{V_s}{2} \sin \omega_s t + \frac{2V}{\pi} \sin \omega_c t + \frac{V_s}{\pi}[\cos(\omega_c - \omega_s)t - \cos(\omega_c + \omega_s)t]$$

$$+ \frac{2V}{3\pi} \sin 3\omega_c t + \frac{V_s}{3\pi}[\cos(3\omega_c - \omega_s)t - \cos(3\omega_c + \omega_s)t] + \cdots \qquad (4)$$

The frequency components of (4) are illustrated graphically in Figure 6.11. We wish to transmit only the frequency ω_c and its sidebands at $\omega_c - \omega_s$ and $\omega_c + \omega_s$, that is, the modulated carrier component $(V + V_s \sin \omega_s t)\frac{2}{\pi} \sin \omega_c t$ as indicated in Figure 6.7. Thus, the purpose of the ideal filter is to short all signals except those in the frequency band $\omega_c - \omega_s \leq \omega \leq \omega_c + \omega_s$.

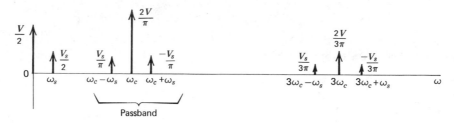

Figure 6.11   Frequency spectrum of $v_o(t)$.

6.3 LOGIC CIRCUITS

In digital systems information is transmitted by bits. A bit is denoted by a 1 or a 0. In positive logic systems the 1 corresponds to a high voltage and the 0 corresponds to a low voltage. In a magnetic memory a 1 corresponds

to a given polarization of the material and a zero to the opposite polarization. Usually bits are grouped into words and the words denote instructions
or data. An n-bit word can represent 2^n different instructions or data. For
example, for a three-bit word there are eight possible instructions or data
as shown below.

$$000$$
$$001$$
$$010$$
$$011$$
$$100$$
$$101$$
$$110$$
$$111$$

The word 101 could tell the digital machine to print the contents in its
memory, read data cards at the input, add two pieces of data, or in control
situations turn off or on a motor. In the binary number system, the bits
represent the coefficients of powers of *two* rather than *ten* as in the
decimal system. For example, the binary word for the decimal number 5 is

$$101 \equiv 1 \times 2^2 + 0 \times 2^1 + 1 \times 2^0$$

The digital machine determines which instruction word was sent and
performs the operation dictated by that instruction by means of logic gates.
Two important logic gates are the AND gate and the OR gate shown
symbolically in Figure 6.12 and 6.13. The truth tables indicate the logical
output for a given combination of logical inputs. In the OR gate note that
the logical output is zero only when all the logical inputs are zero. In the AND
gate the logical output is "1" only when all the logical inputs are "1." The
above logic gate functions can be implemented with diode circuits as shown in
Figure 6.14. However, in practice the diodes are not ideal, and, on cascading
the gates, the output current i_o is no longer zero so that the circuit response
quickly deteriorates. Practical digital gate circuits employ transistors that are
discussed in Chapter Seven.

Let us give a simplified example in order to see how logic gates can be
used in a digital system to make decisions. Suppose that we have a candy
machine with 15c candy. The machine will accept three nickels or a dime
and a nickel. Therefore, assume that we have three slots for nickels in the
machine labeled n_1, n_2, and n_3, and one slot for a dime labeled d_1. If we
assume that the first nickel always falls into slot n_1 then the combinations
that yield a candy bar are $n_1n_2n_3$ or n_1d_1 where the symbolism $n_1n_2n_3$ means
n_1 and n_2 and n_3. A circuit realization of this logic function is shown in
Figure 6.15. Note that the output is 1 if and only if we have three nickels
and/or a dime and one or more nickels.

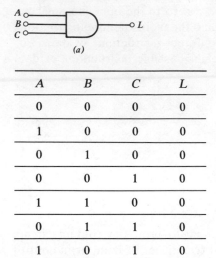

Figure 6.12   AND gate. (a) Symbol. (b) Truth table.

A	B	C	L
0	0	0	0
1	0	0	0
0	1	0	0
0	0	1	0
1	1	0	0
0	1	1	0
1	0	1	0
1	1	1	1

(b)

Figure 6.12   AND gate. (a) Symbol. (b) Truth table.

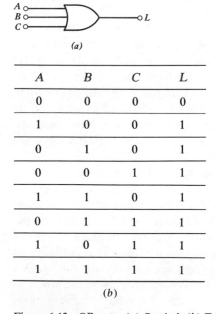

(a)

A	B	C	L
0	0	0	0
1	0	0	1
0	1	0	1
0	0	1	1
1	1	0	1
0	1	1	1
1	0	1	1
1	1	1	1

(b)

Figure 6.13   OR gate. (a) Symbol. (b) Truth table.

164

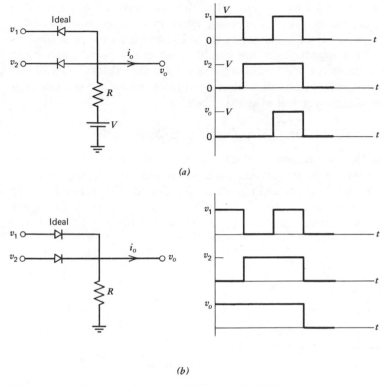

Figure 6.14 Diode logic gates. (*a*) AND gate. (*b*) OR gate.

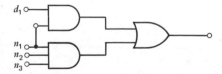

Figure 6.15 Candy machine logic circuit.

6.4 SEMICONDUCTOR DIODE

The original diode was the electron tube discussed in Chapter One. It consists of a pair of metal plates—the anode and cathode—in a vacuum. The cathode is heated to 1000 K to 2500 K by a heater. At these temperatures the electrons in the cathode have sufficient energy to break from their atomic bonds and enter the vacuum. Since only the cathode is a source of free electrons, when the anode is positive with respect to the cathode current flows. But if the anode is negative with respect to the cathode, electrons are repelled from the anode and no current flows.

However, experimenters have found that this property of unidirectional current flow can also be realized across a junction between two dissimilar semiconductor materials. Furthermore, the semiconductor diode is smaller, does not require a heater, does not dissipate as much power, and more closely approximates the ideal diode characteristic than the vacuum diode. For these reasons the vacuum diode has been replaced by the semiconductor diode in modern electrical systems.

6.4.1 DERIVATION OF THE DIODE EQUATION

The construction of the semiconductor diode is shown in Figure 6.16a. It consists of a semiconductor crystal such as silicon with one side doped with acceptor atoms (p type) and the other side doped with donor atoms (n type). A typical i-v characteristic of a silicon diode is shown in Figure 6.16b. The behavior of this pn junction is explained below.

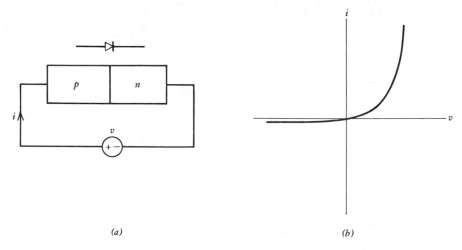

(a)   (b)

Figure 6.16   Semiconductor diode. (a) Circuit. (b) i-v characteristic.

Let us assume that the voltage v is zero as shown in Figure 6.17a. In Figure 6.17a the right side of the semiconductor material (n region) contains more free electrons than the left side of the material (p region). Similarly, the left side has more holes than the right side. Thus electrons diffuse to the left and holes diffuse to the right in Figure 6.17a. However, when an electron moves from the n region to the p region it leaves behind a positively charged atom. Similarly, when a hole moves from the p region to the n region it leaves behind a negatively charged atom. In Figure 6.17 we assume that all free electrons within a distance x_n have diffused into the p region, and all holes within a distance x_p of the junction have diffused

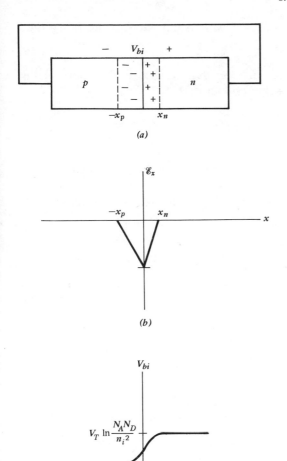

Figure 6.17   (a) Depletion region ($v = 0$). (b) Electric field across pn junction. (c) Built-in voltage.

into the n region. Thus the region bounded by $-x_p \leq x \leq x_n$ is depleted of all mobile charges so that it is called the charge-depletion region. The total charge on the right side of the depletion region is

$$Q_+ = eN_dAx_n \tag{5}$$

where $A(\text{cm}^2)$ is the area of the junction, $x_n(\text{cm})$ is the width of the depletion region in the n-type semiconductor and N_d is the number of donor atoms/cm³. We are assuming that all donor atoms are ionized in the depletion region (no free electrons). If we assume charge neutrality then

$$N_A A x_p = N_D A x_n \tag{6}$$

where N_A is the number of acceptor atoms/cm^3 and x_p is the width of the depletion region in the p-type semiconductor. Thus

$$x_p = \frac{N_D}{N_A} x_n \tag{7}$$

In Figure 6.17a it is presumed that $N_D > N_A$ as the figure shows $x_p > x_n$. *Note that the depletion width is greatest on the more lightly doped side of the semiconductor.*

In order to compute the electric field, from Chapter One recall that

$$\frac{d\mathscr{E}_x}{dx} = \rho/\epsilon \tag{8}$$

With zero volts applied to the semiconductor diode the current flow is zero and the electric field is confined to the depletion region. The integration of (8) yields the function in Figure 6.17b. The maximum value of the field occurs at $x = 0$.

$$|\mathscr{E}_x|_{x=0} = eN_A x_p/\epsilon = eN_D x_n/\epsilon \tag{9}$$

where $\rho = N_A$ on the left side of the junction and $\rho = N_D$ on the right side of the junction. From assumption (6) $N_A x_p = N_D x_n$ so that the electric field goes to zero at $x = x_n$.

The potential difference at the junction can be found as follows. The current flow per unit area across the junction due to electron flow is

$$J_n = e\left(u_n n \mathscr{E}_x + D_n \frac{\partial n}{\partial x} \right) \tag{10}$$

and that due to hole flow is

$$J_p = e\left(u_p p \mathscr{E}_x - D_p \frac{\partial p}{\partial x} \right) \tag{11}$$

where μ denotes the mobility of the carriers cm^2/V-s and D(cm^2/s) is called the diffusion constant. The first term in (10) and (11) is the current flow caused by the electric field in the semiconductor, and the second term is the current flow caused by the diffusion of carriers to a less dense region. We said that when the external voltage is zero the current is zero. This implies that $J_n = 0$ and $J_p = 0$ so that

$$\mu_n n \mathscr{E}_x = - D_n \frac{\partial n}{\partial x} \tag{12}$$

and

$$\mu_p p \mathscr{E}_x = D_p \frac{\partial p}{\partial x}, \tag{13}$$

That is, the diffusion current and field current are equal and opposite. For example, in Figure 6.17a the electrons diffuse to the left, but due to the electric field they experience a force to the right.

In order to solve for the voltage across the depletion region recall that

$$\frac{dv}{dx} = -\mathscr{E}_x \tag{14}$$

Substituting (14) into (12) we obtain

$$\int dv = \frac{D_n}{\mu_n} \int \frac{dn}{n} \tag{15}$$

The density of free electrons in the p region is n_i^2/N_A (see Chapter One) and in the n region it is N_D. Thus (15) becomes

$$V_{bi} = \frac{D_n}{\mu_n} \ln n \bigg|_{n_i^2/N_A}^{N_D} = \frac{D_n}{\mu_n} \ln \frac{N_A N_D}{n_i^2} \tag{16}$$

We write

$$\boxed{V_{bi} = V_T \ln \frac{N_A N_D}{n_i^2}} \tag{17}$$

where $V_T \overset{\Delta}{=} kT/q = D_n/\mu_n$ is approximately 26 mV at room temperature. This voltage is called the *built-in junction potential*. As an example suppose that the semiconductor is silicon in which the intrinsic concentration of the electrons at room temperature in the pure crystal is $n_i = 1.5 \times 10^{10}$ electrons/cm^3. Let $N_D = 10^{18}$ donor atoms/cm^3 and $N_A = 10^{16}$ acceptor atoms/cm^3, then

$$V_{bi} = (26 \text{ mV}) \ln \left(\frac{10^{34}}{2.25 \times 10^{20}} \right) \tag{18}$$

or

$$V_{bi} = 0.82 \text{ V} \tag{19}$$

This built-in voltage is illustrated in Figure 6.17a and does not cause a current to flow and thus does not violate our assumption of zero current, because in equilibrium similar built-in voltages occur at the junctions between the connecting wires and the n and p regions so that the net voltage drop around the loop is zero. This equilibrium can be upset by having the junctions at different temperatures or by radiating one of the regions with light that creates excess minority carriers (principle of the photocell). When equilibrium is upset the diode becomes a source of energy.

The built-in voltage across the pn junction is a potential barrier that resists the flow of electrons into the p region and the flow of holes into the n region. We now apply a forward voltage to the diode as shown in Figure

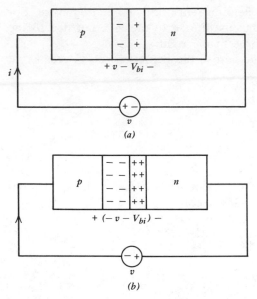

Figure 6.18   Depletion region behavior with a nonzero external voltage source. (*a*) Forward bias reduces the junction voltage. (*b*) Reverse bias increases the junction voltage.

6.18*a*. Assuming that the voltage drop in the *n* and *p* regions is negligible, then the total drop across the junction is $v - V_{bi}$, that is, a forward voltage reduces the potential barrier and allows electrons and holes to diffuse across the junction. In the forward direction the current increases exponentially with the applied voltage. Now if a reverse voltage is applied as shown in Figure 6.18*b* then the potential barrier increases and few carriers have enough energy to surmount the barrier. It can be shown that the diode current is[1]

$$i = I_s(e^{v/\eta V_T} - 1) \tag{20}$$

where I_s is called the saturation current, $V_T \overset{\Delta}{=} kT/q$ and is equal to 26 mV at room temperature, η is an empirical constant typically bounded by $1 \le \eta \le 2$. The saturation current I_s is directly proportional to the area A of the *pn* junction. If we double A, then we double the current I_s.

6.4.2 MEASUREMENT OF DIODE PARAMETERS

The diode equation (20) is plotted in Figure 6.19 for a silicon diode with $I_s = 10^{-14} A$ and $\eta = 1$. To dramatize the exponential increase in current, the value of the current as a function of v is listed in Table 6.1. For a 0.1 V

[1] A. S. Grove, *Physics and Technology of Semiconductor Devices*, Wiley, 1967.

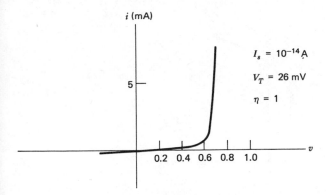

Figure 6.19   Silicon diode characteristic.

Table 6.1   Forward Biased Diode

v	e^{v/V_T}	$I \approx I_s e^{v/V_T}$	$r \approx V_T/I$
0.5	2.2×10^8	0.002 mA	13 kΩ
0.6	1.05×10^{10}	0.105 mA	248 Ω
0.7	5.0×10^{11}	5.0 mA	5.2 Ω
0.8	2.31×10^{13}	231 mA	0.1 Ω

increase in voltage the current changes by approximately a factor of "50." Also e^{v/V_T} is so large in this region that the -1 term in (20) can be neglected. Thus we can write

$$i \approx I_s e^{v/\eta V_T} \qquad v \gg V_T \tag{21}$$

and the incremental conductance

$$r^{-1} = \frac{\partial i}{\partial v} = \frac{1}{V_T} I_s e^{v/V_T} \approx \boxed{I/V_T} \tag{22}$$

which is also listed in Table 6.1. *Note that the incremental conductance is simply the current through the diode divided by V_T.* This is a simple and useful expression to remember.

To determine I_s experimentally let us write

$$v = \frac{\eta V_T}{\log_{10} e} \log_{10} i/I_s \tag{23}$$

Equation 23 is plotted in Figure 6.20 on a semilog graph. Note that the ordinate intercept ($v = 0$) corresponds to $i = I_s$. Thus, experimentally one determines I_s by measuring v for several values of i where i is not too

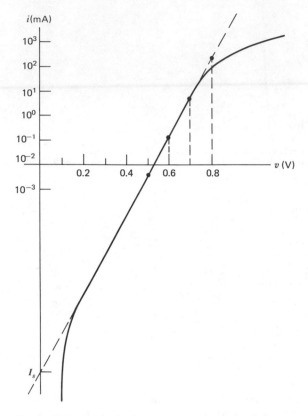

Figure 6.20   Determination of the reverse saturation current I_s.

large nor too small. Then fit the points with a dotted line as shown in Figure 6.20. The solid line gives the actual characteristic. At large forward currents the voltage drop outside the depletion region due to the resistivity of the semiconductor becomes significant so that the current no longer increases exponentially, but linearly. At low currents the -1 term in (20) cannot be neglected in the sketch of (23).

6.4.3 MODELING THE DIODE

Table 6.1 and Figure 6.19 illustrate that the silicon diode closely approximates the ideal diode except that in the forward direction the voltage drop is approximately 0.7 V. Thus, in analysis one frequently replaces the diode by a 0.7-V source when the diode is forward biased, and an open circuit when it is reverse biased.[2]

---

[2] In Germanium diodes the saturation current is much larger, for example $I_s \approx 10^{-9}$ A, so that the knee in the diode characteristic occurs at approximately 0.3 V in Figure 6.19.

However, in some applications the incremental resistance of the diode is important. In such cases the diode is modeled by a line tangent to its i-v characteristic as shown in Figure 6.21a. The circuit model for this linear characteristic is shown in Figure 6.21b. The branch constraint for this linear model is

$$v_\ell = V_{oc} + ri_\ell \qquad (24)$$

Note that V_{oc} is the open circuit voltage and r is the incremental resistance. As an example, consider the diode in Figure 6.19 and suppose that we wish to model the diode with a linear circuit whose constraint is tangent to the point $i = 5$ mA, $v = 0.7$ V. From (22) the incremental resistance is

$$r = V_T/I_o = 26 \text{ mV}/5 \text{ mA} = 5.2 \, \Omega \qquad (25)$$

The substitution of r, i, and v into (24) yields the open circuit voltage for the model.

$$V_{oc} = 0.7 \text{ V} - (5.2 \, \Omega)(5 \text{ mA}) = 0.674 \text{ V} \qquad (26)$$

If we set $r = 0$, then $V_{oc} = 0.7$ V and the model reduces to a voltage source whose constraint is a vertical line passing through the point $i = 5$ mA, and $v = 0.7$ V in Fig. 6.21a.

The tangent approximation is equivalent to taking the first two terms of the Taylor expansion[3] of the diode characteristic about a point (i_o, v_o),

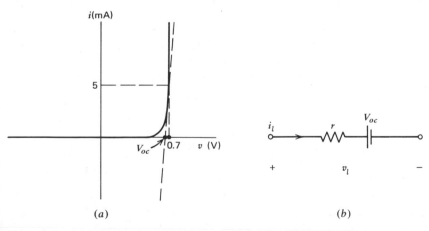

(a) (b)

Figure 6.21 (a) Linearization of the forward diode characteristic. (b) Linear model.

[3] The Taylor series expansion of a function $y = f(x)$ about a point (x_o, y_o) is defined as

$$y = f(x_o) + \frac{df}{dx}\bigg|_{x=x_o}(x - x_o) + \sum_{n=2}^{\infty} \frac{1}{n!}\frac{d^n f}{dx^n}\bigg|_{x=x_o}(x - x_o)^n$$

The terms of order $n \geq 2$ are the nonlinear terms.

that is,

$$i_\ell = I_s(e^{v_0/V_T} - 1) + \frac{I_s}{V_T} e^{v_0/V_T}(v_l - v_0) \tag{27}$$

The factor

$$\frac{I_s e^{v_0/V_T}}{V_T} \approx \frac{i_o}{V_T} \tag{28}$$

is the reciprocal of the incremental resistance so that (27) can be written as

$$i_\ell = \left(i_o - \frac{v_0}{r}\right) + v_\ell/r \tag{29}$$

or

$$v_\ell = ri_\ell + (v_o - ri_o) \tag{30}$$

where $i_o - v_o/r$ is the short circuit current of the model (current axis intercept) and

$$v_o - ri_o = V_{oc} \tag{31}$$

is the open circuit voltage of the model.

6.4.4 REVERSE BREAKDOWN

Finally let us point out one other defect in (20). In the reverse direction the *pn* junction breaks down as shown in Figure 6.22. The breakdown voltage is given by the empirical formula

$$V_B = (2.7)(10^{12})N^{-2/3} \tag{32}$$

where N is the majority carrier concentration of the more lightly doped junction and is valid for $N < 10^{18}$. For example, for $N = 10^{17}$, $V_B = 15$ V and for $N = 10^{15}$, $V_B = 300$ V. Equation 32 models the avalanche break-

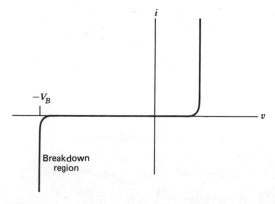

Figure 6.22   Semiconductor diode characteristic including breakdown.

down that results from the resulting field in the depletion region breaking covalent bonds and the resulting mobile electrons colliding with other atoms and releasing additional electrons. When $N > 10^{18}$ a different breakdown mechanism occurs called Zener breakdown. In a heavily doped pn junction a quantum mechanical phenomenon called tunneling occurs in which charges "tunnel" through the potential barrier. Thus, in heavily doped junctions the primary breakdown mechanism is Zener.

The reverse breakdown region closely approximates an ideal voltage source. Thus, the diode can be used for voltage regulation. Diodes used for this purpose are called Zener diodes, although the breakdown mechanism can be avalanche or Zener.

6.5 SMALL-SIGNAL ANALYSIS

The analysis of nonlinear circuits is significantly more difficult than the analysis of linear circuits. If precision is required then iterative techniques must be used, and for practical purposes the aid of a computer is required to carry out all the computational steps. However, frequently a high degree of precision is not required. In such cases approximate analytical techniques can be used. These approximate methods usually give the engineer a better intuitive feel for the relation between the circuit parameters and the response. In this section we discuss a very common analysis technique for electronic circuits which is useful when the nonlinear components can be realistically approximated by linear elements.

Consider the voltage regulator example in Figure 6.23. The 12 V supply has a ripple component v_r. This ripple voltage can cause trouble in digital logic systems and in communication systems. Suppose that the load R_L needs only 6 V, then by placing a 6 V Zener diode across the load and operating the diode in its breakdown region the load voltage v_L is relatively invariant to changes in the supply voltage represented by v_r. In general when a circuit contains dc sources and time-varying sources we will write a response x_A as the sum of two components $x_A = X_A + x_a$ where X_A denotes the response when all the time-varying sources are zero and x_a denotes the

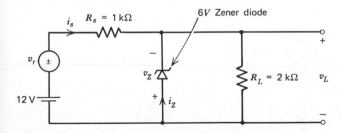

Figure 6.23   Voltage regulator circuit.

variation of the response about the dc response. Thus a capital letter with a capital subscript will denote the dc response, a small letter with a small subscript will denote the variation about the dc response, and a small letter with a capital subscript will denote the total response. To determine how much the ripple is reduced by the Zener diode we proceed as follows.

6.5.1 OPERATING POINT

The first step is to turn off all sources (a voltage source is replaced by a short circuit and a current source is replaced by an open circuit) except for the dc supplies, which are called bias supplies in electronic circuits. Next, the operating point is located for each nonlinear element. This is the dc solution. When only one nonlinear element is present, Thévenin's theorem can be used to determine the load line. The intersection of the load line with the nonlinear characteristic indicates the operating point. For example, at the diode terminals the open circuit voltage is $v_Z = -8$ V and the short circuit current is $i_Z = -12$ mA. These two points determine the load line that has a slope of $-1/667$ where $667\,\Omega$ is the Thévenin circuit resistance ($2\,\mathrm{k}\Omega \| 1\,\mathrm{k}\Omega$) seen by the diode. This load line is plotted in Figure 6.24. It intersects the diode characteristic at $V_Z \approx -6$ V and $I_Z \approx -3$ mA, which is the dc solution. This solution is sometimes called the bias point, Q point, or operating point. In order to find the variation v_ℓ about this operating point due to the ripple v_r, we introduce linear approximations in the next section.

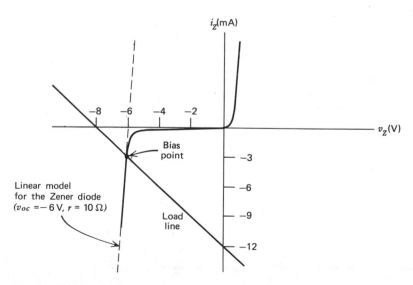

Figure 6.24 Operating point and load line.

6.5.2 LINEARIZATION OF THE NONLINEAR CHARACTERISTICS

In the operation of the regulator the diode should always be operated in its breakdown region. Note that in Figure 6.24 in this breakdown region the diode is very nearly linear; thus the diode can be approximated by the dashed line. The circuit equivalent of this dashed line is a 10-Ω resistor (incremental resistance of the diode in the breakdown region) in series with -6 V (the voltage axis intercept). These values are typically determined by measurement techniques. Figure 6.25 illustrates the linearized circuit. The caret above the responses indicates that since Figure 6.25 is an approximation of Figure 6.23 the responses will not necessarily be identical.

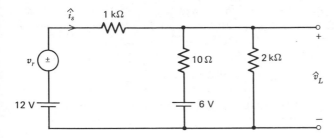

Figure 6.25   Linearized regulator circuit.

6.5.3 SMALL-SIGNAL MODEL

Once the circuit has been linearized, the *superposition principle* can be applied, and the response can be separated into two parts, a *dc response* determined by Figure 6.26a, and a *small signal* response determined by analyzing Figure 6.26b. For example, the ripple voltage across the load is found by analyzing the small signal circuit.

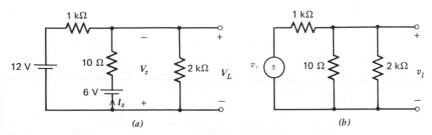

Figure 6.26   Separation of dc bias and small-signal responses. (a) dc bias circuit. (b) Small-signal circuit.

$$v_l = \cfrac{\cfrac{2\,k\Omega \times 10\,\Omega}{2\,k\Omega + 10\,\Omega}}{\cfrac{2\,k\Omega \times 10\,\Omega}{2\,k\Omega + 10\,\Omega} + 1\,k\Omega} \cdot v_r$$

or

$$v_l \approx 10^{-2} v_r \tag{33}$$

The small signal analysis indicates that the supply voltage ripple is reduced by a factor of a hundred at the load terminals. The term small signal is used because our analysis of the linear circuit model is accurate only in the region where the linear model closely approximates the nonlinear device characteristic. If v_r is too large the solution will be driven out of this region. For example, in Figure 6.27 we graphically illustrate the solutions for the circuits in Figures 6.23 and 6.25 when $v_r = -6$ V at some time instant. The solution for the actual circuit is point B whereas the solution for our linear model is point A. Clearly there is a large error at this time instant. In this example if $i_z > 0$ for some values of v_r (i.e., v_r becomes too large) then the linear model is not accurate in this region.

The small-signal circuit concept is important since it finds wide use the analysis of electronic amplifiers that are discussed in the next chapter and

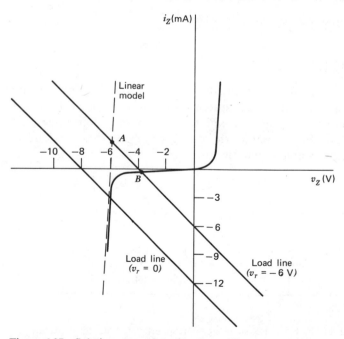

Figure 6.27   Solution comparison for $v_r = -6$ V.

in many other engineering systems in which nonlinear components are operated in a nearly linear mode.

6.6 TUNNEL DIODE

The tunnel diode characteristic in Figure 6.28 is different from the ordinary diode characteristic (dashed line). The tunnel diode is a very heavily doped pn junction in which, in addition to the usual diode conduction mechanism, current flows by means of "tunneling." Tunneling is a quantum mechanical phenomenon in which electrons on one side of a potential barrier have a high probability of reaching the other side of the potential barrier without acquiring sufficient energy to surmount the barrier provided that the barrier width is sufficiently narrow and there exists empty energy states on the other side of the barrier which the tunneling electrons can occupy. Recall that in heavily doped pn junctions the depletion width is indeed narrow.

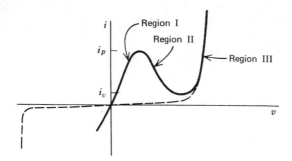

Figure 6.28   Tunnel diode characteristic.

In Figure 6.28 the peak current i_p and valley current i_v are critical parameters. Between these two currents the diode voltage can have three different values. Thus, the three segments of the curve are designated as region I, region II, and region III. In region I tunneling occurs and in region II the tunneling phenomenon decreases. Also, in region II note that the incremental resistance is negative. To show some of the interesting applications of the tunnel diode let us use the simple circuit in Figure 6.29.

Firstly let us select R and V_B to realize the load line in Figure 6.30. Three operating points are possible, A, B, and C. In practice the circuit never operates in state B. Therefore the circuit is called a "bistable circuit" or "flip-flop," because an input as shown in Figure 6.31a can switch the diode from state A to state C and vice versa resulting in the output in Figure 6.31b. A positive pulse switches the diode to state C provided that the load line is shifted above the peak current i_p as shown in Figure 6.30,

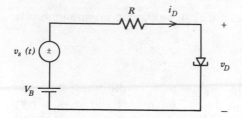

Figure 6.29   Tunnel diode circuit.

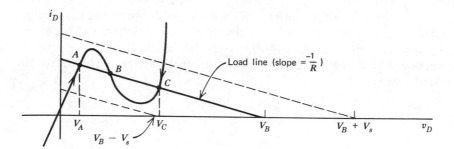

Figure 6.30   Bistable operation.

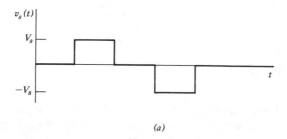

(a)

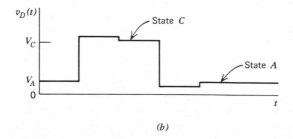

(b)

Figure 6.31   Switching states with input pulses. (a) Input pulse. (b) Output voltage.

180

and a negative pulse switches the diode from C to A provided that the load line drops below the current i_v as shown in Figure 6.30. In digital circuits the flip-flop is made from transistors since it is easier to cascade the transistor flip-flops with other circuits. The tunnel diode bistable circuit is used sometimes if extremely high switching speeds are required.

Finally, if the tunnel diode is biased in the negative resistance region as indicated in Figure 6.32 then a small change in the input voltage causes a larger change in the output voltage. For example, from Figure 6.32 the input change $v_s = AB$ causes the output change $v_d = AC$. Obviously

$$\frac{v_d}{v_s} = \frac{AC}{AB} > 1 \tag{34}$$

The approximate gain can be calculated using small-signal analysis. The small-signal model is illustrated in Figure 6.33 where r_d is the incremental resistance of the tunnel diode at point A in Figure 6.32. Note that r_d is negative in this region.

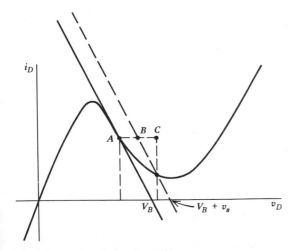

Figure 6.32 Small-signal amplifier.

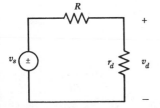

Figure 6.33 Small-signal model of tunnel diode amplifier.

The tunnel diode is seldom used as an amplifier because of its sensitivity to changes in the bias V_B, which is only a few millivolts. A small change in bias can shift the operating point into region I or III where it ceases to amplify. Also if v_s is too large the operation extends into region I or III and these nonlinear effects can severely distort the signal. Finally, again it is difficult to cascade diode stages since they tend to load each other too much by drawing currents that affect the operation of the other stages.

REFERENCES

1.   Gray, P. E. and C. L. Searle. *Electronic Principles: Physics, Models, and Circuits,* Chapters 4, 6, Wiley, 1969.

PROBLEMS

1.   Sketch the transfer characteristic v_2 versus v_1 of the circuit in Figure P6.1. If $v_1 = V \sin \omega_o t$ sketch v_2.

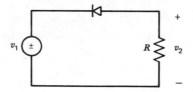

Figure P6.1

2.   Sketch the DP characteristic i versus v for the circuits in Figure P6.2.

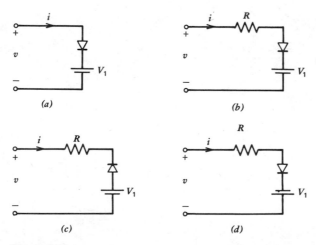

Figure P6.2

3. Repeat Problem 2 for the circuits in Figure P6.3.

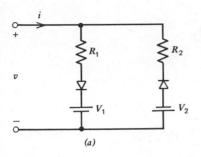 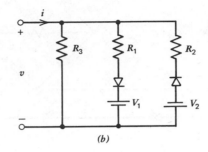

Figure P6.3

4. Repeat Problem 2 for the circuits in Figure P6.4.

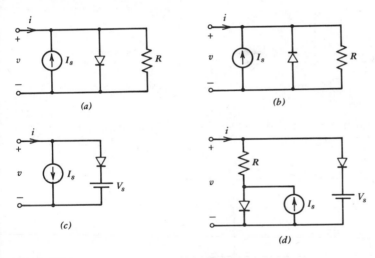

Figure P6.4

5. In the half-wave rectifier power supply circuit, Figure 6.4, what amplitude of the input sinusoid is needed to obtain a 5-V dc ouput? Suppose that the filter is not ideal and allows 1% of the signal at the frequency ω_o through. What is the variation in the dc response at the output? Neglect the other frequency components ($2\omega_o$, $4\omega_o$, etc.).

6. Sketch the transfer characteristics v_2 versus v_1 of the two full-wave rectifier circuits in Figure P6.6. If $v_1 = V \sin \omega_o t$, sketch $v_2(t)$. What is the relation between the dc component of the output of the full-wave rectifier to the dc component of the output of the half-wave rectifier for the same input? Does the output voltage of the full-wave rectifier have a frequency component ω_o as does the half-wave rectifier?

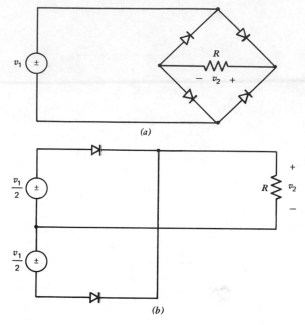

Figure P6.6

7. Sketch the transfer characteristic of the circuit in Figure P6.7. If $v_s = V \sin \omega t$ where $V > V_1, V_2$, sketch $v_2(t)$.

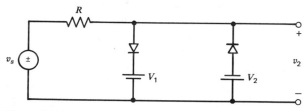

Figure P6.7

8. Verify Equation 4 using trigonometric identities.

9. In Figure 6.11 note that in the passband of the simple amplitude modulator circuit in Figure 6.7 there is a strong spectral component of amplitude $2V/\pi$ at the frequency ω_c. Sometimes it is desirable to conserve power by not transmitting this carrier frequency component, but only the information components at the sidebands $\omega_c - \omega_s$ and $\omega_c + \omega_s$. This is done by means of the balanced amplitude modulator circuit in Figure P6.9. Show that

$$v_o(t) = (V_s \sin \omega_s t) \cdot s(t)$$

and plot the spectrum of $v_o(t)$ as was done in Figure 6.11 for the simple modulator. The signals $g(t)$ and $c(t)$ are as defined in Figure 6.8 where $V > V_s/2$.

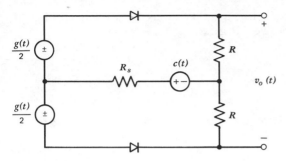

Figure P6.9

10. In the binary number system what is the decimal equivalent of the five-bit binary number 01011? What is the largest decimal that a five-bit binary word can represent? What is the largest decimal number that an n-bit binary word can represent?
11. Write the truth table for the logic circuit in Figure P6.11.

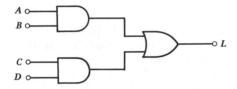

Figure P6.11

12. A useful logic gate is the inverter circuit in Figure P6.12a, which can be realized with a transistor (Chapter Seven). Use the inverter, AND, and OR gates to realize the exclusive OR function in Figure P6.12b.

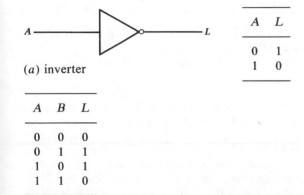

(a) inverter

A	L
0	1
1	0

A	B	L
0	0	0
0	1	1
1	0	1
1	1	0

(b) exclusive or function

Figure P6.12

13. Realize the half-adder function in Figure P6.13 using any combination of inverter, AND, OR, and exclusive OR gates.

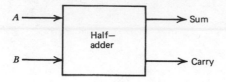

A	B	Sum	Carry
0	0	0	0
0	1	1	0
1	0	1	0
1	1	0	1

Figure P6.13

14. Two semiconductor diodes are connected in parallel. They are identical except that the *pn* junction area of the one diode is twice as large as the *pn* junction area of the other diode. How are their currents related?
15. Make a table similar to Table 6.1 only for (a) $I_s = 10^{-6}$ A, (b) $I_s = 10^{-9}$ A.
16. Find the saturation current I_s of a semiconductor diode if when $v = 0.7$ V, $i = 1$ mA and when $v = 0.5$ V, $i = 100$ nA. The temperature of the diode junction is unknown. Assume that $\eta = 1$.
17. Repeat Problem 16 for the measurements $v = 0.5$ V, $i = 1$ mA and $v = 0.7$, $i = 100$ mA.
18. Given $I_s = 10^{-12}$ A and $V_T = 26$ mV, how much power does a semiconductor diode dissipate if $i = 10$ mA?
19. Give that $i = 10^{-9}(e^{v/V_T} - 1)$ where $V_T = 26$ mV. Assume that $i = 5$ mA and determine r and V_{oc} of the linear model in Figure 6.21*b*.
20. Repeat Problem 19 for (a) $i = 100$ mA, (b) 1 mA, and (c) $i = 10\ \mu$A.
21. If $N = 10^{16}$ where N is the majority carrier concentration of the more lightly doped side of a *pn* junction, then what is the reverse breakdown voltage of this *pn* junction?
22. In the voltage regulator circuit in Figure 6.23 let $R_s = 100\ \Omega$ and $R_L = 2\ k\Omega$. With the other parameters unchanged compute the ripple at the output. What is the smallest value that the resistance R_L can have such that the circuit still functions as a regulator?
23. Repeat Problem 22 with $R_S = 1\ k\Omega$.
24. In Figure 6.28 suppose that the tunnel diode is biased in region II. Let $R = 20\ \Omega$ and suppose that the incremental resistance in region II of the diode is $-25\ \Omega$. Let v_d be the small signal voltage across the diode. Find the small signal gain v_d/v_s.

Chapter Seven
The Controlled Source and Its Applications in Electronic Circuit Modeling and Analysis

Frequently in the analysis of circuits we are dealing with N-terminal devices ($N > 2$) where the response at one terminal pair is dependent on the voltage or current at another terminal pair. This is particularly true in electronic circuits in which a small signal is used to control a large current. As a result the small signal is amplified if the device is operated in a linear mode, or the small signal can switch on or off the large current if the device is operated in a nonlinear mode.

In this chapter we introduce the dependent or controlled source. It is used to model the effect that a signal at one terminal pair has on a signal at another terminal pair. As examples, the modeling of the operational amplifier, the bipolar transistor, and the field-effect transistor with dependent sources is discussed. First let us discuss the concept of linear modeling for two-ports.

7.1 LINEAR MODELS FOR TWO-PORTS

Consider the circuit in Figure 7.1. This N-port could be a phonograph amplifier in which the load resistor R_L at port B (output port or terminal pair) represents the resistance of the speaker, and the source v_s and its resistance R_s at port A (input port or terminal pair) represent the Thévenin equivalent circuit of the pickup cartridge that converts mechanical vibrations to an electric signal by means of a piezoelectric crystal or magnetic coil. The N-port itself is a network of electrical components such

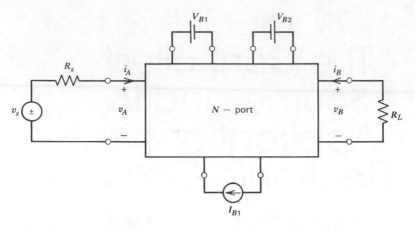

Figure 7.1 N-port circuit.

as resistors and transistors. The N-port could be an integrated circuit, that
is, all the components are fabricated on a single silicon chip and interconnected by photolithographic techniques. In some circuits, particularly electronic circuits, dc sources are required to bias the components
at a desired operating point. In Figure 7.1 we illustrate two dc voltage
sources and a dc current source for generality, but the dc bias source could
be a single dc voltage source, such as the typical 9-V source used in pocket
transistor radios.

   An analysis of the input-output behavior of this N-port proceeds as
follows. The external branches place the following constraints on the input
and output port voltages and current, respectively.

$$i_A R_s + v_A = v_s \tag{1}$$

$$i_B R_L + v_B = 0 \tag{2}$$

Since we have four unknowns we need two more equations. These equations are the port constraints of the N-port at the input and output ports.
For complete generality let us assume that the N-port constraint equations
are nonlinear, that is, let us assume that

$$v_A = f_A(i_A, i_B) \tag{3}$$

and

$$v_B = f_B(i_A, i_B) \tag{4}$$

Note that the port voltage v_A could depend not only on the input current i_A,
but also on the output current i_B. We need not include v_B in (3) since it can
be expressed in terms of i_B by means of (2). Equations 3 and 4 can be
determined by taking measurements of v_A and v_B versus i_A and i_B, or

sometimes one can represent the circuit and components inside the N-port by their mathematical constraints and reduce the equations to (3) and (4) by the elimination of internal variables similar to what we did in finding Thévenin and Norton equivalents in Chapter Five.

In Figure 7.2 we illustrate that when the input signal v_s is zero, the port

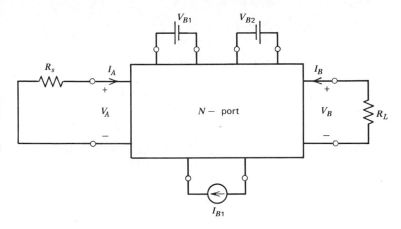

Figure 7.2 Port voltages and currents for $v_s = 0$.

voltage-current pairs that satisfy (1) to (4) will be the dc operating point values that we represent as $\{V_A, I_A\}$ and $\{V_B, I_B\}$. The above equations become

$$I_A R_s + V_A = 0 \tag{1'}$$

$$I_B R_L + V_B = 0 \tag{2'}$$

$$V_A = f_A(I_A, I_B) \tag{3'}$$

$$V_B = f_B(I_A, I_B) \tag{4'}$$

In analysis we are not only interested in obtaining the proper dc operating point values, but we are also interested in finding the relationship between the input signal v_s and the deviation in the input and output voltages and currents from the dc operating point caused by v_s, that is, the voltages and currents

$$v_a \overset{\Delta}{=} v_A - V_A \tag{5}$$

$$i_a \overset{\Delta}{=} i_A - I_A \tag{6}$$

$$v_b \overset{\Delta}{=} v_B - V_B \tag{7}$$

and

$$i_b \overset{\Delta}{=} i_B - I_B \tag{8}$$

In this book we will only be interested in the case when the input signal v_s is sufficiently small such that the relationship between v_s and the voltages and currents defined in (5) to (8) is approximately linear. The linear circuit that conveniently models this relationship is derived below.

We begin by expressing (3) and (4) by their respective Taylor series about the operating point V_A, I_A, V_B, I_B.

$$v_A = f_A(I_A, I_B) + \frac{\partial f_A}{\partial i_A}\bigg|_{I_A, I_B} (i_A - I_A) + \frac{\partial f_A}{\partial i_B}\bigg|_{I_A, I_B} (i_B - I_B)$$

$$+ \text{higher-order terms in } (i_A - I_A)^n \text{ and } (i_B - I_B)^n, \ n \geqslant 2. \tag{9}$$

$$v_B = f_B(I_A, I_B) + \frac{\partial f_B}{\partial i_A}\bigg|_{I_A, I_B} (i_A - I_A) + \frac{\partial f_B}{\partial i_B}\bigg|_{I_A, I_B} (i_B - I_B)$$

$$+ \text{higher-order terms in } (i_A - I_A)^n \text{ and } (i_B - I_B)^n, \ n \geqslant 2 \tag{10}$$

Let us assume that the higher-order terms in (9) and (10) are negligible (linearization of the nonlinear function), and let us define

$$z_{ab} \triangleq \frac{\partial f_A}{\partial i_B}\bigg|_{I_A, I_B} \tag{11}$$

and so on, so that (9) and (10) become

$$\hat{v}_a + V_A = f_A(I_A, I_B) + z_{aa}\hat{i}_a + z_{ab}\hat{i}_b \tag{12}$$

and

$$\hat{v}_b + V_B = f_B(I_A, I_B) + z_{ba}\hat{i}_a + z_{bb}\hat{i}_b \tag{13}$$

where the caret above the voltages and currents denotes that, since we neglected the higher-order terms in (9) and (10), these terms are no longer the exact deviation from the dc operating point, but only an approximation. We call them the *small-signal* voltages and currents.

Next we substitute (12) and (13) for (3) and (4) and substract (1') to (4') from (1) to (4) to obtain the following equations for our linear small-signal model.

$$\hat{i}_a R_s + \hat{v}_a = v_s \tag{14}$$

$$\hat{i}_b R_L + \hat{v}_b = 0 \tag{15}$$

$$\hat{v}_a = z_{aa}\hat{i}_a + z_{ab}\hat{i}_b \tag{16}$$

$$\hat{v}_b = z_{ba}\hat{i}_a + z_{bb}\hat{i}_b \tag{17}$$

The equivalent circuit represented by these equations is shown in Figure 7.3. Note that the dependence of the voltage and current at port A on the current $\hat{i}_b$ is modeled by a *dependent* or *controlled* voltage source. The source is said to be dependent or controlled because its value depends on another circuit variable. The dependent source is represented by a diamond symbol to distinguish it from an independent source. The dependent

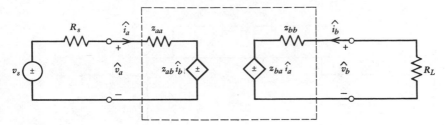

Figure 7.3   Small-signal model.

voltage sources in Figure 7.3 are called *current controlled voltage sources* since a branch current is the controlling variable.

The relationship between any voltage or current in the circuit and v_s is found by solving (14) to (17). For example, the transfer function

$$\frac{\hat{v}_b}{v_s} = \frac{z_{ba}R_L}{(z_{aa} + R_s)(z_{bb} + R_L) - z_{ab}z_{ba}} \tag{18}$$

is found by eliminating $\hat{i}_a$, $\hat{i}_b$ and $\hat{v}_a$ from (14) to (17). In the next section we show how to determine the z parameters and other linear circuit N-port parameters by measurement techniques.

7.2 MEASUREMENT OF LINEAR TWO-PORT PARAMETERS

In order to perform a small signal analysis of the N-port in Figure 7.1 we need the z parameters in (16) and (17). These z parameters can be determined as follows. From (16) and (17) note that

$$z_{aa} = \left.\frac{v_a}{i_a}\right|_{i_b=0} \tag{19}$$

and

$$z_{ba} = \left.\frac{v_b}{i_a}\right|_{i_b=0} \tag{20}$$

that is, z_{aa} is the small-signal driving-point resistance at port A with $i_b = 0$, and z_{ba} is the small-signal transfer resistance with $i_b = 0$. The caret above the small-signal variables is omitted. Henceforth, it will be understood that the small-signal variables associated with the linear model are only approximations to the actual deviation of the voltage and current from the dc operating point. Of course, if the N-port is linear, that is, the higher order Taylor series terms are zero, then our model is exact. The z parameters are measured as illustrated in Figure 7.4. Note that the current at port B must

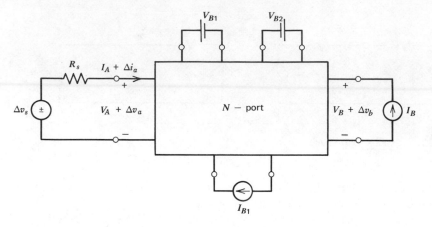

Figure 7.4   Measurement of z_{aa} and z_{ba}.

be held at its dc operating point value. From this circuit

$$z_{aa} = \lim_{\Delta v_s \to 0} \left(\frac{\Delta v_a}{\Delta i_a}\right) \tag{21}$$

and

$$z_{ba} = \lim_{\Delta v_s \to 0} \left(\frac{\Delta v_b}{\Delta i_a}\right) \tag{22}$$

Similarly, the parameters z_{ab} and z_{bb} can be measured by setting $i_a = 0$, that is, a current source I_A is attached to port A as shown in Figure 7.5 and

$$z_{ab} = \lim_{\Delta v_b \to 0} \left(\frac{\Delta v_a}{\Delta i_b}\right) \tag{23}$$

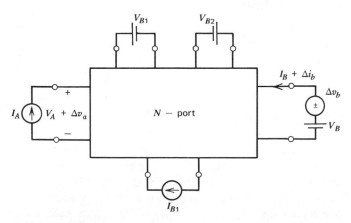

Figure 7.5   Measurement of z_{ab} and z_{bb}.

and

$$z_{bb} = \lim_{\Delta v_b \to 0} \left(\frac{\Delta v_b}{\Delta i_b} \right) \tag{24}$$

The z parameters are often called the *open-circuit resistance parameters*[1] because they are measured with the *small-signal current* at the other port equal to zero.

Finally we should point out that one is not merely limited to the z parameters in the linear characterization of an N-port. For example, sometimes it is more convenient to express the port currents as functions of the port voltages in (3) and (4). Such a characterization leads to the y or *short-circuit conductance parameters*[1] defined in Table 7.1. Note that the

Table 7.1   Linear Two-Port Models

Parameters	Circuit Equations	Circuit Model
Open-circuit resistance	$v_a = z_{aa}i_a + z_{ab}i_b$ $v_b = z_{ba}i_a + z_{bb}i_b$	
Short-circuit conductance	$i_a = y_{aa}v_a + y_{ab}v_b$ $i_b = y_{ba}v_a + y_{bb}v_b$	
Hybrid	$v_a = h_{aa}i_a + h_{ab}v_b$ $i_b = h_{ba}i_a + h_{bb}v_b$	
Inverse hybrid	$i_a = g_{aa}v_a + g_{ab}i_b$ $v_b = g_{ba}v_a + g_{bb}i_b$	

[1] In general these parameters are called the open-circuit impedance parameters where z denotes impedance. Impedance is defined in Chapter Ten and essentially it is the equivalent of resistance but includes the effect of inductors and capacitors in sinusoidal steady state analysis. Similarly, the short-circuit conductance parameters (discussed next) are usually called the short-circuit admittance parameters where y denotes admittance.

y parameter circuit model requires voltage controlled current sources. The y parameters are measured with the small-signal port voltages set to zero, that is, the port voltage is held to its dc value. This can often be accomplished by placing a large capacitor across the port terminals. Often the y parameters are easier to measure, and thus used more frequently in the characterization of N-ports.

Other permutations of voltage and current lead to the hybrid and inverse hybrid parameters also listed in Table 7.1. These equations and models can be easily extended to N-ports where $N > 2$. The parameters in Table 7.1 that one chooses to model the small-signal behavior of an N-port are usually based on the ease with which the parameters can be measured or calculated. Also, sometimes one or more parameter characterizations do not exist for a circuit (one or more of the parameter values is infinite).

In Table 7.1 we can convert one set of parameters to another. For example, suppose that one has determined the hybrid parameters for a two-port, but now it is desired to use the y parameter model. Since in the y parameters the current is a function of the voltage we write the h parameters in the form

$$\begin{bmatrix} -h_{aa} & 0 \\ -h_{ba} & 1 \end{bmatrix} \begin{bmatrix} i_a \\ i_b \end{bmatrix} = \begin{bmatrix} -1 & h_{ab} \\ 0 & h_{bb} \end{bmatrix} \begin{bmatrix} v_a \\ v_b \end{bmatrix} \tag{25}$$

and solve for i_a and i_b to obtain

$$\begin{bmatrix} i_a \\ i_b \end{bmatrix} = \frac{1}{h_{aa}} \begin{bmatrix} 1 & -h_{ab} \\ h_{ba} & h_{aa}h_{bb} - h_{ab}h_{ba} \end{bmatrix} \begin{bmatrix} v_a \\ v_b \end{bmatrix} \tag{26}$$

From the above equation we conclude that

$$y_{aa} = \frac{1}{h_{aa}} \qquad y_{ab} = -h_{ab}/h_{aa}$$

$$y_{ba} = \frac{h_{ba}}{h_{aa}} \quad \text{and} \quad y_{bb} = \frac{h_{aa}h_{bb} - h_{ab}h_{ba}}{h_{aa}}$$

Thus, the y parameters for the small-signal circuit model exist if $h_{aa} \neq 0$. Note that the above result can also be achieved by converting the input port of the hybrid circuit model to a Thévenin equivalent, and by converting the current controlled source at the output port to a voltage controlled current source by expressing i_a in terms of v_a and v_b.

The remainder of this chapter emphasizes the application of these models in circuit analysis and design.

7.3 MODELING THE VOLTAGE AMPLIFIER AND THE OPERATIONAL AMPLIFIER

Typically, the voltage amplifier consists of a package that contains transistors and resistors, and it has signal input and output terminals and at least two other terminals for biasing the electronic devices. One of these bias terminals is a (+) or (−) dc source terminal and the other is a ground terminal, or in many cases two dc supplies are required, one of the bias terminals is a positive (+) dc source terminal and the other a negative (−) dc source terminal as shown in Figure 7.6. The symbol μ denotes the gain of the amplifier, although frequently A, G, or K are used in place of μ. After the dc operating point has been established by the dc sources, one is interested in the small signal response at the input and output terminals of the amplifier. Ideally the input current i_s should be zero so that there is no loading on the input source, and the output voltage should be $v_o = \mu v_s$ for all i_o. Thus we define *an ideal voltage amplifier as one in which*

$$i_s = 0 \qquad \text{(27a)}$$

and

$$v_o = \mu v_s \qquad \text{(27b)}$$

for any connection to the input and output terminals. This definition allows us to model the *ideal* voltage amplifier at the input port with an open circuit and at the output port with a voltage source whose voltage is *controlled* by the input voltage. This model is illustrated in Figure 7.7 where $v_i = v_s$ and $v_o = \mu v_i$. Note that this is the inverse hybrid model in Table 7.1 where $g_{ba} = \mu$, $g_{bb} = 0$ since the output port voltage is independent of the output

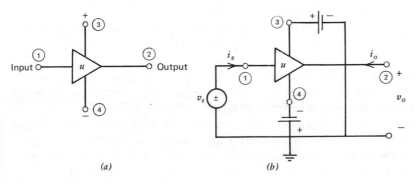

(a)   (b)

Figure 7.6   Ideal voltage amplifier.

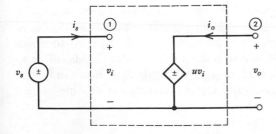

Figure 7.7   Small-signal model of the ideal voltage amplifier.

port current, $g_{aa} = 0$ since the input resistance is infinite, and $g_{ab} = 0$ since the input port current does not depend on the output port current. A model which includes the small-signal input resistance and output resistance of the voltage amplifier is shown in Figure 7.8.

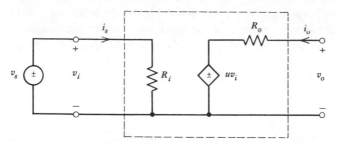

Figure 7.8   Small-signal model of a nonideal voltage amplifier.

Remember that both of the above models are only small-signal models and not large signal models. The small-signal model implies that the amplifier can deliver any current, voltage, or power to an external load provided that the input voltage is large enough. However, in reality, electronic amplifiers cannot deliver unlimited current, voltage, or power to a load. Both i_0 and v_0 are limited. In fact, some amplifiers have built-in protection circuitry at the output to limit the output current so that if the output terminal of the amplifier is shorted the device will not burn out. Next we model the operational amplifier.

The operational amplifier (op amp) is a very high gain amplifier with a differential input as illustrated in Figure 7.9a. It is used as a universal building block in many electronic circuits. Ideally it draws zero current at the input, its output resistance is zero, and its gain A is infinite. The op amp was the first and is the most widely used analog integrated circuit. Some important characteristics of the third generation 741 operational amplifier are listed in Table 7.2. Note that in Figure 7.9a the operational amplifier has an output terminal ⓞ, an inverting terminal ⓘ, a noninverting terminal ⓝ, and also terminals for a positive and a negative dc supply voltage V_+ and V_-. Not shown are additional terminals for nulling the ouput when the input voltage is zero, that is, in the actual amplifier the output voltage may not be zero when $e_n = e_i$. The voltage difference $e_n - e_i$ required to make $e_o = 0$ is called the offset voltage and it is desirable to reduce this offset to zero. A small-signal model for the ideal op amp with gain A is shown in Figure 7.9b. Note that in this simple model the input resistance is assumed to be infinite and the output resistance is zero.

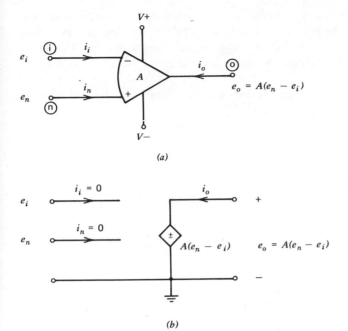

Figure 7.9 Operational amplifier. (a) Op amp symbol. (b) Ideal op amp model.

Table 7.2 Typical 741 Parameters

Parameters	Typical Values
Input bias current	80 nA
Input resistance	2 MΩ
Voltage gain	200,000
Output resistance	75 Ω
Supply voltage	+15 V and −15 V
Power consumption	50 mW

7.4 CIRCUIT ANALYSIS WITH DEPENDENT SOURCES

In the analysis of circuits with dependent sources our previous analysis methods are still valid. However, some caution is advised in certain operations. For example, in the parallel-series reduction of a circuit one must not lose the branch whose current or voltage, or the node whose voltage controls the dependent source. Also, in calculating the equivalent resistance seen at a port set all *independent* source values to zero inside the circuit, but *not the dependent sources*. Dependent sources are not zero unless their controlling variable is zero, which is usually not the case. Therefore, dependent sources can have a significant effect on the resistance measured at a port.

As an example consider the feedback amplifier in Figure 7.10a. (It is

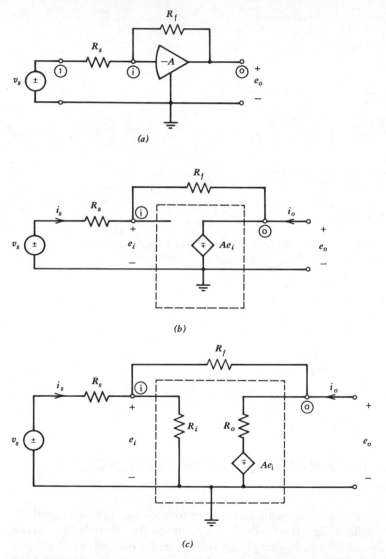

Figure 7.10 Negative feedback amplifier. (*a*) Circuit. (*b*) Ideal amplifier model. (*c*) Nonideal amplifier model.

common practice to omit the power supply terminals.) If we assume that the amplifier is ideal and has gain $-A$, the small signal model is as shown in Figure 7.10b. The mesh equation for this circuit is

$$-v_s + R_s i_s + R_f i_s - Ae_i = 0 \qquad (28)$$

Since the controlled voltage source introduces the variable e_i, it must be expressed in terms of the mesh current.

$$e_i = -R_s i_s + v_s \qquad (29)$$

Hence the circuit equations in the matrix form are

$$\begin{bmatrix} R_s + R_f & -A \\ R_s & 1 \end{bmatrix} \begin{bmatrix} i_s \\ e_i \end{bmatrix} = \begin{bmatrix} v_s \\ v_s \end{bmatrix} \qquad (30)$$

From (30) we obtain the gain of the feedback amplifier.

$$e_o = -Ae_i = \frac{-AR_f}{(1+A)R_s + R_f} \cdot v_s$$

or

$$\boxed{\frac{e_o}{v_s} = \frac{-AR_f}{(1+A)R_s + R_f}} \qquad (31)$$

Note that for sufficiently large A

$$\boxed{\frac{e_o}{v_s} \approx \frac{-R_f}{R_s}} \qquad (32)$$

This is a very interesting result that says that the gain is independent of A for a sufficiently large A and depends only on the ratio of two external resistors.

From (30) we can also calculate the input resistance $v_s/i_s|_{i_o=0}$. We obtain

$$\boxed{\frac{v_s}{i_s} = R_{\text{in}} = R_s + \frac{R_f}{1+A}} \qquad (33)$$

The input resistance depends on A and approaches R_s for very large values of A. One cannot set controlled sources to zero when computing the input resistance; if we had, we would have obtained the incorrect result that $R_{\text{in}} = R_s + R_f$ from Figure 7.10b. Note that setting the controlled source to zero in Figure 7.10b when calculating v_s/i_s also implies that $e_i = 0$, which is not the case for finite A. We should expect the equivalent resistance to depend on A since the parameter A appears in the square matrix in (30) along with the resistors. The response of a circuit with dependent sources but no nonzero independent sources is zero. *At least one nonzero independent source is needed to excite a circuit.*

As a second example let us assume that the amplifier in Figure 7.10a is *not* ideal. Let us assume that the amplifier's input voltage and current are independent of its output voltage and current and that the input resistance is R_i. At the output suppose that $e_o/e_i|_{i_o=0} = -A$ and $e_o/i_o|_{e_i=0} = R_o$. We say that the gain of the amplifier is $-A$ and its output resistance is R_o. Since the output voltage is controlled by the input voltage it makes sense to use the inverse hybrid parameters to characterize the input-output behavior of the voltage amplifier. It follows from Table 7.1 that $g_{aa} = 1/R_i$, $g_{ab} = 0$, $g_{ba} = -A$ and $g_{bb} = R_o$. The small signal circuit model is shown in Figure 7.10c. Excluding the ground node this circuit has four nodes, but two of the nodes have voltage sources connected to ground, hence in the node method one only need sum currents at nodes ⓘ and ⓞ. We obtain

$$\frac{e_i - v_s}{R_s} + \frac{e_i}{R_i} + \frac{e_i - e_o}{R_f} = 0$$

$$\frac{e_o - e_i}{R_f} + \frac{e_o - e_2}{R_o} = 0$$

(34a)

Note that we have two equations and three unknowns. The problem is that we have neglected to include the controlled voltage source constraint $e_2 = -Ae_i$. Upon substitution of this constraint into (34a) the equations can be written in the form

$$\begin{bmatrix} \dfrac{1}{R_s} + \dfrac{1}{R_f} + \dfrac{1}{R_i} & -\dfrac{1}{R_f} \\ -\dfrac{1}{R_f} + \dfrac{A}{R_o} & \dfrac{1}{R_o} + \dfrac{1}{R_f} \end{bmatrix} \begin{bmatrix} e_i \\ e_o \end{bmatrix} = \begin{bmatrix} v_s/R_s \\ 0 \end{bmatrix}$$

(34b)

The solution is

$$e_o = \frac{-\left(-\dfrac{1}{R_f} + \dfrac{A}{R_o}\right)(v_s/R_s)}{\left(\dfrac{1}{R_s} + \dfrac{1}{R_f} + \dfrac{1}{R_i}\right)\left(\dfrac{1}{R_o} + \dfrac{1}{R_f}\right) + \left(\dfrac{1}{R_f}\right)\left(-\dfrac{1}{R_f} + \dfrac{A}{R_o}\right)}$$

(35)

and the voltage gain

$$\frac{e_o}{v_s} = \frac{-(AR_f - R_o)}{R_o + (1 + A)R_s + R_f + \dfrac{R_s R_f + R_s R_o}{R_i}}$$

(36)

Besides computing the gain of an amplifier circuit, it is usually important to know the resistance seen at the input port and the resistance seen at the output port. In Figure 7.10c the input port resistance is

$$R_{in} = v_s/i_s \tag{37}$$

and the output port resistance is computed by connecting a current source i_o or voltage source e_o to the output port and calculating

$$R_{out} = e_o/i_o|_{v_s=0}. \tag{38}$$

Now the input resistance will usually depend on the load current i_o. We will assume that $i_o = 0$, although one could easily perform the calculation with a load resistor R_L in shunt at the output port. The input resistance obtained will not necessarily be the same for each of the above loading conditions.

To calculate R_{in} note that

$$i_s = \frac{v_s - e_i}{R_s} \tag{39}$$

but from (34)

$$e_i = \frac{(R_o + R_f)v_s}{R_o + (1 + A)R_s + R_f + \dfrac{R_s R_f + R_s R_o}{R_i}} \tag{40}$$

Substitute (40) into (39), then

$$i_s = \left[1 - \frac{(R_o + R_f)}{R_o + (1 + A)R_s + R_f + \dfrac{R_s R_f + R_s R_o}{R_i}} \right] \frac{v_s}{R_s}$$

or

$$i_s = \left[\frac{(1 + A)R_s + \dfrac{R_s R_f + R_s R_o}{R_i}}{R_o + (1 + A)R_s + R_f + \dfrac{R_s R_f + R_s R_o}{R_i}} \right] \frac{v_s}{R_s} \tag{41}$$

If R_i is very large, then (41) becomes

$$R_{in} = \frac{e_s}{i_s} \approx \frac{(R_o + R_f + (1 + A)R_s)}{(1 + A)} \tag{42}$$

Furthermore, if A is sufficiently large

$$\boxed{R_{in} \simeq R_s} \tag{43}$$

To compute the equivalent resistance at the output port in Figure 7.10c let us set $v_s = 0$, inject a current i_o at the output port (current source), and compute e_o. Equation 34 becomes

$$\begin{bmatrix} \dfrac{1}{R_s} + \dfrac{1}{R_f} + \dfrac{1}{R_i} & -\dfrac{1}{R_f} \\[2ex] -\dfrac{1}{R_f} + \dfrac{A}{R_o} & \dfrac{1}{R_o} + \dfrac{1}{R_f} \end{bmatrix} \begin{bmatrix} e_i \\[1ex] e_o \end{bmatrix} = \begin{bmatrix} 0 \\[1ex] i_o \end{bmatrix} \tag{44}$$

and

$$e_o = \frac{R_o\left(R_s + R_f + \dfrac{R_s R_f}{R_i}\right)}{R_o + (1 + A)R_s + R_f + \dfrac{R_s R_f + R_s R_o}{R_i}} i_o \tag{45}$$

Again, if R_i is very large, then

$$\boxed{R_{\text{out}} = \frac{e_o}{i_o} \approx \frac{R_o(R_s + R_f)}{R_o + R_f + (1 + A)R_s}} \tag{46}$$

The output resistance goes to zero as A goes to infinity or if R_o is zero.

   The above example of a negative feedback amplifier clearly indicates the role of the dependent source in determining the Thévenin equivalent resistance at a port.

   As a final example for this section consider the op amp circuit in Figure 7.11. In order to analyze the op amp circuit we use the model in Figure 7.12. We have assumed that the output resistance is zero and the input resistance is infinite. This circuit has five nodes, but nodes ① and ② have grounded voltage sources connected to them as does the output node. Hence, using the node method we write

$$\text{node } \textcircled{i} \quad : \quad \frac{e_i - e_2}{R_1} + \frac{e_i - e_o}{R_f} = 0 \tag{47}$$

$$\text{node } \textcircled{n} \quad : \quad \frac{e_n - e_1}{R_2} = 0 \tag{48}$$

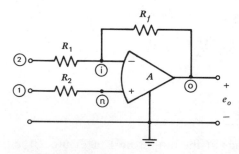

Figure 7.11 Negative feedback amplifier configuration.

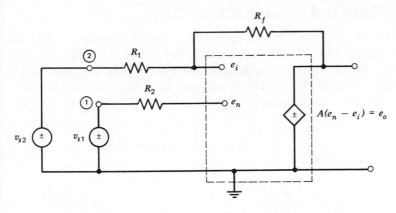

Figure 7.12   Negative feedback amplifier circuit model.

and since a dependent voltage source is connected to the output node we must write its constraint.

$$e_o = A(e_n - e_i) \tag{49}$$

Note also that $e_1 = v_{s1}$ and $e_2 = v_{s2}$. Solving (47) to (49) the result is that

$$e_o\left(1 + \frac{R_f + R_1}{AR_1}\right) = \frac{R_f + R_1}{R_1} v_{s1} - \frac{R_f}{R_1} v_{s2} \tag{50}$$

In the ideal op amp A is infinite, hence

$$\lim_{A \to \infty} e_o\left(1 + \frac{R_f + R_1}{AR_1}\right) = e_o$$

yields

$$e_o = \frac{R_f + R_1}{R_1} v_{s1} - \frac{R_f}{R_1} v_{s2} \tag{51}$$

From (51) we note that the circuit in Figure 7.12 is a very versatile amplifier whose gain is easily set by a couple of resistors, and the output is the difference between two input signals in the above configuration. In actual op amps the gain A varies drastically with temperature and from lot to lot in the integrated circuit manufacturing process. However, the negative feedback resistor R_f has a stabilizing effect on the entire amplifier by making the overall gain dependent on the ratio of resistors.

Even though the op amp contains typically 20 or more diodes and transistors, due to the integrated circuit process some op amps can be bought in very large quantities for less than 25¢ each.

7.5 THE BIPOLAR TRANSISTOR

The bipolar transistor is the three-terminal semiconductor device illustrated in Figure 7.13. The three terminals are called the collector, the base, and the emitter. Note that the emitter-base region and the base-collector region form a diode. There are basically two types of bipolar transistors, the npn and the pnp as shown in Figure 7.13. In the forward

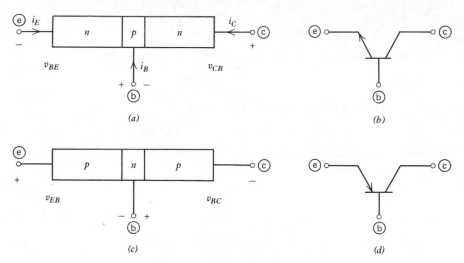

Figure 7.13 Bipolar transistors. (*a*) *npn* transistor. (*b*) Symbol. (*c*) *pnp* transistor. (*d*) Symbol.

active operation of the transistor the emitter-base junction is forward biased ($v_{BE} > 0$ for *npn*, $v_{BE} < 0$ for *pnp*), and the collector-base junction is reversed biased ($v_{BC} < 0$ for *npn*, $v_{BC} > 0$ for *pnp*). Recall from our discussion of diodes that the current carriers in the base region are the minority carriers (electrons in the *npn* transistor and holes in the *pnp* transistor) that have diffused across a forward-biased junction. In the transistor the base region is made very narrow so that the minority carriers traverse the base region before they can recombine with majority carriers. When they reach the base-collector junction they are captured by the electric field in the depletion region of the reversed-biased collector-base junction and are accelerated into the collector region. This is illustrated graphically in Figure 7.14 for the *npn* transistor. In transistor design the goal is to make $i_C \approx -i_E$. Remember that current flow is in the opposite direction of electron flow. Thus, in Figure 7.14 the emitter region is much more heavily doped than the base region. This means that most of the current flow across the emitter-base junction is due to electron flow. Since the base region is very narrow, most of the electrons that enter the base

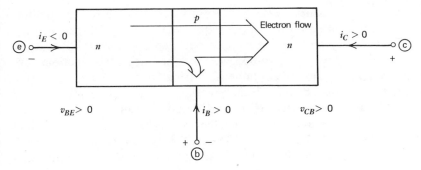

Figure 7.14   Forward active mode of operation.

region will be collected by the collector region where they recombine with holes. The electron-hole recombination in the base region will be much less than that in the collector region so that i_B will be much smaller than i_C. Recall from Chapter Six that a small diode voltage change (v_{BE}) can cause a large change in the current. Thus, we have an amplifier since small changes in v_{BE} cause large changes in i_C. Also, the collector region is virtually isolated from the base by means of the reverse-biased base-collector junction.

7.5.1 IC BIPOLAR TRANSISTORS

The integrated circuit (IC) structure of the *npn* transistor is shown in Figure 7.15. The structure is similar to that for making a diffused resistor except that two extra diffusions are added. First, before the *n* epitaxial layer is grown a buried layer n^+ diffusion is made to reduce the resistance of the collector. The *n* epitaxial layer is grown ($\approx 10^{16}$ dopant atoms/cm^3), the *p* isolation diffusions are made, next the *p* base diffusion is made ($\approx 10^{18}$ dopant atoms/cm^3). Note that the base diffusion is frequently the

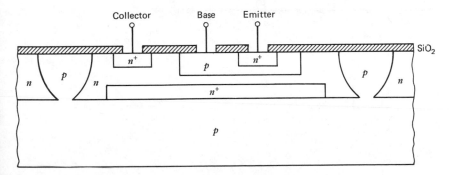

Figure 7.15   *npn* integrated circuit transistor structure.

one used to make resistors. Next an n^+ diffusion is made for the emitter and the collector contact ($\approx 10^{20}$ dopant atoms/cm³). The collector contact diffusion is necessary so that the contact between the n material and the metal does not behave as a diode. Lastly the metal connections are deposited. Typically a diode is obtained by using the base-emitter junction of a transistor whose collector-base junction is shorted. Thousands of transistors, diodes, and resistors can be made simultaneously on the three-inch silicon wafer and interconnected by means of this diffusion and photolithographic mask IC process. The initial circuit design and mask layout steps are expensive. But once these costs are paid, the circuits can be manufactured at a ridiculously low cost.

7.6 TRANSISTOR MODELS

To a first-order approximation the operation of the npn transistor can be modeled by the circuit in Figure 7.16 where the two diodes represent the pn junctions and the controlled source $\alpha_F i_F$ represents the collector current component due to minority carrier flow across the emitter-base junction, and $\alpha_R i_R$ represents the component of the emitter current due to minority carrier flow across the collector-base junction. For the npn transistor the equations for this model are

$$i_E = -i_F + \alpha_R i_R \tag{52}$$

$$i_C = \alpha_F i_F - i_R \tag{53}$$

and

$$i_B + i_C + i_E = 0 \tag{54}$$

where

$$i_F = I_{ES}(e^{v_{BE}/V_T} - 1) \tag{55a}$$

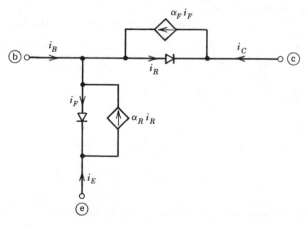

Figure 7.16 *npn* circuit model.

and

$$i_R = I_{CS}(e^{v_{BC}/V_T} - 1) \tag{55b}$$

These equations are called the Ebers-Moll equations. The factors I_{CS} and I_{ES} represent the saturation current of the b-c and b-e junctions, respectively, and the coefficients α_F and α_R are current gain constants.

Typically the transistor is operated in its forward active mode in which the b-e junction is forward biased and the b-c junction is reversed biased. In this case $i_R \approx 0$ and the model in Figure 7.16 simplifies to that shown in Figure 7.17a. By means of current source splitting as shown in Figure 7.17b the model reduces to that shown in Figure 7.17c which is more commonly used to model the transistor in the forward active mode. Note that

$$i_B = (1 - \alpha_F)i_F \tag{56}$$

so that

$$\beta_F = \frac{\alpha_F}{1 - \alpha_F} \tag{57}$$

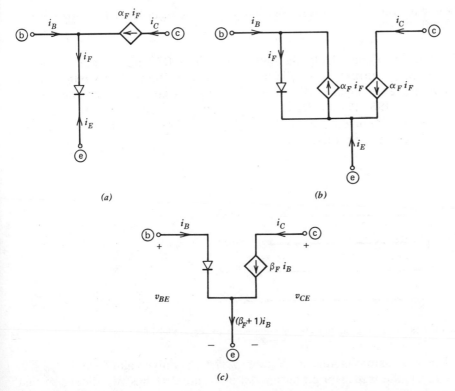

(a)

(b)

(c)

Figure 7.17   Forward active transistor models. (a) Basic model. (b) Source splitting. (c) Typical model.

Typically $\alpha_F \approx 0.98$ or 0.99 in which case the dc forward current gain from the base of the transistor to the collector is $\beta_F \approx 50$ to 100. Since β_F is the reciprocal of the difference between two nearly equal numbers it is difficult to control in the manufacturing process.

The DP characteristics of a silicon transitor between base-emitter and collector-emitter in the forward active mode are shown in Figure 7.18.

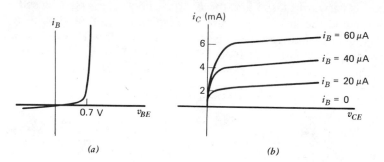

(a)   (b)

Figure 7.18   DP characteristics of *npn* transistor. (*a*) Base-emitter port. (*b*) Collector-emitter port.

Note that in Figure 7.18*b* the collector-emitter DP characteristic depends on the base current i_B. In Figure 7.19 we note that indeed our forward active model in Figure 7.17*c* does closely approximate the characteristics in Figure 7.18*b*. The model can be improved by adding a conductance in shunt with the current source $\beta_F i_B$ in Figure 7.17*c* whose slope matches the slope of the DP characteristic in Figure 7.18*b*.

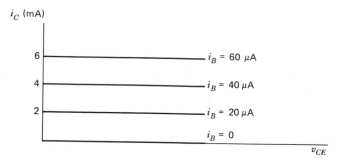

Figure 7.19   Collector-emitter DP characteristics of the model in Figure 7.17*c*.

The transistor model in Figure 7.17*c* is only valid when $v_{BC} < 0$. However, the transistor circuit model in Figure 7.16 has two diodes so that there are four possible states of these diodes. We have already considered the case when the base-emitter diode is forward biased and the base-collector diode is reversed biased which led to the simple model in Figure 7.17c. A second possibility is both diodes reversed biased that yields the simple open circuit model in Figure 7.20a. This is called the cutoff state. A third possibility is both diodes forward biased, which means that $v_{BE} \approx$ 0.7 V and $v_{BC} \approx 0.7$ V (for silicon transistors) so that $v_{CE} \approx 0$ V. This yields the simple model in Figure 7.20b. The transistor is said to be saturated. Finally, the fourth possible state occurs when $v_{BE} < 0$ and $v_{BC} > 0$. This is called the reverse active mode and the derivation of the model in Figure 7.20c is similar to that in Figure 7.17. However, since the collector region is usually more lightly doped than the base region, the electron current flow across the base-collector junction is no longer significantly greater than the hole flow. Typically $\alpha_R \approx 0.5$ so that $\beta_R = \alpha_R/(1 - \alpha_R) \approx 1$ compared to a typical β_F of 50 to 100. Thus, in the reverse active mode the current gain of a typical integrated circuit transistor is very poor.

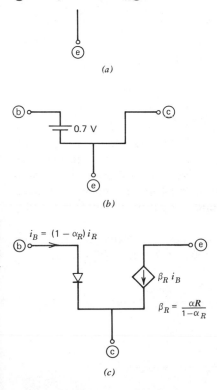

Figure 7.20 Remaining transistor models. (a) Open-circuit model. (b) Saturated model (both diodes on). (c) Reverse active model.

7.7 THE TRANSISTOR INVERTER

The basic transistor inverter circuit is illustrated in Figure 7.21a. The collector-emitter DP characteristics of the transistor as a function of the base current i_B are plotted in Figure 7.21b along with the load line

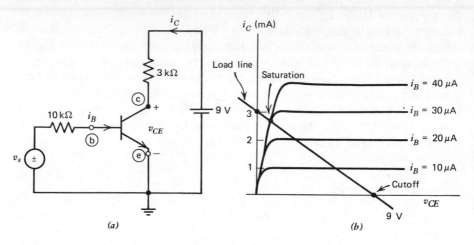

(a) (b)

Figure 7.21 Transistor inverter. (a) Inverter circuit. (b) Graphical analysis.

determined by the mesh equation

$$-9\text{ V} + (3\text{ k}\Omega)i_C + v_{CE} = 0 \tag{58}$$

For any base current i_B the solution must lie on the load line. Furthermore, with the voltage sources connected as shown, it is not possible for both $v_{BE} < 0$ and $v_{BC} > 0$ so that the reverse active state of operation is impossible. The transistor can only operate in the forward active state, the cutoff state, or the saturation state. The state of operation is determined by the input source v_s.

Let us begin our analysis by assuming that the transistor is in the forward active or cutoff state so that the model in Figure 7.22 can be used.

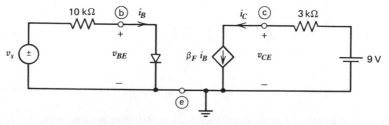

Figure 7.22 Forward active model of inverter circuit.

Note that in Figure 7.21b when $i_B = 10\,\mu A$, $I_C = 1\,mA$ so that for this transistor

$$\beta_F = \frac{1\,mA}{10\,\mu A} = 100$$

This model is valid as long as the base-collector junction of the transistor is reversed biased. Furthermore, let us assume that the transistor is made from silicon. Recall from Chapter Six that in the silicon transistor $i_B \approx 0$ for $v_{BE} < 0.5\,V$ and $V_{BE}(on) \approx 0.7\,V$. Thus, for $v_s < 0.5\,V$, $i_B \approx 0$ which, from our model in Figure 7.22, implies that $i_C \approx 0$ so that $v_{CE} = 9\,V$. This cutoff point is illustrated in Figure 7.21b.

As v_s increases so does i_B and i_C, which means that v_{CE} decreases. We are moving up the load line in Figure 7.21b toward the saturation point. For example, to find the value of v_s necessary to yield a collector current $i_C = 1\,mA$ we write

$$\frac{v_s - V_{BE}(on)}{10\,k\Omega} = i_B \tag{59}$$

but $i_B = 1\,mA/\beta_F = 10\,\mu A$ and $V_{BE}(on) \approx 0.7\,V$ so that

$$v_s \approx 0.8\,V$$

The transistor goes into saturation when the base-collector junction begins to conduct current. This occurs when $v_{BC} \approx 0.5\,V$ so that

$$v_{CE} = v_{BE} - v_{BC} \approx 0.7\,V - 0.5\,V = 0.2\,V$$

The substitution of $v_{CE} = 0.2\,V$ into (58) yields

$$i_C = \frac{9\,V - 0.2\,V}{3\,k\Omega} = 2.93\,mA$$

or

$$i_B = \frac{i_C}{\beta_F} = 29.3\,\mu A$$

Thus, for input voltages v_s for which $i_B \geqslant 29.3\,\mu A$ the transistor is in saturation and $v_{CE} \approx 0.2\,V$.

The response of this circuit to a sinusoidal input voltage v_s, which drives the transistor through all three states, is shown in Figure 7.23. Note that v_{CE} can only swing between the limits of the 9-V supply voltage and the 0.2-V saturation voltage. Also, $i_B \geqslant 0$ and $0 \leqslant i_C < 3\,mA$. The inverter circuit is the basic element in digital logic circuits in which the transistor is switched between cutoff and saturation by means of the base current i_B. An example follows.

Let us conclude this section with an analysis of the resistor-transistor logic gate (RTL gate) in Figure 7.24. The two input nodes are A and B and the output node is C. When both input nodes are low ($< 0.5\,V$) both

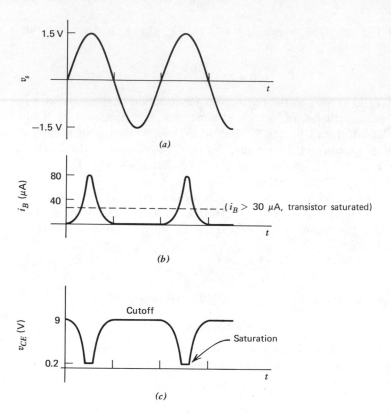

Figure 7.23   Inverter response. (*a*) Input. (*b*) Base current. (*c*) Output voltage.

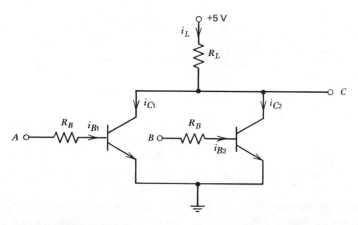

Figure 7.24   RTL gate.

212

transistors are off and the current $i_L = 0$ so that $v_C \approx 5$ V. If one or both of the inputs are high (≈ 5 V), then, assuming that the base current is sufficiently large to cause saturation, the output will be low (≈ 0.2 V). In a positive logic system a high voltage is a 1 and a low voltage a 0. The truth table in Table 7.3 indicates that this circuit is a NOR gate (not OR).

Table 7.3 Truth Table for NOR gate

A	B	C
0	0	1
1	0	0
0	1	0
1	1	0

Suppose that it is desired to restrict i_C of the transistors to less than 5 mA. Then

$$R_L = \frac{5\text{ V}}{5\text{ mA}} = 1\text{ k}\Omega$$

Next let us assume that $20 \leqslant \beta_F \leqslant 100$, then for the worst case in order to saturate the transistor for a high input voltage we find that

$$i_B \geqslant \frac{5\text{ mA}}{20} = 250\ \mu\text{A}$$

so that

$$R_B = \frac{5\text{ V} - 0.7\text{ V}}{250\ \mu\text{A}} = 17.2\text{ k}\Omega$$

In the actual design one might choose $R_B = 10$ kΩ since the input voltage may be less than 5 V in the high state.

7.8 THE TRANSISTOR AMPLIFIER

In the previous section the output signal was distorted primarily due to the base-emitter diode, which clipped the negative half of the input signal. This clipping can be eliminated by injecting a bias current I_B at the base as shown in Figure 7.25a. The input signal is injected as a current source i_s so as not to disturb the bias I_B. The graphical picture in Figure 7.25b gives us a good visual image of the operation of the circuit, but to obtain numerical results we must return to our linear and small signal models. In Figure 7.26 we replace the transistor by its forward active model, and in Figure 7.27a the diode is linearized about the bias point I_B. Finally, by means of super-

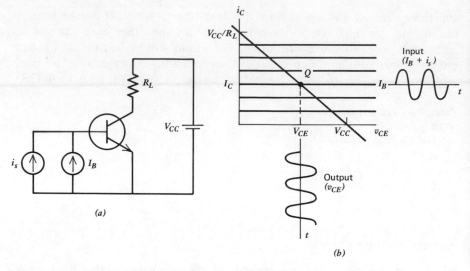

Figure 7.25 Linear amplifier. (*a*) Circuit. (*b*) Response.

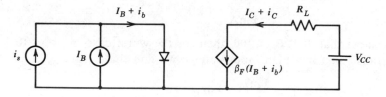

Figure 7.26 Forward active model.

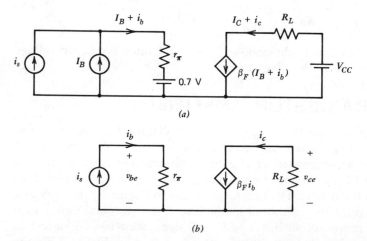

Figure 7.27 Linearization of the active model. (*a*) Linear model. (*b*) Small-signal model.

214

position, we separate the dc bias response and the response due to the input i_s, which is determined from the small-signal model in Figure 7.27b. To calculate the small-signal resistance of the diode recall that in the forward active region

$$i_B = -(1 - \alpha_F)i_E \tag{60}$$

where from (52)

$$i_E \approx -I_{ES}(e^{v_{BE}/V_T} - 1)$$

so that the reciprocal of the small-signal resistance at the dc operating point is

$$\frac{1}{r_\pi} = \frac{\partial i_B}{\partial v_{BE}}\bigg|_{V_{BE}} = (1 - \alpha_F)\frac{1}{V_T}I_{ES}\, e^{V_{BE}/V_T} \tag{61}$$

However, $i_C = -\alpha_F i_E$ so that at the bias point

$$\frac{1}{r_\pi} = \frac{(1 \quad \alpha_F)}{\alpha_F}\frac{1}{V_T}I_C$$

Therefore

$$\boxed{r_\pi = \frac{\beta_F V_T}{I_C}} \tag{62}$$

where $V_T \approx 26$ mV at room temperature.

In Figure 7.27b, since

$$v_{be} = r_\pi i_b \tag{63}$$

we can generate a second equivalent small-signal model as shown in Figure 7.28. It follows that

$$\boxed{g_m = \frac{\beta_F}{r_\pi} = \frac{I_C}{V_T}} \tag{64}$$

The model in Figure 7.28 is usually preferred if nodal equations are to be

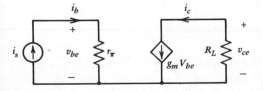

Figure 7.28   Voltage-controlled, current-source model.

written to analyze the circuit. An analysis of this circuit yields the voltage gain

$$\boxed{\frac{v_{ce}}{v_{be}} = -g_m R_L} \tag{65}$$

Substituting (64) into (65) we obtain

$$\frac{v_{ce}}{v_{be}} = -\frac{I_C R_L}{V_T} \tag{66}$$

Thus, the voltage gain can be increased by increasing R_L or the dc bias current I_C. These changes also affect the dc operating point.

A transistor is limited to the amount of power it can dissipate. The average power dissipated by the transistor is

$$P_{\text{avg}} = \frac{1}{T} \int_0^T (V_{CE} + v_{ce})(I_C + i_c) \, dt \tag{67}$$

If we assume that the average values of the v_{ce} and i_c are zero, then

$$P_{\text{avg}} = V_{CE} I_C + \frac{1}{T} \int_0^T v_{ce} i_c \, dt \tag{68}$$

It is interesting to note from Figure 7.25b that v_{ce} and i_c are 180° out of phase so that the second term in (68) is negative, that is, the average power dissipation decreases as the small-signal variation increases. For example, in Figure 7.25 let $R_L = 3 \text{ k}\Omega$, $V_{CC} = 10 \text{ V}$, $I_B = 20 \ \mu\text{A}$ and $\beta_F = 100$ so that

$$I_C = \beta_F I_B = 2 \text{ mA}$$

$$V_{CE} = V_{CC} - R_L I_C = 10 \text{ V} - (3 \text{ k}\Omega)(2 \text{ mA}) = 4 \text{ V}$$

Furthermore, let the input current $i_s = 10^{-5} \sin \omega t$ so that from Figure 7.27b

$$i_c = 10^{-3} \sin \omega t$$

$$v_{ce} = -3 \sin \omega t$$

From (68) we obtain the average power dissipated by the transistor.

$$P_{\text{avg}} = (2 \text{ mA})(4 \text{ V}) - \frac{1}{T} \int_0^T (10^{-3} \sin \omega t)(3 \sin \omega t) \, dt$$

$$= 8 \text{ mW} - \frac{1}{2}(3 \text{ mW}) = 6.5 \text{ mW}$$

The average power delivered by the dc voltage source is constant, that is,

$$P_{\text{avg}} = \frac{1}{T} \int_0^T V_{CC}(I_C + i_c) \, dt = V_{CC} I_C = 20 \text{ mW}$$

Thus, as the input signal increases, less power is dissipated by the transistor and more power is dissipated in the load R_L at the signal frequency ω. The average power dissipated in the load resistor is

$$P_L = \frac{1}{T} \int_0^T (I_C + i_c)^2 R_L \, dt = R_L[I_C^2 + i_c^2(\text{rms})]$$

The small-signal current swing i_c about the bias point is restricted to the range $0 \leq |i_c| \leq i_C(\text{sat})$.

7.9 PRACTICAL TRANSISTOR AMPLIFIER CIRCUITS

As an amplifier the transistor is operated in its forward active region. Therefore, the first step in designing a transistor amplifier is to bias it in its forward active region. A simple bias circuit that uses only one battery is shown in Figure 7.29a. We can use voltage source splitting and Thévenin's theorem to arrive at the simple equivalent circuit in Figure 7.29b. Let us suppose that $\beta_F = 50$, $V_{CC} = 10$ V, and we desire $I_C = 1$ mA. Typically one designs the circuit so that $V_{CE} = V_{CC}/2 = 5$ V. This requires that $R_L = 5$ V/1 mA $= 5$ kΩ. Since $\beta_F = 50$, then at the dc operating point

$$I_B = \frac{I_C}{\beta_F} = 20 \ \mu\text{A} \tag{69}$$

We can write the following loop equation for the base of the transistor.

$$\frac{10R_{B2}}{R_{B1} + R_{B2}} - V_{BE} = \frac{R_{B1}R_{B2}}{R_{B1} + R_{B2}} (20 \ \mu\text{A}) \tag{70}$$

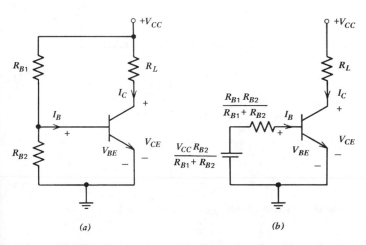

(a)   (b)

Figure 7.29   Practical transistor bias circuit.

If we assume that $V_{BE} \approx 0.7$ V and $R_{B1} = 200$ kΩ, then from (70)

$$R_{B2} = 26.4 \text{ k}\Omega \tag{71}$$

Now that the transistor is properly biased we can apply our small signal input $i_s(t)$ as shown in Figure 7.30a. The incremental model is shown in Figure 7.30b.

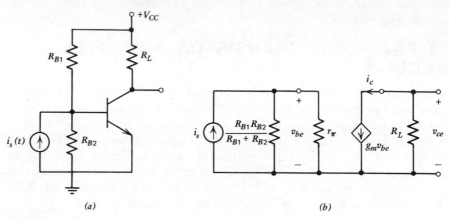

(a)   (b)

Figure 7.30   Linear amplifier. (a) Amplifier. (b) Incremental model.

Recall that the incremental parameters

$$g_m = \frac{I_C}{V_T} = \frac{1 \text{ mA}}{25.9 \text{ mV}} \approx 0.04 \text{ mho} \tag{72}$$

and

$$r_\pi = \frac{\beta_F}{g_m} = 1300 \ \Omega \tag{73}$$

A small-signal analysis of the circuit in Figure 7.30b yields

$$v_{be} = \frac{r_\pi R_B}{r_\pi + R_B} i_s \tag{74}$$

where

$$R_B = \frac{R_{B1} R_{B2}}{R_{B1} + R_{B2}}$$

$$i_c = g_m \cdot v_{be} \tag{75}$$

and

$$v_{ce} = -R_L g_m v_{be} \tag{76}$$

Therefore from (74) and (75) the small-signal current gain is

$$A_i \triangleq \frac{i_c}{i_s} = g_m \frac{r_\pi R_B}{r_\pi + R_B} \tag{77}$$

but since $R_B \gg r_\pi$

$$A_i \approx g_m r_\pi = \beta_F = 50$$

and from (76) the small-signal voltage gain is

$$A_v \triangleq \frac{v_{ce}}{v_{be}} = -g_m R_L = -200 \tag{78}$$

We must point out that the bias circuit in Figure 7.29 is not very good since β_F can easily vary by a factor of 2 in a lot of supposedly identical transistors. Also, β_F has a temperature coefficient of $+1\%/°C$ so that in the range 25 to 125°C β_F can increase by a factor of 2. Returning to Figure 7.29 with $\beta_F = 100$ we note that $I_B = 20\ \mu A$ (unchanged), but for $\beta_F = 100$ $I_C = 100 \cdot 20\ \mu A = 2\ mA$, and so $V_{CE} = -R_L I_C + 10\ V = 0\ V$, the transistor is now *saturated*.

This problem can be remedied by means of negative feedback. This is accomplished by placing a resistor R_E in the emitter of the transistor as shown in Figure 7.31a. Now any increase in I_C raises the emitter voltage and reduces V_{BE} and hence I_B. Unfortunately, the emitter feedback resistor complicates our bias equations. To determine this bias equation we use the relations

$$I_C = \beta_F I_B \tag{79}$$

and

$$I_E = I_C + I_B = (\beta_F + 1)I_B \tag{80}$$

The input mesh equation in Figure 7.31b is

$$V_B - V_{BE} = R_B I_B + R_E(\beta_F + 1)I_B \tag{81}$$

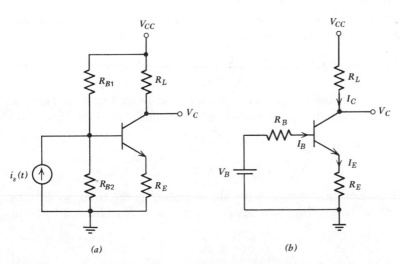

(a)   (b)

Figure 7.31   Stable bias circuit. (a) Circuit. (b) Equivalent bias circuit.

where we have defined

$$R_B = R_{B1} \| R_{B2} \tag{82}$$

and

$$V_B = \frac{R_{B2}}{R_{B1} + R_{B2}} V_{CC} \tag{83}$$

Also note that

$$V_C = V_{CC} - R_L I_C \tag{84}$$

Let us choose $V_{CC} = 10$ V, $V_B = 2$ V, $R_L = 5$ kΩ, $R_E = 1$ kΩ, $\beta_F = 50$, and $I_C = 1$ mA, then (79) to (84) yield

$$I_B = I_C / \beta_F = 20 \ \mu A \tag{85}$$

$$I_E = I_B + I_C = 1.02 \text{ mA} \tag{86}$$

$$V_C = V_{CC} - R_L I_C = 5 \text{ V} \tag{87}$$

$$V_{CE} = V_C - R_E I_E \approx 4 \text{ V} \tag{88}$$

and from (81)

$$R_B = \frac{V_B - V_{BE} - R_E(\beta_F + 1)I_B}{I_B} = 14 \text{ k}\Omega \tag{89}$$

From (82) and (83) we find that

$$R_{B1} = 70 \text{ k}\Omega \tag{90}$$

$$R_{B2} = 17.5 \text{ k}\Omega \tag{91}$$

which are the values of the bias resistors that we need to operate at $I_C = 1$ mA and $V_{CE} \approx 4$ V.

Now if β_F increases to 100 due to a temperature change or replacement of the transistor with another transistor, then from (81)

$$I_B = \frac{V_B - V_{BE}}{R_B + (\beta_F + 1)R_E} = 11.3 \ \mu A \tag{92}$$

Thus, the feedback resistor causes I_B to decrease from 20 μA to 11.3 μA when β_F doubles, whereas I_B remains unchanged in Figure 7.29. From (84)

$$V_C = 4.35 \text{ V}$$

and

$$V_{CE} = 3.2 \text{ V}$$

so that the resistor R_E prevents the dc operating point from going into saturation when β_F doubles. This result can easily be seen from (81). Note that

$$I_B = \frac{V_B - V_{BE}}{R_B + (\beta_F + 1)R_E} \tag{93}$$

Thus, if $(\beta_F + 1) \gg R_B$, then if β_F doubles, I_B will be reduced by approximately a factor of 2. Since

$$I_C = \beta_F I_B \tag{94}$$

the current I_C will remain approximately constant as will the bias point.

However, the resistor R_E not only stabilizes the bias, but also reduces the gain of the circuit. For example, in Figure 7.32 we present a small-signal model of the amplifier circuit in Figure 7.31 in which the input

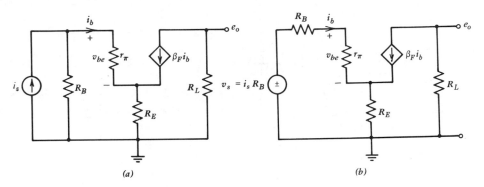

(a)   (b)

Figure 7.32 Small-signal circuits.

current source has been converted to an equivalent voltage source. The input mesh equation is

$$(R_B + r_\pi + (\beta_F + 1)R_E)i_b = R_B i_s = v_s \tag{95}$$

Thus, the input resistance of the amplifier is

$$R_{in} = \frac{v_s}{i_b} = R_s + r_\pi + (\beta_F + 1)R_E \tag{96}$$

Note the large increase due to R_E. The voltage gain of this circuit is

$$e_o = -\beta_F i_b R_L \tag{97}$$

but the substitution of (95) into (97) for i_b yields

$$\frac{e_o}{v_s} = \frac{-\beta_F R_L}{R_B + r_\pi + (\beta_F + 1)R_E} = \frac{-g_m R_L}{1 + \dfrac{R_B}{r_\pi} + (\beta_F + 1)\dfrac{R_E}{r_\pi}} \tag{98}$$

Compare (98) to (78) and note the severe reduction in gain due to R_E. If $(\beta_F + 1)R_E \gg r_\pi + R_B$ then

$$\frac{e_o}{v_s} \approx -\frac{R_L}{R_E} \tag{99}$$

7.9.1 THE EMITTER-FOLLOWER CIRCUIT

The emitter-follower (sometimes called the common-collector circuit) is illustrated in Figure 7.33. In this case the dc voltage source V_{EE} sets the bias current. To find the dc operating point set $v_s = 0$ and if we assume that

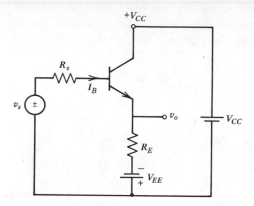

Figure 7.33 Emitter-follower circuit.

$R_s I_B$ is negligible, then

$$I_E \approx \frac{V_{EE} - 0.7}{R_E} \tag{100}$$

and the V_{CC} supply reverse biases the b-c diode. This circuit has a high input resistance, a very low output resistance, and a voltage gain of approximately one. It is used as a buffer between circuits since its high input resistance will not load the circuit connected to its input, and its low output resistance makes it a good voltage source supply to the circuit connected to its output. This example also gives us good practice in analysis with dependent sources.

The small signal model is shown in Figure 7.34. Note that we can easily write KVL for the outer loop in terms of i_b.

$$[R_s + r_\pi + (\beta_F + 1)R_E]i_b = v_s \tag{101}$$

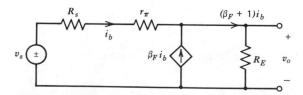

Figure 7.34 Small-signal model.

Therefore, the input resistance seen by the source v_s whose resistance is R_s is

$$R_{in} = \frac{v_s}{i_b} - R_s = \boxed{r_\pi + (\beta_F + 1)R_E}$$ (102)

To compute the circuit gain note that

$$v_o = (\beta_F + 1)i_b R_E$$ (103)

From (101) and (103) the gain is

$$\frac{v_o}{v_s} = \frac{(\beta_F + 1)R_E}{R_s + r_\pi + (\beta_F + 1)R_E} < 1$$ (104)

Finally to compute the resistance seen at the output port we use Figure 7.35. Note that

$$i = \frac{v}{R_E} - \beta_F i_b + \frac{v}{r_\pi + R_s}$$

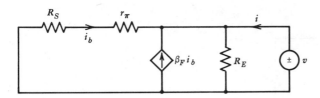

Figure 7.35   Small-signal model to compute output port resistance.

and

$$i_b = -\frac{v}{r_\pi + R_s}$$

so that

$$i = \left[\frac{1}{R_E} + \frac{\beta_F + 1}{r_\pi + R_s}\right]v$$

The output resistance is

$$R_o = \frac{v}{i} = \left(\frac{r_\pi + R_s}{\beta_F + 1}\right) \| (R_E)$$ (105)

Since (104) yields the open circuit output voltage and (105) is the resistance seen at the output port, Figure 7.36 is the Thévenin equivalent circuit of the emitter-follower at the output port for small-signal operation.

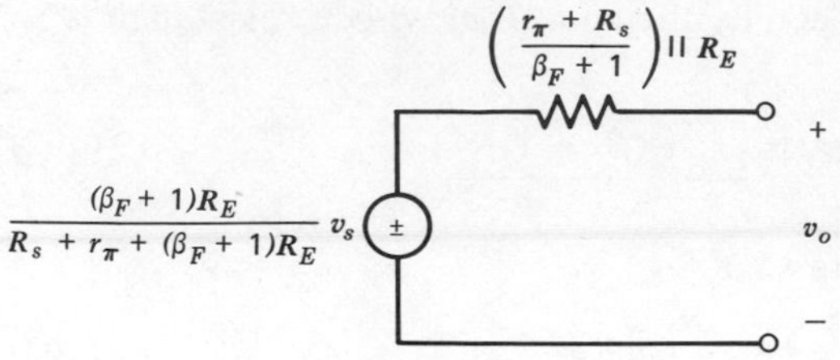

$\left(\frac{r_\pi + R_s}{\beta_F + 1}\right) \| R_E$
$\frac{(\beta_F + 1)R_E}{R_s + r_\pi + (\beta_F + 1)R_E} v_s$
±
+
v_o
−

Figure 7.36 Thévenin equivalent of the emitter-follower circuit at the output port terminals.

7.10 DIFFERENTIAL AMPLIFIER

The differential amplifier is the symmetrical circuit in Figure 7.37. It requires almost perfect matching of components which is relatively easy in integrated circuits but difficult in discrete circuits. Therefore, the differential amplifier is an ideal circuit for analog integrated circuits and frequent use is made of it, particularly in the operational amplifier.

$+V_{CC}$
R_C
R_C
v_o
v_1
v_2
I_E
I_E
R_E
$-V_{EE}$

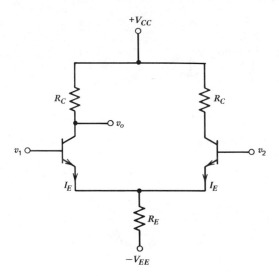

Figure 7.37 Differential amplifier.

The bias equation ($v_1 = v_2 = 0$) is

$$V_{BE} + 2I_E R_E - V_{EE} = 0. \tag{106}$$

Once the transistors are biased in their active region we can replace them by their incremental models, and split the input signals into their common mode and a differential mode component, that is,

$$v_d = \frac{v_1 - v_2}{2} \tag{107}$$

and

$$v_c = \frac{v_1 + v_2}{2} \tag{108}$$

Since the circuit is symmetrical we need only analyze the half-circuits in Figure 7.38.

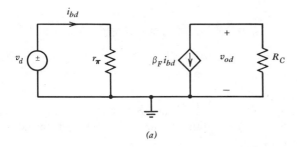

(a)

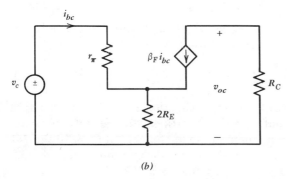

(b)

Figure 7.38   Half circuits for the differential amplifier. (a) Differential mode. (b) Common mode.

Our analysis of the half-circuits gives

$$v_{od} = -g_m R_C v_d \tag{109}$$

and

$$v_{oc} = -g_m R_C \frac{1}{1 + (\beta_F + 1)\dfrac{2R_E}{r_\pi}} v_c \tag{110}$$

Superposition yields

$$v_o = v_{od} + v_{oc}$$

$$= -g_m R_C \left[\frac{(v_1 - v_2)}{2} + \frac{1}{1 + (\beta_F + 1)\dfrac{2R_E}{r_\pi}} \frac{v_1 + v_2}{2} \right] \tag{111}$$

Thus, the output signal

$$v_o = A_d(v_2 - v_1) + A_c \frac{v_1 + v_2}{2} \tag{112}$$

where the differential mode gain

$$A_d \triangleq \frac{g_m R_c}{2}$$

the common mode gain

$$A_c \triangleq \frac{-g_m R_c}{1 + (\beta_F + 1)2R_E/r_\pi}$$

and $(v_2 - v_1)$ is called the differential input voltage. In the differential amplifier the desired response is

$$v_o \approx A_d(v_1 - v_2) \tag{113}$$

that is, we want $A_c = 0$ so that the response depends only on the differential input. The common mode rejection ratio

$$\text{CMRR} \triangleq \left| \frac{A_d}{A_c} \right| = \frac{1 + (\beta_F + 1)2R_E/r_\pi}{2} \tag{114}$$

is a measure of the goodness of the approximation (113). Thus, in a true operational amplifier the output is not proportional to the difference between the inputs, but contains a common mode component also. In the 741 operational amplifier $\text{CMRR} \approx 10^5$, so that a 1-V common mode signal yields the same output as a 10-μV differential signal.

The typical op amp consists of a cascade of two gain stages and an output stage. The first gain stage is always a differential amplifier. The output stage must have a low output resistance and must be capable of delivering a fair amount of power to a load connected to the output of the op amp. In some of the first op amps the emitter-follower was used in the output stage, but now more elaborate output stages are employed.

7.11 THE MOSFET TRANSISTOR

We conclude this chapter with a discussion of a very important device that is the basic building block of large-scale integrated circuits used in solid-state memories, and logic circuits for calculators, watches, and cameras. It is the *metal-oxide-semiconductor-field-effect transistor* (MOS-FET). In the early development of large-scale-integrated circuits the MOS device required fewer diffusions, less area, and less power than the bipolar transistor, but bipolar circuits were faster and achieved higher gains when operated as a linear amplifier. In recent developments the MOS transistor is closely approaching the switching speed of bipolar circuits, and on the bipolar side the invention of integrated injection logic (I^2L) has significantly reduced the area and power of bipolar circuits. Therefore, at present the choice between the bipolar or the MOS technology is no longer obvious.

The physical construction of the n-channel MOS transistor (NMOS) is shown in Figure 7.39. It consists of a p-type substrate with two n^+ diffusions separated by a distance L, which is called the channel length. The two metal contacts to the n^+ regions form the source and drain

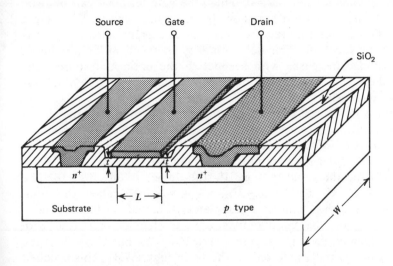

Figure 7.39 *n*-channel MOSFET.

terminals. The third terminal is formed by a metal contact over the channel but separated from the channel by a dielectric (S_iO_2) of thickness t. When the gate voltage is sufficiently large so that $V_{GS} > V_T$ and $V_{GD} > V_T$ an inversion layer is formed in the channel, that is, the majority carriers are now electrons rather than holes in the p-type substrate. Thus, a conducting channel is formed between the source and drain as shown in Figure 7.40. The threshold voltage V_T for MOS devices *is not the same V_T used in bipolar transistors* $(V_T \neq kT/q)$. Instead, it is a threshold that must be exceeded in order to invert the surface of the channel and it depends on

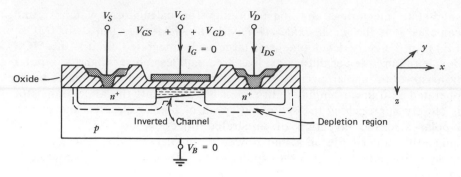

Figure 7.40   NMOS in the nonsaturation region ($V_{GS} > V_T$, $V_{GD} > V_T$).

the concentration of holes in the p-type substrate, the thickness of the gate oxide, plus several other parameters. If $V_{GS} < V_T$ and $V_{GD} < V_T$ then there is no conducting channel and $I_{DS} = 0$. Thus, the gate terminal can be used to switch on or off the current between the drain and source. We have a voltage controlled switch that requires zero input current since the gate is insulated from the substrate. This device makes an excellent logic element. As the gate voltage increases with respect to the drain and source, the resistance in the channel decreases because the number of free electrons increases with increasing V_{GS} and V_{GD}.

Typically the gate-source terminals form the input port and the drain-source terminals form the output port. Thus, it is desirable to express the characteristics of the MOS devices in terms of V_{DS}, I_{DS}, and V_{GS}. Recall that $I_G = 0$. The characteristics are plotted in Figure 7.42 for $V_{DS} > 0$. When V_{GS} is less than the threshold voltage V_T the device is off ($I_{DS} = 0$). The region where $V_{GS} > V_T$ and $V_{GS} - V_{DS} = V_{GD} > V_T$ is called the non-saturation region and the channel is completely inverted. In this region in Figure 7.42 note that the conductance of the channel increases with increasing V_{GS}. Finally, there is a region where $V_{GS} > V_T$ but $V_{GD} < V_T$, then the channel is only partially inverted as shown in Figure 7.41. This is called the saturation region (see Figure 7.42). In this region the current I_{DS} is constant and equal to the value when $V_{GD} = V_T$, which forms the boundary between the saturation region and the nonsaturation region. In this region the voltage across the inverted channel remains constant and further increases in V_{DS} simply increase the voltage across the depletion region. Therefore the drain-source current is constant.

The NMOS characteristics are approximated to a first order by the following derivation. The voltage drop between the gate and the channel is $V_G - V(x)$. These voltages are measured with respect to the substrate. The voltage of the substrate is sometimes referred to as the body voltage V_B. We assume that $V_B = 0$ in this discussion and that both V_S and V_D are

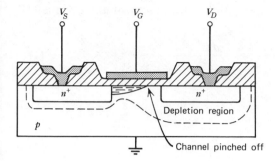

Figure 7.41   NMOS in the saturation region ($V_{GS} > V_T$, $V_{GD} < V_T$).

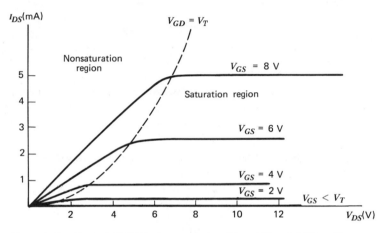

Figure 7.42   Typical NMOS characteristics ($V_T = 1$ V and $W/L = 10$).

greater than or equal to zero. Note that the substrate voltage must always be less than the voltage of the source and drain so that the diodes formed by the p-type substrate and the n^+ regions are reversed biased. The voltage $V(x)$ is the result of the electric field in the channel due to the applied voltages. From electric field theory one can show that the charge induced in the channel at point x is

$$Q(x) = -\frac{\epsilon_{ox}}{t}[V_G - V_T - V(x)]C/cm^2, \text{ for } V_G - V_T - V(x) > 0 \quad (115)$$

where ϵ_{ox} is the permittivity of the oxide and t is the thickness of the oxide between the gate and channel. The incremental resistance at point x is proportional to the incremental channel length dx and inversely proportional to the charge per unit area at point x, the width of the channel W, and the mobility of the charge μ_n.

$$dR(x) = \frac{dx}{\mu_n W \frac{\epsilon_{ox}}{t}(V_G - V_T - V(x))} \tag{116}$$

This equation is only valid for $V_G - V_T > V(x)$, where $V_G > V_T$ and all voltages are assumed positive. Now we write that

$$dV = dR I_{DS} \tag{117}$$

where I_{DS} is the drain to source current. We write

$$\int_{V_S}^{V_D} \mu_n W \frac{\epsilon_{ox}}{t}(V_G - V_T - V)\, dV = \int_0^L I_{DS}\, dx \tag{118}$$

and obtain

$$I_{DS} = \mu_n \left(\frac{W}{L}\right) \frac{\epsilon_{ox}}{t}\left[(V_G - V_T)(V_D - V_S) - \frac{V_D^2}{2} + \frac{V_S^2}{2}\right] \tag{119}$$

for V_{GS} and $V_{GD} > V_T$. However, $V_D = V_{DS} + V_S$ so that (119) becomes

$$I_{DS} = \mu_n \left(\frac{W}{L}\right) \frac{\epsilon_{ox}}{t}[(V_{GS} - V_T)V_{DS} - V_{DS}^2/2] \tag{120}$$

Equation 120 describes the behavior of the NMOS transistor in the *nonsaturation* region.

Recall that saturation occurs when the channel is pinched off. Assuming $V_{DS} > 0$, this occurs when $V_{GD} = V_T$ or equivalently when

$$V_{DS} = (V_{GS} - V_T) \tag{121}$$

Upon substitution of (121) into (120) we obtain the current I_{DS} in the saturation region that depends only on V_{GS} and not V_{DS}.

$$I_{DS} = \mu_n \left(\frac{W}{L}\right) \frac{\epsilon_{ox}}{t} \frac{1}{2}(V_{GS} - V_T)^2 \tag{122}$$

for

$$V_{GS} > V_T \quad \text{and} \quad V_{GD} < V_T$$

Typical parameters for the NMOS transistor are listed below.

$$\mu_n \approx 500 \text{ cm}^2/\text{V-s}$$
$$t \approx 0.1 \ \mu m$$
$$\epsilon_{ox} = \epsilon_r \epsilon_o \approx 4 \times 8.85 \times 10^{-14} \text{ C/V-cm}$$

so that

$$I_{DS} = 1.77 \times 10^{-5} \left(\frac{W}{L}\right)\left[(V_{GS} - V_T)V_{DS} - \frac{V_{DS}^2}{2}\right] \tag{123}$$

$$\text{(nonsaturation)}$$

and

$$I_D = 1.77 \times 10^{-5} \left(\frac{W}{L}\right) \frac{1}{2} (V_{GS} - V_T)^2 \qquad (124)$$

(saturation)

Equations 123 and 124 are sketched in Figure 7.42 for $V_T = 1$ V, $L = 5$ μm, and $W = 50$ μm so that $W/L = 10$. The ratio W/L is an important design parameter in MOSFET design. Recall that in the nonsaturation region the MOSFET is a variable resistor. The incremental conductance in this region is

$$g_{DS} \overset{\Delta}{=} \frac{\partial I_{DS}}{\partial V_{DS}} \approx 1.77 \times 10^{-5} \left(\frac{W}{L}\right) (V_{GS} - V_T) \qquad (125)$$

so that for $0 \leqslant V_{GS} \leqslant 12$ V the drain to source resistance r_{DS} lies approximately in the range $500 \, \Omega < r_{DS} < \infty$. In the saturation region the small-signal model for the MOSFET is shown in Figure 7.43 where

$$g_m \overset{\Delta}{=} \frac{\partial I_{DS}}{\partial V_{GS}} = 1.77 \times 10^{-5} \frac{W}{L} (V_{GS} - V_T) \qquad (126)$$

The gate current is essentially zero so that the model is quite simple. If $V_{GS} = 5$ V, and $W/L = 10$, then $g_m \approx 0.7 \times 10^{-3}$. This value of g_m is roughly 100 times smaller than the value of g_m for the bipolar transistor so that lower gains are achieved with the MOSFET. This can be corrected by making $W/L = 10^3$, but then the device is extremely large.

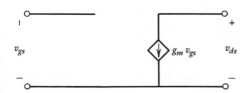

Figure 7.43   Small-signal model.

There is also a p-channel MOSFET (PMOS) that is obtained by interchanging the p and the n regions in Figure 7.39. The channel carriers are now holes rather than electrons and holes have a mobility about three times smaller than the mobility of electrons so that the PMOS transistor is slower than the NMOS transistor. The characteristics of the PMOS transistor are similar to those in Figure 7.42 except that the polarities of I_{DS}, V_D, and V_G are reversed. In the early days of MOS development it was easier to control V_T for the PMOS device than for the NMOS device. Thus many of the first MOS integrated circuits used the PMOS process. The circuit symbols for the MOSFET are given in Figure 7.44. Frequently the body (substrate) terminal is omitted in circuit diagrams.

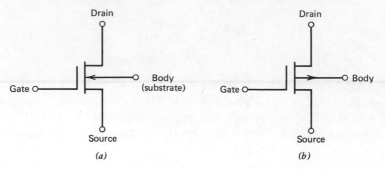

Figure 7.44 MOSFET symbols. (*a*) *n*-channel. (*b*) *p*-channel.

Mostly MOS devices are used in digital logic and memory circuit applications. An example of an NAND gate is given in Figure 7.45. The transistor T_1 is used as a resistor load. The gate of T_1 is held at a much higher voltage than the drain to hold T_1 in the nonsaturation region. A MOS

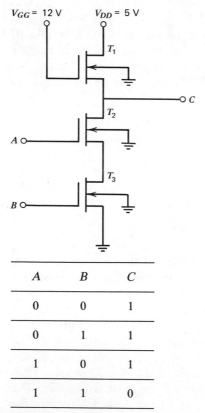

A	B	C
0	0	1
0	1	1
1	0	1
1	1	0

Figure 7.45 NAND gate (positive logic).

device requires much less area than a resistor, therefore it is cheaper to use T_1 as a load. In order for the voltage at terminal C to be low both transistors T_2 and T_3 must be switched on, that is, the voltage at terminals A and B must be high (≈ 5 V). Also, in this state we want most of the voltage drop to be across T_1 and not T_2 and T_3. Thus, W_2/L_1 and $W_3/L_3 \gg W_1/L_1$ so that the resistance of T_2 and T_3 is much less than the resistance of T_1 when both A and B are high. In this state it is important that $V_C < V_T$ so that the following stage is not incorrectly turned on.

REFERENCES

1. Millman, J. and C. C. Halkias. *Electronic Fundamentals and Applications for Engineers and Scientists*, McGraw-Hill, 1976.
2. Grinich, V. H. and H. G. Jackson. *Introduction to Integrated Circuits*, McGraw-Hill, 1975.

PROBLEMS

1. Small-signal measurements are made on the two-port in Figure 7.1 and it is found that the input resistance at port A is 2.5 kΩ and is independent of i_b and v_b. Also $i_b/i_a|_{v_b=0} = 100$ and $i_b/v_b|_{i_a=0} = 10^{-4}$ mho. Find the hybrid parameters for this circuit and draw the equivalent circuit model.
2. Convert the hybrid parameters in Problem 1 to short-circuit conductance parameters and draw the equivalent circuit.
3. Small-signal measurements on the two-port in Figure 7.1 yield the following data: (a) The input resistance at port A is infinite for all i_b, v_b, (b) at port B $i_b/v_a|_{v_b=0} = 0.04$ mho, and (c) $i_b/v_b|_{v_a=0} = 2 \times 10^{-5}$ mho. Find the short-circuit conductance parameters and draw the equivalent circuit for the two-port.
4. In Table 7.1 given one set of two-port parameters one can always convert to another set of parameters provided those parameters exist. Express the open-circuit resistance and inverse hybrid parameters in terms of the hybrid parameters and state under what conditions these parameters exist.
5. Repeat Problem 4 only express the other three two-port parameters as a function of the open-circuit resistance parameters.
6. Repeat Problem 5 for the short-circuit conductance parameters.
7. Repeat Problem 5 for the inverse hybrid parameters.
8. In the negative feedback amplifier circuit in Figure 7.10a, assume that the amplifier is ideal and that A is infinite. Design the amplifier to have a gain of 10 and an input resistance of 1 kΩ.
9. Repeat Problem 8 only let (a) $A = 10^4$, (b) $A = 10^2$.
10. What is the output resistance of the amplifier in Figure 7.10b?
11. For the nonideal amplifier in Figure 7.10c, suppose that $A = 5000$, $R_i = 10$ kΩ, $R_o = 10$ Ω, $R_s = 1$ kΩ, and $R_f = 10$ kΩ. Calculate the gain from (36) and compare it with the ideal gain (32). Also calculate the input resistance from (42) and compare it to the approximation (43). Finally compute the output resistance from (45).
12. Repeat Problem 11 for $A = 100$.

13. The circuit in Figure P7.13 is an ideal current amplifier with negative feedback

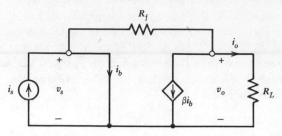

Figure P7.13

through the resistor R_f. show that the current gain

$$\frac{i_o}{i_s} = \frac{-\beta R_f}{(\beta + 1)R_L + R_f} \approx -\frac{R_f}{R_L}$$

for large β. Also show that the input resistance v_s/i_s is zero and that the output resistance is $-v_o/i_o = R_f/(\beta + 1)$. Draw the Thévenin and Norton equivalent circuits seen by the load R_L.

14. Find the current gain i_o/i_s, the input resistance v_s/i_s, and the output resistance $-v_o/i_o|_{i_s=0}$ of the nonideal current amplifier in Figure P7.14. Draw the Thévenin and Norton equivalent circuits seen by the load R_L.

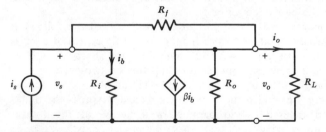

Figure P7.14

15. Find the voltage gain v_o/v_s, the input resistance v_s/i_b, and the output resistance $v_o/i_o|_{v_s=0}$ of the circuit in Figure P7.15. Draw the Thévenin and Norton equivalent circuits seen by the load resistor R_L.

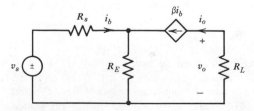

Figure P7.15

16. Repeat Problem 15 for the circuit in Figure P7.16.

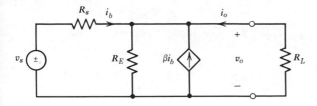

Figure P7.16

17. A gyrator is defined as a two-port with short-circuit conductance parameters $y_{aa} = y_{bb} = 0$ and $- y_{ab} = y_{ba} = g_r$. A gyrator model with the output port terminated in a resistor R is shown in Figure P7.17. Find the input resistance v_a/i_a.

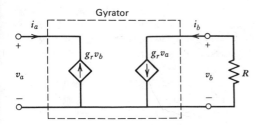

Figure P7.17

18. Design the operational amplifier circuit in Figure 7.11 so that $e_o = 10v_{s1}$. Let $R_1 = 1 \text{ k}\Omega$ and $v_{s2} = 0$. Use Equation 51 and draw your circuit. What is the input resistance?

19. Repeat Problem 18 only let $e_o = - 10v_{s2}$. Assume $v_{s1} = 0$ and that A is very large.

20. For the ideal op amp circuit in Figure P7.20 show that

$$e_o = - \left(\frac{R_f}{R_1} e_1 + \frac{R_f}{R_2} e_2 \right)$$

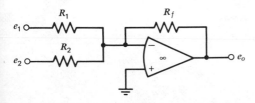

Figure P7.20

21. Design the ideal op amp circuit in Figure P7.21 so that $e_0 = 10(v_{s1} - v_{s2})$ and the input resistance at each input terminal (1 and 2) is $2\,k\Omega$.

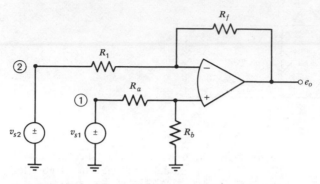

Figure P7.21

22. In the inverter circuit, Figure 7.21, replace the $3\,k\Omega$ resistor with a $10\,k\Omega$ resistor. What base current i_B is required to saturate the transistor? Assuming that $V_{BE} \approx 0.7\,V$ in saturation, what voltage v_s is required to saturate the transistor? What base current i_B is required to bias the transistor in the forward active region with $v_{CE} = 5\,V$?

23. The circuit in Figure P7.23 is a diode-transistor logic (DTL) NAND gate (positive logic). When diodes D_A and D_B are off (terminals A and B at 5 V), the transistor should be saturated. Assume that $\beta = 20$ and that the voltage across the base-emitter diode and diodes D_1 and D_2 is approximately 0.7 V each when they are on, and calculate the maximum value of R_B allowed to insure saturation. The purpose of diodes D_1 and D_2 is to prevent the transistor from being turned on by the forward voltage across D_A or D_B when one of the inputs is low. Give the positive logic truth table for this circuit with A and B as inputs and C the output.

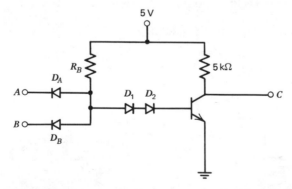

Figure P7.23

24. A bipolar npn transistor is biased in its forward active region. Assume $V_T = 25$ mV, $\beta_F = 100$ and $I_C = 2$ mA. Compute the small-signal model parameters r_π and g_m.

25. Repeat Problem 24 for $I_c = 200$ μA.

26. Let $V_{CC} = 20$ V, $R_L = 1$ kΩ, and $\beta_F = 200$ for the linear amplifier in Figure 7.25. Furthermore, suppose that the transistor is rated for a maximum power dissipation of 150 mW. Calculate the bias current I_B such that $V_{CE} = 10$ V. Calculate the voltage gain from (65). At the dc operating point what power is dissipated by the transistor? Suppose that the small-signal voltage swing v_{ce} is $-10 \sin \omega t$ volts and the small-signal current swing i_c is $10 \sin \omega t$ milliamperes. Calculate the average power dissipated (a) by the transistor, (b) the 1-kΩ load resistor, and (c) the power supplied by the 20-V source.

27. In Figure 7.29 let $V_{CC} = 15$ V and choose R_{B1} and R_{B2} such that the Thévenin equivalent voltage is 3 V and the bias current I_B is 25 μA. Assume $V_{BE} = 0.7$ V. Calculate I_C and R_L if $\beta_F = 80$ and we want $V_{CE} = 7$ V. Calculate the voltage gain from (78). What is the new dc operating point if $\beta_F = 160$?

28. Repeat Problem 27 only this time use the bias circuit in Figure 7.31. Assume the same value of R_L and let $R_E = 500$ Ω. Calculate the voltage gain from (98) and compare it to (99).

29. Repeat Problems 28 only this time choose R_{B1} and R_{B2} such that the Thévenin equivalent voltage is 2.0 V and $I_B = 25$ μA.

30. In Figure 7.33 let $R_s = 1$ kΩ, $V_{CC} = V_{EE} = 5$ V, $\beta_F = 100$, $R_E = 1$ kΩ. Calculate the (a) dc operating point, (b) the input resistance, (c) the output resistance, and (d) the voltage gain.

31. Given the differential amplifier in Figure P7.31, assume $\beta_F = 100$ and

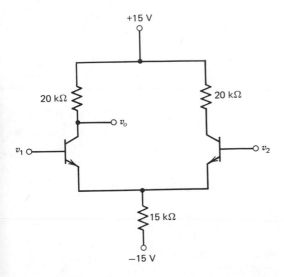

Figure P7.31

(a) Compute the dc operating point.
(b) Determine the small signal parameters of the transistors.
(c) Compute the differential and common-mode gains.
(d) Compute the common-mode-rejection ratio.
32. The drain-source characteristics of an NMOS transistor are given by the equations

$$I_{DS} = \begin{cases} 0, & \text{for } V_{GS} \text{ and } V_{GD} < V_T \text{ (off)} \\ (5 \ \mu A/V^2)\left(\dfrac{W}{L}\right)[2(V_{GS} - V_T)V_{DS} - V_{DS}^2], & \text{for } V_{GS} \text{ and } V_{GD} > V_T \\ & \text{(nonsaturation)} \\ (5 \ \mu A/V^2)\left(\dfrac{W}{L}\right)(V_{GS} - V_T)^2 & \text{for } V_{GS} > V_T \text{ and } V_{GD} < V_T \text{ (saturation)} \end{cases}$$

Plot I_{DS} versus V_{DS} for $V_{DS} > 0$ assuming that the threshold voltage $V_T = 1$ V, $V_{GS} = 3$ V and $W/L = 10$. Note the value of V_{DS} when the transistor goes into the saturation region. What is the incremental resistance ($\partial I_{DS}/\partial V_{DS}$) when (a) $V_{DS} = 0$ V, (b) $V_{DS} = 1$ V, and (c) $V_{DS} = 2$ V?
33. Repeat Problem 32 for $V_{GS} = 5$ V.
34. Repeat Problem 32 for $W/L = 100$.
35. Using the equation in Problem 32 calculate the small-signal gain (v_o/v_s) of the linear amplifier in Figure P7.35. Assume $V_T = 1$ V and $W/L = 10$.

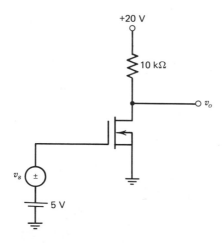

Figure P7.35

36. Assuming positive logic draw the truth table for the NMOS logic gate in Figure P7.36.

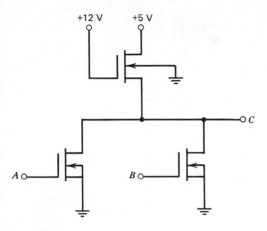

Figure P7.36

Chapter Eight
Energy Storing Elements

The past several chapters have been devoted to the study of resistive circuit models. In these circuits the response at any instant of time depends only on the present value of the independent sources in the circuit. In this chapter we introduce two new ideal elements, the capacitor and the inductor, which are capable of storing energy in their electric and magnetic fields, respectively. In circuit models that contain these devices the response at any given time depends not only on the independent source values, but also on the energy stored in the capacitors and the inductors and this stored energy depends on the past history of the circuit. This property complicates the analysis of these circuits. Now analysis requires the solution of differential-algebraic equations rather than just algebraic equations. However, in return the capacitor and inductor yield some very interesting and useful circuit characteristics that cannot be obtained with resistive elements alone.

8.1 CAPACITOR

The *ideal capacitor* is a two-terminal element whose characteristic can be mapped into the q-v plane as indicated in Figure 8.1a. The characteristic can be linear or nonlinear, and if it changes with time then we say that the capacitor is *time varying*. Typically we can mathematically express the charge of the ideal nonlinear capacitor characteristic as a function of the

240

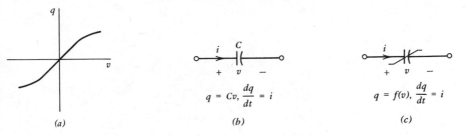

Figure 8.1   Ideal capacitor. (*a*) Characteristic. (*b*) Linear time-invariant capacitor symbol. (*c*) Nonlinear time-invariant capacitor symbol.

voltage across the capacitor terminals

$$q = f(v) \tag{1a}$$

and the current flow is

$$i = \frac{dq}{dt} \tag{2}$$

If the capacitor is linear time invariant we write (1a) as

$$q = Cv \tag{1b}$$

If the capacitor is linear but time varying we write

$$q = C(t)v \tag{1c}$$

and if it is nonlinear and time varying

$$q = f(v, t) \tag{1d}$$

In Figure 8.1*b* and 8.1*c* we give the typical symbols and branch constraints for the linear and nonlinear capacitor whose characteristics are time invariant. The voltage polarity is assigned according to our standard reference system.

8.1.1 PRACTICAL CAPACITORS

The most simple capacitor is the parallel plate capacitor in Figure 8.2. In Chapter One we pointed out that if the dimension d is small compared to the dimensions of the plate, then the electric field between these plates is approximately

$$\mathscr{E}_x = \frac{Q}{\epsilon A} \tag{3}$$

therefore

$$v = \mathscr{E}_x d = \frac{Qd}{\epsilon A} \tag{4}$$

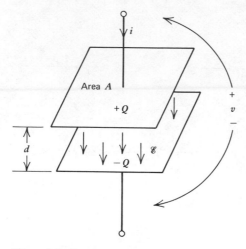

Figure 8.2 Parallel plate capacitor.

We write

$$Q = Cv \qquad (5a)$$

where $C = \epsilon A/d$. Equation 5a is linear and the quantity C is called the capacitance with the unit of measurement called the farad (F) in honor of Michael Faraday. Sometimes (5a) is expressed in the form

$$i = \frac{d}{dt}(Cv) \qquad (5b)$$

Thus, in this physical capacitor $C > 0$ and $i > 0$ implies that the charge and the voltage are increasing as is the case in Figure 8.2. In practice, numerous other geometries are used for the capacitors, but (5a) and (5b) are still appropriate.

A typical metal-insulator semiconductor (MIS) capacitor used in integrated circuits is shown in Figure 8.3. The n epitaxial layer is heavily

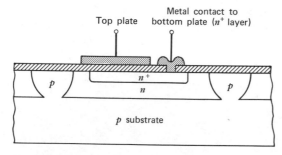

Figure 8.3 Integrated circuit capacitor.

doped with donor atoms to form the bottom plate. The dielectric is silicon dioxide (SiO_2). Since it is important that the capacitor occupy little area in an integrated circuit, typically these capacitors are constrained to values less than 100 picofarads (pF).

8.1.1 *Semiconductor Junction Capacitance.* An important nonlinear capacitor is the semiconductor *pn* junction capacitance associated with the depletion region. Recall that in Chapter Six we assumed a rectangular charge distribution in the depletion region as shown in Figure 8.4a, and the

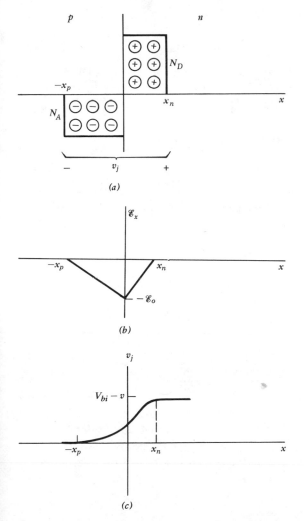

Figure 8.4 Electrical characteristics of the depletion region. (a) Charge density in the depletion region. (b) Electric field in the depletion region. (c) Voltage drop across depletion region.

voltage drop across this region from the n side to the p side of the crystal is $v_j = V_{bi} - v$ where V_{bi} is the "built-in" voltage and v is the applied voltage in the forward direction. The electric field in this depletion region is found from the equation

$$\frac{d\mathscr{E}_x}{dx} = \frac{\rho(x)}{\epsilon}. \tag{6}$$

where ϵ is the permittivity of the crystal and $\rho(x)$ is the charge density. In Chapter Six we said that

$$\rho(x) = -eN_A \qquad -x_p < x < 0 \tag{7}$$

and

$$\rho(x) = eN_D \qquad 0 \leq x < x_n \tag{8}$$

where $e = 1.6 \times 10^{-19}$ C. Also we assumed that the total charge in the depletion region was zero, that is,

$$Q_p + Q_n = 0 \tag{9}$$

where

$$Q_p = -eN_A x_p A \tag{10}$$

$$Q_n = eN_D x_n A \tag{11}$$

and A is the area of the junction.

Upon the integration of (6) under the above assumptions we obtain the field distribution in Figure 8.4b where

$$\mathscr{E}_o = \frac{eN_D x_n}{\epsilon} = \frac{eN_A x_p}{\epsilon} \tag{12}$$

Since $dv/dx = -\mathscr{E}_x$, we obtain, from Figure 8.4b,

$$V_{bi} - v = \frac{1}{2}\mathscr{E}_o(x_n + x_p) \tag{13}$$

and using (9) to (12) we write

$$V_{bi} - v = \frac{1}{2}\frac{eN_D x_n}{\epsilon}\left(x_n + \frac{N_D}{N_A}x_n\right) \tag{14}$$

From (14) note that the externally applied voltage v controls the width of the depletion region $x_n + x_p$, but changing the width of the depletion region changes the charge Q_p and Q_n. Thus, the charge in the depletion region is controlled by the "built-in" voltage V_{bi}, which is a constant determined by V_T, N_A, and N_D (see Equation (16), Chapter Six) and by the applied voltage v. From (9), (10), and (11) we write (14) as

$$Q_p = -A\left[\frac{2\epsilon N_A N_D e}{N_A + N_D}(V_{bi} - v)\right]^{1/2} \tag{15}$$

The incremental capacitance of this nonlinear junction capacitor is frequently expressed as

$$\boxed{\frac{dQ_p}{dv} = C_j(v) = \frac{C_{jo}}{(1 - v/V_{bi})^{1/2}}}$$ (16)

where the capacitance at zero bias

$$C_{jo} = \frac{1}{2} A \left[\frac{2\epsilon N_A N_D e}{(N_A + N_D)V_{bi}}\right]^{1/2}$$ (17)

Equations 15 and 16 are plotted in Figure 8.5. Typically, in integrated circuits, C_{jo} is approximately 0.5 to 2 pF.

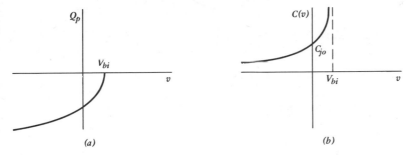

Figure 8.5 *pn* junction depletion capacitance. (*a*) *q-v* characteristic. (*b*) Nonlinear junction capacitance.

In the 1940s, experimenters working with semiconductor diodes in high frequency mixer circuits found that amplification could be achived with the semiconductor diode. This phenomenon was later explained to result from the fact that one of the input signals varies the voltage and charge in the depletion region. It is said that this signal "pumps" the depletion capacitance. Under this condition, and with proper circuit design, the diode can look like a negative resistor in a certain frequency band that is dependent on the pump frequency. Thus, it is possible to achieve amplification for a second input signal in this frequency band. These amplifiers are called *parametric amplifiers* and semiconductor diodes sold for their depletion capacitance characteristics are called *varactor diodes*. Parametric amplifiers were used for many years as the first amplification stage in high frequency radio communication receivers because of their low noise properties. However, now bipolar and FET technology has advanced to the state where they are replacing the parametric amplifier. Presently, the varactor diode is used in circuits where an electronically adjustable capacitor is needed; for example, the mechanical tuner in television is being replaced by a varactor tuner.

8.2 ENERGY STORAGE IN THE CAPACITOR

In order to determine the energy storage capability of the capacitor, consider the circuit in Figure 8.6. Let us suppose that initially the capacitor has a charge Q_1 corresponding to a voltage V_1 before the switch is closed.

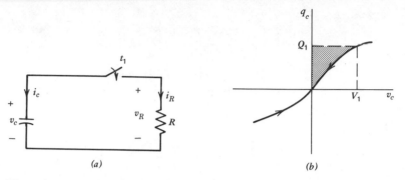

Figure 8.6   Capacitor circuit. (a) Circuit. (b) Capacitor characteristic.

When the switch is closed at time t_1, note that

$$v_R = V_1 > 0 \tag{18}$$

therefore

$$i_R = \frac{V_1}{R} = -i_c > 0 \tag{19}$$

so that

$$\frac{dq_c}{dt} = i_c < 0 \tag{20}$$

Equation 20 implies that the charge on the capacitor is decreasing. The charge decreases as long as $v_c > 0$. If $v_c < 0$, then $i_c > 0$ and the charge increases as shown in Figure 8.6b. The origin of this characteristic is a stable equilibrium point. When the voltage is zero all of the stored energy in the capacitor has been dissipated in the linear resistor R.

The total energy dissipated in R must come from the capacitor. The maximum possible energy that can be dissipated in R is

$$w_R(\infty) - w_R(t_1) = \int_{t_1}^{\infty} v_R i_R \, dt = - \int_{t_1}^{\infty} v_c(q_c) \frac{dq_c}{dt} \, dt \tag{21}$$

If the capacitor is time invariant, then

$$w_R(\infty) - w_R(t_1) = - \int_{Q_1}^{0} v_c(q_c) \, dq_c$$

$$= \int_{0}^{Q_1} v_c(q_c) \, dq_c \tag{22}$$

At $t = \infty$ the capacitor is completely discharged so that we say that the total energy stored in C at time t_1 is

$$w_c(t_1) = \int_0^{Q_1} v_c(q_c)\,dq_c \qquad (23)$$

The above expression simply represents the shaded area in Figure 8.6b. In the case of the *linear time-invariant capacitor*,

$$q_c = Cv_c \qquad (24)$$

so that (23) becomes

$$w_c(t_1) = \frac{1}{2C}\,Q^2 = \frac{1}{2}\,CV_1^2 \qquad (25)$$

Neither (23) nor (25) depends on R, therefore the stored energy is independent of R.

In the case of the *linear time-invariant capacitor*, the differential equation that governs the behavior of the response in Figure 8.6a is

$$C\frac{dv_c}{dt} + \frac{v_c}{R} = 0 \qquad (26)$$

The solution to this equation is

$$\frac{1}{v_c}\frac{dv_c}{dt} = -\frac{1}{RC} \qquad (27)$$

or

$$\frac{d}{dt}\ln v_c = -\frac{1}{RC} \qquad (28)$$

The integration of (28) yields

$$v_c(t) = k\,e^{-t/RC} \qquad (29)$$

The initial energy stored in the capacitor when the switch closes determines the constant k. For example, if $v_c(t_1) = V_1$, then

$$V_1 = k\,e^{-t_1/RC} \qquad (30)$$

Solve (30) for k and substitute the answer into (29) to obtain

$$v_c(t) = V_1\,e^{-(t-t_1)/RC}, \quad t \geq t_1 \qquad (31)$$

This response is plotted in Figure 8.7. This is the trajectory that the voltage $v_c(t)$ follows as it decays to the origin in Figure 8.6b. The quantity $\tau = RC$ is called the time constant of the circuit. When $t = t_1 + \tau$, that is, one time

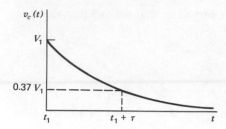

Figure 8.7   Capacitor voltage response.

constant, then

$$v_c(t_1 + \tau) = V_1 e^{-1} = 0.37 V_1 \qquad (32)$$

The capacitor voltage has lost 63% of its initial amplitude and 86% of its stored energy. When $t = t_1 + 4\tau$ (four time constants) the capacitor voltage is approximately $0.02 V_1$ and it has lost approximately 99.96% of its stored energy. For convenience it is usually assumed that the reference point $t_1 = 0$.

   In the *time-varying* capacitor the energy supplied to the resistance R in the time interval $[t_1, \infty]$ is not necessarily equal to the energy stored in the capacitor at time t_1. Consider Figure 8.8. Suppose that the current in the capacitor is zero so that Q is constant. If at t_1 the capacitance is C_1 and at time t_2 the capacitance is C_2, then

$$\frac{1}{2C_2} Q^2 > \frac{1}{2C_1} Q^2 \qquad (33)$$

that is, the stored energy has been increased by decreasing the capacitance. For example, the capacitance could be decreased by an external force moving the capacitor plates further apart. *By changing the q-v characteristic we can change the stored energy in the capacitor.* Usually this is referred to as pumping the capacitor. In a time-varying capacitor the energy delivered to the resistor R in Figure 8.6 not only depends on the stored energy at time t_1, but also on the variation of the q-v characteristic.

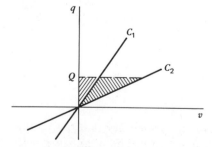

Figure 8.8   Time-varying linear capacitor characteristic.

Let us now compute the power delivered to a resistor from a *time-varying* linear capacitor, $q = C(t)v$. The energy dissipated in the time interval $[t_1, \infty]$ is

$$w_R(\infty) - w(t_1) = -\int_{t_1}^{\infty} v_c \frac{d}{dt}(C(t)v_c)dt = -\int_{t_1}^{\infty} \left[\frac{d}{dt}\left(\frac{1}{2}Cv_c^2\right) + \frac{1}{2}v_c^2\frac{dC}{dt}\right]dt \tag{34}$$

Thus,

$$w_R(\infty) - w_R(t_1) = \frac{1}{2}[C(t_1)v_c^2(t_1) - C(\infty)v_c^2(\infty)] - \frac{1}{2}\int_{t_1}^{\infty} v_c^2\frac{dC}{dt}dt \tag{35}$$

If $C(t) > 0$, then from our previous discussions $q \to 0$, hence,

$$w_R(\infty) - w_R(t_1) = \underbrace{\frac{1}{2}C(t_1)v_c^2(t_1)}_{\substack{\text{stored energy} \\ \text{at time } t_1}} - \underbrace{\frac{1}{2}\int_{t_1}^{\infty} v_c^2\frac{dC}{dt}dt}_{\substack{\text{energy delivered} \\ \text{by the "pump"}}} \tag{36}$$

Note that the first term in (36) is the energy delivered to the resistor provided the capacitance remains constant. The second term is the energy delivered to the resistor due to moving the capacitor plates, that is, the energy supplied by the pump or external source. If dC/dt is negative, then the integral is negative and the external source is supplying energy to the circuit. If dC/dt is positive, then the integral is positive, and the capacitor is supplying energy to the external source.

Finally, consider the q-v characteristic in Figure 8.9, which does not intersect the origin. In this case, we can show that the equilibrium point is at B where $v = 0$, so that if the capacitor is biased at A, then the stored energy is the shaded area indicated in Figure 8.9.

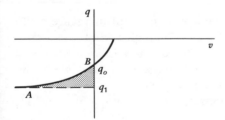

Figure 8.9   Varactor diode characteristic.

Hence, *we now define the energy stored in the time-varying capacitor (linear or nonlinear) at time t_1 as*

$$w_c(t_1) = \int_{q_0}^{q_1} v_c(q_c, t_1)dq_c \tag{37}$$

where q_o is the charge at which $v = 0$. Remember that if the capacitor is time varying, then this is not necessarily the energy that can be delivered to a load.

8.3 RESPONSE OF THE CAPACITOR TO SIMPLE WAVEFORMS

First, let us consider the response of a linear constant capacitor to a current step input (see Chapter Two) of the form $ku(t)$ as shown in Figure 8.10a. The integral form of (5b) is

$$v_c(t) = v_c(0) + \frac{1}{C} \int_0^t i(\tau) d\tau \tag{38}$$

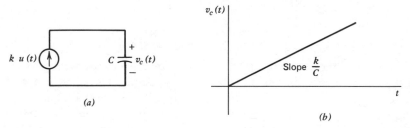

Figure 8.10 Capacitor voltage response to a constant current. (a) Circuit. (b) Response.

where τ is a dummy variable. Since $i(t)$ is constant in the interval $[0, t]$ we obtain

$$v_c(t) = v_c(0) + \frac{kt}{C} \qquad t > 0 \tag{39}$$

This response is called a ramp function and is illustrated in Figure 8.10b for $v_c(0) = 0$. This type of circuit is used to sweep an electron beam across the face of an oscilloscope display. The voltage ramp is applied to the deflection plates of the cathode-ray tube. At the end of the sweep the capacitor is reset (shorted) before the next sweep begins.

As a second example, consider the current waveform in Figure 8.11. Assume that $C = 100 \text{ pF}$ and $v_c(0) = 0$, then

$$v_c(t) = 10^{10} \int_0^t i(\tau) d\tau \tag{40}$$

In the interval $[0, 1 \,\mu\text{s}]$ the current is a constant 1 mA. Therefore, the current increases linearly. At $t = 1 \,\mu\text{s}$ the total area is $1 \text{ mA} \times 1 \,\mu\text{s} = 10^{-9} \text{ A-s}$. From (40) we obtain $v_c(1 \,\mu\text{s}) = 10 \text{ V}$. In the interval $[1 \,\mu\text{s}, 3 \,\mu\text{s}]$

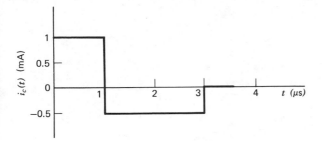

Figure 8.11 Current input.

$i(t) = -0.5$ mA, therefore

$$v_c(t) = 10 \text{ V} - 10^{10} \int_{10^{-6}}^{t} (0.5)d\tau \qquad 1\,\mu s \leqslant t \leqslant 3\,\mu s \tag{41}$$

The area in the interval $[1\,\mu s, 3\,\mu s]$ is 0.5 mA $\times 2\,\mu s = 10^{-9}$ A-s. Thus the change in voltage in this interval is -10 V so that

$$v_c(3\,\mu s) = 0 \text{ V} \tag{42}$$

Since $i(t) = 0$ for $t > 3\,\mu s$, $v_c(t) = 0$ V for $t > 3\,\mu s$. The complete voltage waveform is shown in Figure 8.12.

Note that since the charge and voltage are proportional to the integral of the current, they cannot change instantaneously for finite values of current even though the current may be discontinuous.

Next let us consider the response of the capacitor to a voltage step. See Figure 8.13. Kirchhoff's voltage law demands that $v_c = 0$ for $t < 0$ and $v_c = k$ for $t > 0$. This means that the energy in the capacitor jumps from zero to $\frac{1}{2}Ck^2$ J instantly at $t = 0$, however

$$i_c = C\frac{dv_c}{dt} = \begin{cases} 0, t < 0 \\ 0, t > 0 \\ ?, t = 0 \end{cases} \tag{43}$$

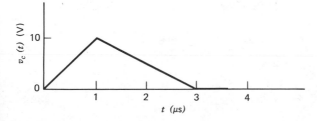

Figure 8.12 Capacitor voltage response.

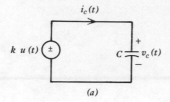

(a)

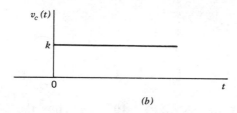

(b)

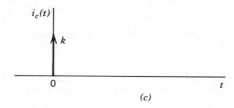

(c)

Figure 8.13 Capacitor response to a voltage step. (a) Circuit. (b) Voltage input. (c) Current response.

What is the derivative of the step function at the discontinuity? Mathematically we write

$$\frac{d}{dt} u(t - a) = \delta(t - a) \tag{44}$$

where $\delta(t - a)$ is called the unit impulse. It is not a mathematical function in the usual sense, but the following properties can be ascribed to it.

$$\int_{a-\epsilon}^{a+\epsilon} \delta(t - a)dt = 1 \qquad \epsilon > 0 \tag{45}$$

and

$$\int_{a-\epsilon}^{a+\epsilon} f(t)\delta(t - a)dt = f(a) \tag{46}$$

if $f(t)$ is continuous at $t = a$; $\epsilon > 0$.

The unit impulse is usually represented by a pulse of infinite height and unit area. For example, in Figure 8.13c the capacitor current is a unit impulse with area k, since the voltage jumped by k volts. The unit impulse is an extremely useful tool for keeping track of simple discontinuities in functions that are differentiated.

8.4 THE INDUCTOR

In 1820 Oersted made the important discovery that a compass needle was deflected in the presence of a current-carrying conductor and concluded that a magnetic field is produced by a flow of charge. The experimental law of Biot and Savart, colleagues of Ampere, states that the *magnitude* of the magnetic field intensity vector **H** (amperes/m) at any point P due to the current i in an element of differential length dl, Figure 8.14, is proportional to the product of i with dl and with the sine of the angle between dl and the line from dl to P; the magnitude is also inversely proportional to the square of the distance r, that is,

$$H = \int \frac{i\,dl\,\sin\theta}{4\pi r^2} \tag{47}$$

The magnetic field intensity vector **H** is normal to the plane formed by dl and r and is in the direction that a right-hand screw would progress in turning dl toward r. For example, the magnitude of the magnetic field at the point P produced by a current i in an infinitely long wire is

$$H = \int_{-\infty}^{\infty} \frac{i\,dl\,\sin\phi}{4\pi r^2} = \frac{ir_o}{4\pi} \int_{-\infty}^{\infty} \frac{dl}{(r_o^2 + l^2)^{3/2}}$$

$$H = \frac{i}{2\pi r_0} \tag{48}$$

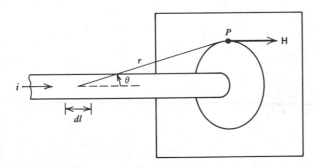

Figure 8.14   Magnetic field intensity at P.

As the point P is rotated about the wire at a constant radius r_o, the value of H does not change, that is, lines of constant magnetic field intensity close on themselves. This is true for any geometry.

The magnetic flux density vector **B** (webers/m²) is related to the magnetic field intensity vector by the expression

$$\mathbf{B} = \mu \mathbf{H} \tag{49}$$

and μ is the permeability of the material in henrys/m. In air the B-H relationship is linear and $\mu_o = 4\pi \times 10^{-7}$ henrys/m. However, the B-H relationship for ferromagnetic materials is highly nonlinear as illustrated in Figure 8.15. At point A the ferromagnetic material begins to saturate and the slope of the B-H curve approaches that of air (a factor of

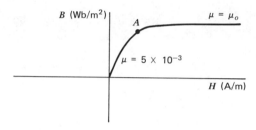

Figure 8.15 B-H curve for a ferromagnetic material.

approximately 10^4 smaller). Ferromagnetic materials also exhibit a hysteresis loop (Figure 8.16). At point a material is completely demagnetized. The current is increased to point b and then decreased to zero, point c. Note that at point c the ferromagnetic material retains some of its magnetism. A negative current must be applied to demagnetize the material, point d. Points e, f, and g indicate the path followed if i is decreased further and then increased.

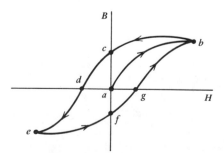

Figure 8.16   Hysteresis loop.

The magnetic flux is the integral of the magnetic flux density

$$\phi = \int B \cos \theta \, dA \tag{50}$$

where θ is the angle between $\mathbf{B}$ and the surface area dA. If the magnetic flux surrounds N conductors, then we define

$$\Psi = N\phi \tag{51}$$

as the total number of flux linkages. For example, the magnitude of the magnetic flux density within the space enclosed by the toroidal winding, Figure 8.17, is

$$B = \frac{\mu N i}{l} \tag{52}$$

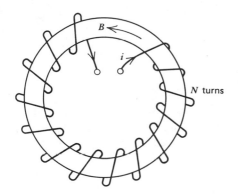

Figure 8.17   Toroidal winding.

where N is the number of turns and l is the length of the core. The flux density is

$$\phi = \frac{\mu N i A}{l} \tag{53}$$

where A is the cross-sectional area of the winding. Thus, the flux linkage is

$$\Psi = \frac{\mu N^2 i A}{l} \tag{54}$$

Although we have picked the toroid as an example, in general one can find a relationship between the flux linkages and the current for any configuration.

An *ideal inductor* is a branch whose constraints can be mapped in the Ψ-i plane, Figure 8.18. This relationship can be linear or nonlinear, time

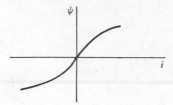

Figure 8.18   Inductor characteristic.

invariant or time varying. *The incremental slope of the Ψ-i characteristic is defined as the self-inductance or inductance of the device*, thus

$$L = \frac{d\Psi}{di} \tag{55}$$

The unit of measurement for the inductor is the henry (H). In the toroidal winding if the core is air, then $\mu = \mu_o$, and the self-inductance is linear

$$L = \frac{\mu_o N^2 A}{l} \tag{56}$$

If the core is a ferromagnetic material, then μ is a function of i, and hence, the inductance is nonlinear. The inductance can be made time varying by changing any one of the parameters with respect to time, for example, μ, N, A, or l.

Realizing that a current produced a magnetic field, Michael Faraday worked for 10 years trying to show that a magnetic field could produce a current. In 1831 he was successful in showing that if the flux linkages enclosing a conductor varied with time, then a voltage was induced in the conductor. This experimental law

$$v = \frac{d\Psi}{dt} \tag{57}$$

is now known as Faraday's law. A German scientist by the name of Lenz established a rule for determining the direction of the induced voltage. Lenz's law states:

The direction of the induced voltage is such as to oppose the cause producing it.

This law affirms the conservation-of-energy principle. Faraday's law (57) led to the development of the electric generator and the electric motor.

The symbol for the ideal linear inductor is given in Figure 8.19. The

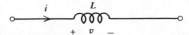

Figure 8.19   Inductor symbol.

branch constraint is

$$\Psi = Li \tag{58a}$$

or

$$\boxed{v = \frac{d\Psi}{dt} = \frac{d}{dt}(Li)} \tag{58b}$$

Note that our standard polarity assignment is in accordance with Lenz's law, that is, for a positive physical inductor an increasing current i produce a voltage v to oppose the increase in current. This becomes more obvious when we discuss energy storage in inductors in the next section.

8.5 ENERGY STORAGE IN THE INDUCTOR

Consider the circuit in Figure 8.20. Before the switch is opened at time t_1 we have $v_L = 0$ so that $d\Psi_L/dt = 0$, which implies a constant flux. Suppose that the initial operating point is $i_L = I_1$ and $\Psi_L = \Psi_1$ in Figure 8.20b. After the switch is opened

$$i_L = -\frac{v_L}{R} > 0 \tag{59}$$

so that

$$v_L = d\Psi_L/dt < 0 \tag{60}$$

which implies that the flux always moves along the characteristic toward the origin. The equilibrium point is at $i_L = 0$. The total energy stored in the magnetic field of the inductor at time t_1, $w_L(t_1)$, is the maximum energy that

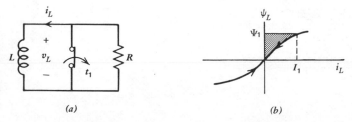

(a)   (b)

Figure 8.20   Inductor circuit. (a) Circuit. (b) Inductor characteristic.

can be delivered to R, which is,

$$w_L(t_1) = w_R(\infty) - w_R(t_1) = -\int_{t_1}^{\infty} v_L i_L \, dt = \int_0^{\Psi_1} i_L(\Psi_L) d\Psi_L \qquad (61)$$

or the shaded area in Figure 8.20b. Again, this derivation is only valid if the characteristic is *constant*.

In the case of a *linear time-invariant inductor*, (61) becomes

$$w_L(t_1) = \frac{1}{2L} \Psi_1^2 = \frac{1}{2} L I_1^2 \qquad (62)$$

Let us now consider the effect of hysteresis on the stored energy in an inductor. Suppose that the current is increased from zero to the value at point A along the path OA in Figure 8.21. Then, the total energy supplied to the inductor is equal to the sum of the area S and D. Now suppose that resistor is connected to the inductor as in Figure 8.20a. Due to the hysteresis effect, the current decreases along path AB. Thus the energy delivered to the resistor is only equal to area S. An energy of magnitude D has been lost due to hysteresis. Note that at point B, $i = 0$; hence, $v = 0$ and $d\Psi/dt = 0$ so that point B is an equilibrium point.

Finally, let us discuss the energy in a *time-varying inductor*. For the sake of simplicity we consider only a linear inductor. In this case (61) becomes

$$w_R(\infty) - w_R(t_1) = -\int_{t_1}^{\infty} i_L \frac{d}{dt}(Li_L) dt = -\int_{t_1}^{\infty} \left[\frac{1}{2}\frac{d}{dt}(Li_L^2) + \frac{1}{2} i_L^2 \frac{dL}{dt}\right] dt \qquad (63)$$

Hence the energy supplied to R is

$$w_R(t) = \frac{1}{2}[L(t_1)i_L^2(t_1) - L(\infty)i_L^2(\infty)] - \frac{1}{2}\int_{t_1}^{\infty} i_L^2 \frac{dL}{dt} dt \qquad (64)$$

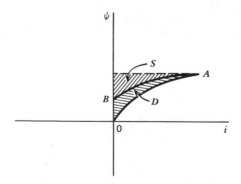

Figure 8.21 Energy stored in an inductance with hysteresis.

If $i_L(\infty) = 0$, which is the case for $L(t) > 0$, then

$$w_R(t) = \underbrace{\frac{1}{2}L(t_1)i_L^2(t_1)}_{\text{stored energy}} - \underbrace{\frac{1}{2}\int_{t_1}^{\infty} i_L^2 \frac{dL}{dt}\,dt}_{\substack{\text{energy supplied} \\ \text{by the pump}}} \tag{65}$$

Thus, decreasing the inductance increases the energy supplied to R. In an iron core inductor, this is equivalent to removing the core while the coil is energized.

We are now in a position to define the *stored energy* in an inductor at time t_1. It is

$$w_L(t_1) = \int_{\Psi_1}^{\Psi_2} i_L(\Psi, t_1)\,d\Psi_L \tag{66}$$

where Ψ_1 is the flux when the current is zero, for example, point B in Figure 8.21. Remember that point B is the equilibrium point when a positive R is connected across the inductor and the incremental slope in the Ψ_L–i_L plane is positive.

8.6 RESPONSE OF THE INDUCTOR TO SIMPLE WAVEFORMS

The integral form of (58b) for a linear constant inductor is

$$i(t) = i(0) + \frac{1}{L}\int_0^t v(\tau)\,d\tau \tag{67}$$

Thus, if the voltage applied to the inductor is a step function $ku(t)$, then

$$i(t) = i(0) + \frac{1}{L}\int_0^t k\,d\tau \qquad t \geqslant 0 \tag{68}$$

or

$$i(t) = i(0) + \frac{kt}{L} \qquad t \geqslant 0 \tag{69}$$

Equation 69 is illustrated in Figure 8.22 for a zero initial condition [$i(0) = 0$].

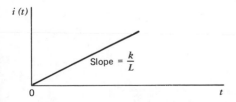

Figure 8.22   Response of an inductor to a constant voltage.

Since the flux and current are proportional to the integral of the voltage, they cannot change instantaneously for finite values of voltage even though the voltage may be discontinuous.

A discontinuity in the current in an inductor implies an infinite voltage. As an example suppose that the current in Figure 8.23 is 5 A for $t < 0$ and

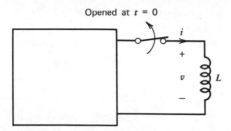

Figure 8.23   Interruption of inductor current.

the switch is opened at $t = 0$. Let $L = 2\,\mathrm{H}$. Then the current $i(t) = 5[1 - u(t)]$, and the voltage across the inductor is

$$v(t) = (2\ \mathrm{H})\frac{d}{dt}[5 - 5u(t)]$$

$$= -10\,\delta(t) \tag{70}$$

that is, the voltage is infinite at $t = 0$ and zero otherwise as shown in Figure 8.24. The inductive effect induces extremely high voltages in a circuit when current flow is interrupted. This is the reason arcing occurs across switch contacts when a switch is opened. In mechanical ignition systems of automobiles, thousands of volts are generated across the spark plug gap of an automobile from a 12 V battery simply by breaking the flow of current in an ignition coil (see Chapter Nine).

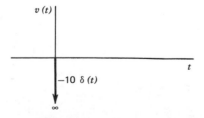

Figure 8.24   Inductor voltage when the flow of current is interrupted.

PROBLEMS

1. Calculate the area needed to make a 100-pF parallel-plate capacitor if the dielectric medium is air and the separation of the plates is 1 mm.
2. Repeat Problem 1 for a MIS capacitor in which the thickness of the silicon dioxide is 0.2 μm and its relative dielectric constant is 4.
3. Calculate the capacitance C_{jo}/cm^2 of a pn junction capacitor given that $N_A = 10^{18}$ and $N_D = 10^{16}$. The relative dielectric constant of silicon is 12 and assume $V_{bi} = 1$ V. Also calculate x_n for $v = 0$ in (14).
4. Repeat Problem 3 for $N_A = 10^{18}$ and $N_D = 10^{15}$.
5. Repeat Problem 3 for $N_A = 10^{19}$ and $N_D = 10^{16}$.
6. The capacitance of a pn junction is measured at -1 V and -5 V bias, respectively. The measured values are $C_j(-1\,V) = 6.67\,pF$ and $C_j(-5\,V) = 3.71\,pF$. Use (16) and find C_{jo} and V_{bi}.
7. In Figure P8.7 $v_c(t) = 10$ V for $t < 0$. Find the energy stored in the capacitor for $t < 0$. The switch is closed at $t = 0$. Sketch $i(t)$ and $v_c(t)$ for $t \geqslant 0$. Find the energy stored in the capacitor at $t = 1$ ms. How much energy has been dissipated in the resistor?

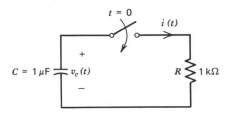

Figure P8.7

8. Repeat Problem 7, only with $C = 0.001\ \mu$F.
9. Repeat Problem 7, only with $R = 1\,\Omega$.
10. Repeat Problem 7, only let $v_c(t) = -5$ V for $t < 0$.
11. A nonlinear capacitor is described by the equations below. What is the incremental capacitance and the energy stored in the capacitor at $v = 1/2$ V?

$$q = \begin{cases} 10^{-20}\,e^{40v}, & v \geqslant 0 \\ 0, & v < 0 \end{cases}$$

12. Repeat Problem 11 for $v = 0.7$ V.
13. For the parallel plate capacitor in Figure 8.2 sketch the current waveforms ($t > 0$) for each of the given charge waveforms in Figure P8.13.

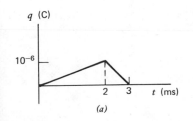

(a)

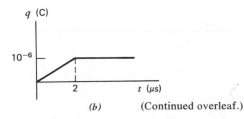

(b)   (Continued overleaf.)

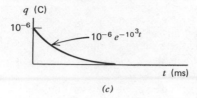

(c)

Figure P8.13   (Continued).

14. Sketch the charge q for $t > 0$ on the plate of a capacitor for each of the current waveforms in Figure P8.14. Assume that $q(0) = 0$.

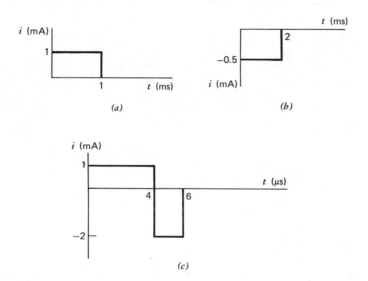

(a)

(b)

(c)

Figure P8.14

15. Given a linear constant 0.1-F capacitor find the voltage $v(t)$, $t > 0$, for each of the current waveforms in Figure P8.14. Assume that $v(0) = 0$.
16. Repeat Problem 15 for a 0.001-F capacitor.
17. Repeat Problem 15 for the current waveform in Figure P8.17.

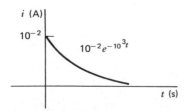

Figure P8.17

18. Given a 1-μF capacitor, sketch the capacitor current ($t > 0$) for each of the voltage waveforms in Figure P8.18.

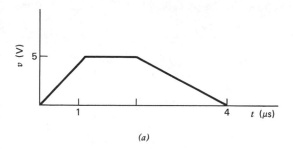

(a)

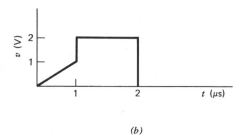

(b)

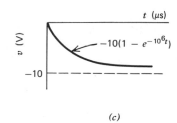

(c)

Figure P8.18

19. A toroidal winding has an area of 0.2 cm^2 and a length of 3 cm. If the permeability of the core is μ_o find the number of turns N to yield an inductance of 1 mH.
20. Repeat Problem 19, only with an iron core whose permeability $\mu = 10^4\,\mu_o$.
21. In Figure P8.21 if $i(t) = 10$ mA for $t < 0$, find the energy stored in the inductor for $t < 0$. The switch is opened at $t = 0$. Sketch $i(t)$ and $v(t)$ for $t \geq 0$. What is the energy stored in the inductor at $t = 10\,\mu$s, and how much energy has been dissipated in the resistor?

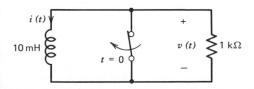

Figure P8.21

22. If the branch voltage of a capacitor is $v_c(t_o) = 100$ V and $v_c(t_1) = 25$ V, find the energy stored or delivered in the time interval (t_o, t_1) where $t_1 > t_o$ for (a) $C = 1\,\mu$F, and (b) $C = 100\,\mu$F.

23. Repeat Problem 22 for an inductor in which $i_L(t_o) = 10$ mA and $i_L(t_1) = 20$ mA and (a) $L = 1$ mH, and (b) $L = 100$ mH.

24. Given a linear constant 1μH inductor, sketch the current waveforms for $t > 0$ corresponding to each of the voltage waveforms given in Figure P8.24. Assume that $i(0) = 0$.

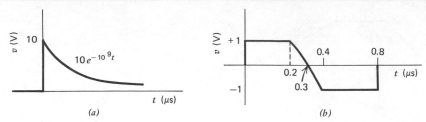

Figure P8.24

25. Repeat Problem 24 with a 100μH inductor.

26. Given a 10-mH inductor sketch the inductor voltage $(t > 0)$ for each of the current waveforms in Figure P8.26.

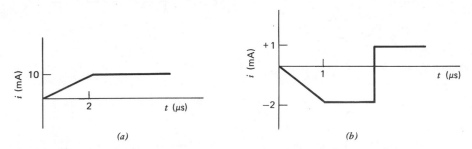

Figure P8.26

27. Repeat Problem 26 for a 1-mH inductor.

28. Repeat Problem 26 for a 100-mH inductor.

29. The capacitance of a time-varying linear capacitor is $C(1 + \sin \omega_1 t)$. If the voltage across the capacitor is $V \sin \omega_2 t$, what is the current $i(t)$ in the capacitor?

30. The inductance of a linear time-varying inductor is $L(1 + \cos \omega_1 t)$. If the current through the inductor is $I \sin \omega_2 t$, what is the voltage across the inductor?

31. In Figure P8.7 suppose that the capacitor is time varying and $C(t) = 10^{-6}(1 + 10^3 t)$. Find the voltage $v_c(t)$ for $t > 0$ if $v_c(0) = 10$ V. Find the energy dissipated in the 1-kΩ resistor in the interval $(0, \infty)$. How does this compare with the stored energy at $t = 0$?

32. The branch constraint of nonlinear inductor is $i = 10^{-2}\Psi + 10^{-3}\Psi^3$. If $i = 1.1$ A at $t = 0$ ($\Psi = 10$ Wb), find the stored energy at $t = 0$. If the linear inductor in Figure P8.21 is replaced with this nonlinear inductor, find $v(t)$ for $t > 0$. (Hint: solve for Ψ first.)

Chapter Nine
Response of Circuits with Energy Storage Elements

The determination of the complete response of circuits containing energy storage elements requires the solution of a set of differential and algebraic equations. If the circuit elements are nonlinear then, as in the case of nonlinear resistive circuits, it is impossible to obtain exact solutions, except in very special cases. In the case of circuits with energy storage elements, time-varying parameters also complicate the problem such that closed form solutions are difficult to obtain. In these cases approximate analytical techniques or computer-aided analyses techniques are required. However, when all the parameters in the circuit are linear and constant, there exist well-established methods for obtaining the solution. These methods are described in this chapter. First, let us discuss an easy technique for determining the differential equation that describes the relationship between a particular response and the inputs including energy stored in the capacitors and inductors.

9.1 DETERMINATION OF THE DIFFERENTIAL EQUATION

Let us write the branch constraint for the linear constant inductor as,

$$v_L = DLi_L \tag{1a}$$

and the branch constraint for the linear constant capacitor as

$$v_C = \frac{1}{DC} i_C \tag{1b}$$

where D denotes the linear operation d/dt and $1/D$ denotes the linear operation $\int dt$. Likewise, we can write

$$i_L = \frac{1}{DL} v_L \tag{2a}$$

and

$$i_C = DC v_C \tag{2b}$$

As long as L and C are constants the operator D and L or C can be interchanged. Henceforth, we will assume that L and C are constants.

Let us now see how the above notation simplifies the determination of our circuit equations, in particular, the elimination of unwanted variables. For example, to determine the differential equation for the circuit in Figure 9.1 we use the model in Figure 9.2. The nodal equations are

$$DC(e_1 - v_s) + \frac{e_1}{R_1} + \frac{1}{DL}(e_1 - e_2) = 0$$

$$\frac{1}{DL}(e_2 - e_1) + \frac{e_2}{R_2} = 0 \tag{3}$$

where $DC(e_1 - v_s) = i_C$ the current through the capacitor leaving node 1 and $(1/DL)(e_1 - e_2) = i_L$ the current through the inductor from node 1 to node 2.

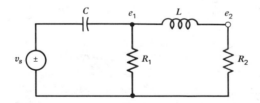

Figure 9.1 Circuit example.

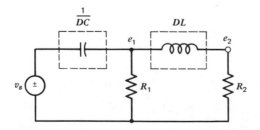

Figure 9.2 Model for finding the differential equations that describe the circuit behavior.

We write (3) as

$$\left(DC + \frac{1}{R_1} + \frac{1}{DL}\right)e_1 - \frac{1}{DL}e_2 = DCv_s \tag{4a}$$

$$-\frac{1}{DL}e_1 + \left(\frac{1}{DL} + \frac{1}{R_2}\right)e_2 = 0 \tag{4b}$$

Let us eliminate the variable e_1 from (4). This can be done by multiplying (4b) by the operation $(DC + 1/R_1 + 1/DL)DL$ and adding the two equations to obtain a differential equation in e_2, or we can use determinant notation and solve for e_2 as follows.

$$e_2 = \cfrac{\begin{vmatrix} DC + \dfrac{1}{R_1} + \dfrac{1}{DL} & DCv_s \\[2ex] -\dfrac{1}{DL} & 0 \end{vmatrix}}{\begin{vmatrix} DC + \dfrac{1}{R_1} + \dfrac{1}{DL} & -\dfrac{1}{DL} \\[2ex] -\dfrac{1}{DL} & \dfrac{1}{DL} + \dfrac{1}{R_2} \end{vmatrix}} \tag{5}$$

The denominator of (5) operates on e_2, therefore, we write

$$\left[D\frac{C}{R_2} + \left(\frac{C}{L} + \frac{1}{R_1 R_2}\right) + \frac{1}{DLR_2}\left(1 + \frac{R_2}{R_1}\right)\right]e_2 = \frac{C}{L}v_s \tag{6}$$

If we operate on both sides of the equation (6) with the operator DR_2/C, then the differential equation for the circuit in terms of the variable e_2 has the form

$$\frac{d^2e_2}{dt^2} + \left(\frac{R_2}{L} + \frac{1}{R_1 C}\right)\frac{de_2}{dt} + \frac{1}{LC}\left(1 + \frac{R_2}{R_1}\right)e_2 = \frac{R_2}{L}\frac{dv_s}{dt} \tag{7}$$

Note that the left side of (7) is determined by the circuit determinant, that is the denominator of (5) and the independent sources have no effect on left side of (7), or in other words *independent* voltage sources and *independent* current sources can be set to zero (shorted and opened, respectively) and the left side of (7) will not change. Recall that their equivalent resistances are zero and infinite, respectively, so this result should be expected. This is not true for dependent sources, since the parameter values of dependent sources will appear in determinant in the denominator of (5) and thus appear as coefficients of the unknown variables.

9.2 ZERO INPUT RESPONSE

Now that we can determine the differential equation for a circuit let us discuss methods for finding the solution of this equation. We will confine our attention to the zero input (unforced) case in this section. Let $y(t)$ denote the response. From the circuit model derived from (1) and (2) it follows that the differential equation for any linear constant parameter circuit will have the form

$$\frac{d^n y}{dt^n} + b_{n-1} \frac{d^{n-1} y}{dt^{n-1}} + \cdots + b_1 \frac{dy}{dt} + b_o y = 0 \tag{8}$$

The integer n is said to be the order of the differential equation, and from our circuit model and our circuit determinant it can be shown that n can never exceed the total number of inductors and capacitors in the circuit. Remember that the left side of (8) is simply the determinant of our circuit model operating on the unknown variable. For example, in the previous section the circuit determinant is the denominator of (5). Of course, this determinant may be multiplied by the operator D several times to remove the $1/D$ coefficients (integral operations). Finally, the coefficients b_i are real numbers in (8) since the parameters (R, L, C, etc.) in the circuit are real.

9.2.1 NATURAL FREQUENCIES OF A CIRCUIT

Note that

$$y(t) = k e^{st} \tag{9}$$

where k is a constant, is a solution to (8) provided that s is a root of the polynomial

$$s^n + b_{n-1} s^{n-1} + \cdots b_1 s + b_o = 0 \tag{10}$$

Observe that the coefficients of (10) are simply the coefficients of (8) and the corresponding power of the complex variable s is the order of the derivative corresponding to that term. This equation is usually referred to as the characteristic equation of the circuit. Note that this equation can be determined simply by replacing the operator D by the complex number s in our circuit determinant. In factored form (10) is written as

$$(s - p_1)(s - p_2) \cdots (s - p_n) = 0 \tag{11}$$

There are n roots of (10). Since (8) is linear we write

$$y(t) = k_1 e^{p_1 t} + k_2 e^{p_2 t} + \cdots k_n e^{p_n t} \tag{12}$$

The roots of (10) are called the *natural frequencies* of the circuit. *These natural frequencies are the roots of a polynomial obtained from the circuit determinant by replacing the operator D with the complex number s. Thus, any response in the circuit has the same natural frequencies.* This result makes sense since the KCL and KVL equations must be satisfied for all t, which implies that all current and voltage responses must have the same waveshape, but their magnitudes and phases as determined by the constants $k_1, \ldots, k_n$ may differ. Shortly we will discuss the evaluation of these constants.

If the roots of (10) are distinct then we have n degrees of freedom in (12). In order to determine $y(t)$ uniquely for $t > a$ we need n independent conditions at $t = a$. These n conditions are the energy stored in the capacitors and inductors at $t = a$ seconds. Remember that n cannot exceed the number of capacitors and inductors in the circuit. The stored energy in a capacitor depends on the voltage across the capacitor, and the stored energy in an inductor depends on the current through the inductor. *The capacitor voltages and inductor currents in the circuit at a given time are called the state of the circuit at that time.* In some cases n is less than the number of capacitors and inductors in the circuit. This occurs when a capacitor voltage or inductor current is dependent on the other capacitor voltages and inductor currents. In such a case the energy stored in that element is *not independent* so that the number of degrees of freedom is less than the number of inductors and capacitors in the circuit. In a loop of capacitors the voltage on one of the capacitors is dependent on the others. Also an inductor cutset (e.g., a node on which only inductor branches terminate) is clearly another case in which one of the inductor currents is dependent on the others so that its energy is not independent.

Now suppose that (10) has a root of multiplicity α, that is, (10) contains a factor $(s - p_i)^\alpha$. In the case of multiple roots it can be shown that the solution has the form

$$y(t) = \sum_{l=1}^{\alpha} k_l t^{l-1} e^{p_i t} + \text{terms for remaining } p_j, \, j \neq i \tag{13}$$

so that we still maintain n degrees of freedom.

Finally, the roots of the characteristic equation (10) can be both real and complex. However, complex roots must occur in conjugate pairs since the coefficients of (10) are real. We express the natural frequencies as

$$p_i = \sigma_i + j\omega_i \tag{14}$$

If p_i is real, then the ith term of (12) is

$$y_i(t) = k_i e^{\sigma_i t} = k_i e^{-t/\tau_i} \tag{15}$$

where $\tau_i = -1/\sigma_i$ is called a time constant. Equation 15 is plotted in Figure 9.3. Note that if $\sigma_i < 0$, in one time constant this term has decayed to $0.37 k_i$ or lost 63% of its original amplitude. In $3\tau_i$ seconds the amplitude has been reduced to $0.05 k_i$, and in $4\tau_i$ seconds the amplitude is $0.02 k_i$. Usually

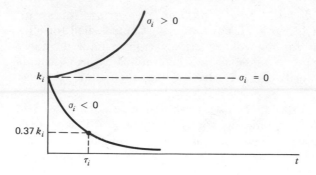

Figure 9.3 Real root contribution to the response.

beyond four time constants the contribution of this term to (12) is neglected. If $\sigma_i > 0$ we say the circuit is *unstable*.

If p_i is complex, then

$$y_i(t) = k_i\, e^{(\sigma_i + j\omega_i)t} + k_i^*\, e^{(\sigma_i - j\omega_i)t} \qquad (16)$$

We have included the conjugate term in (16) because a real circuit has a real solution which implies that the solution must always have the form (16) for complex roots. Let $k_i = k_R + jk_I$ where k_R and k_I denote the real and imaginary parts of k_i, then (16) becomes

$$y_i(t) = e^{\sigma_i t}[k_R(e^{j\omega_i t} + e^{-j\omega_i t}) + jk_I(e^{j\omega_i t} - e^{-j\omega_i t})]$$

or

$$y_i(t) = e^{\sigma_i t}(2k_R \cos \omega_i t - 2k_I \sin \omega_i t)$$

or

$$y_i(t) = A\, e^{\sigma_i t} \cos(\omega_i t + \theta) \qquad (17)$$

where $\theta = \tan^{-1} k_I/k_R$, and $A = 2\sqrt{k_R^2 + k_I^2}$. Usually the contribution to the solution from the complex root terms is expressed by (17) as opposed to (16). Equation 17 is plotted in Figure 9.4. Note that when $\sigma_i < 0$ we have a decaying sinusoid.

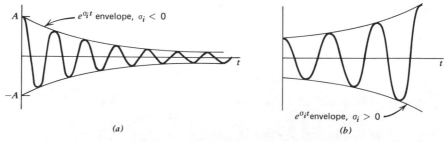

Figure 9.4 Complex root contribution to the response. (*a*) Stable term. (*b*) Unstable term.

From the above discussion we arrive at the following definition. *The circuit is said to be absolutely stable if* $\sigma_i < 0$ *for* $i = 1, 2, \ldots n$. This means that with zero input the energy stored in the capacitors and inductors will decay exponentially to zero.

To illustrate the above principles let us consider the circuit in Figure 9.5.

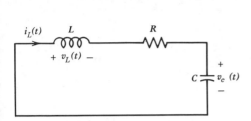

Figure 9.5   Series RLC circuit.

The differential equation for the circuit is

$$\left(DL + R + \frac{1}{DC}\right)i_L = 0 \tag{18}$$

and the characteristic equation is (simply replace the operator D by the complex variable s)

$$L\left(s^2 + \frac{R}{L}s + \frac{1}{LC}\right) = 0 \tag{19}$$

It is common practice to write second-order factors of the characteristic equation in the form

$$s^2 + \frac{\omega_r}{Q}s + \omega_r^2 = 0 \tag{20}$$

where ω_r is called a resonant frequency of the circuit and Q is referred to as the Q of that resonant frequency. The natural frequencies of (20) are

$$s = \left(\frac{-1}{2Q} \pm \sqrt{\frac{1}{4Q^2} - 1}\right)\omega_r \tag{21}$$

If $Q > 1/2$, then the roots of (21) are complex and we write

$$s = \left(-\frac{1}{2Q} \pm j\sqrt{1 - \frac{1}{4Q^2}}\right)\omega_r \tag{22}$$

The loci of these roots is shown in Figure 9.6 with ω_r normalized to one. For $Q > 1/2$ the roots of the characteristic equation lie on the unit circle. As Q decreases the roots move down to the real axis where they split, one moving to the left and the other moving to the origin.

In our particular example $\omega_r^2 = 1/LC$ and $Q = \omega_r L/R$. Let $L = 1$ H, $C =$

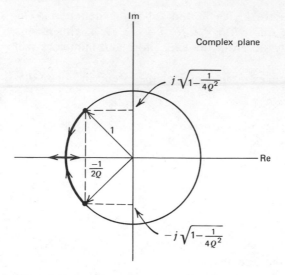

Figure 9.6   Complex root loci.

1 F and $R = 10\ \Omega$. Then $\omega_r = 1$ and $Q = 1/10$ and the natural frequencies are

$$p_i = -5 \pm \sqrt{24} = \begin{cases} -0.1 \\ -9.9 \end{cases} \qquad (23)$$

so that in this circuit any voltage or current response has the form

$$y(t) = k_1\, e^{-t/10} + k_2\, e^{-t/0.1} \qquad (24)$$

Both of the above terms are plotted in Figure 9.7 under the assumption that k_1 is positive and k_2 is negative. The time constants are spread by a factor of 100 so that the time scale must be split. The solution is the sum of these two terms. *We say that a second-order system is over-damped when its natural frequencies are real and distinct.*

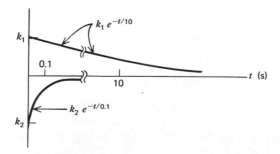

Figure 9.7   Overdamped response.

Next change R to $R = 2\,\Omega$, then

$$p_i = -1 \pm 0 = \begin{cases} -1 \\ -1 \end{cases} \tag{25}$$

We have a multiple root, which means the solution has the form

$$y(t) = k_i\,e^{-t} + k_2 t\,e^{-t} \tag{26}$$

Each of these terms is plotted in Figure 9.8 under the assumption that k_2 is positive and k_1 is negative. The second term peaks when t is equal to one time constant and its value is $0.37\tau_i k_2$. *This time the circuit is said to be critically damped.* Again, the complete solution is the sum of these two waveforms.

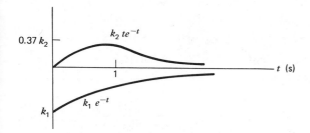

Figure 9.8 Critically damped response plotted for $k_1 < 0$ and $k_2 > 0$.

To conclude this example let $R = 1/5\,\Omega$, then, $Q = 5$ and

$$p_i = -\frac{1}{10} \pm j\sqrt{1 - 0.01} = \begin{cases} -0.1 + j0.995 \\ -0.1 - j0.995 \end{cases} \tag{27}$$

Since the roots are complex the solution has the form of (17).

$$y(t) = A\,e^{-t/10} \cos(0.995t + \theta) \tag{28}$$

The period of the above sinusoid is $2\pi/0.995 = 6.3$ s, and the time constant of the envelope is 10 s. This response is sketched in Figure 9.9, assuming

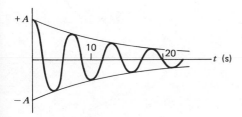

Figure 9.9 Underdamped resonse.

$\theta = 0°$. *In this case the unforced response oscillates and the circuit is said to be underdamped.*

Before we leave this section consider the circuit in Figure 9.10. If we assume that the voltage amplifier is ideal with gain μ, then

$$\left(DC_1 + \frac{1}{R_1}\right)e_1 - \left(\frac{1}{R_1}\mu + DC_1\right)e_2 = 0$$

$$-DC_1e_1 + \left(DC_1 + DC_2 + \frac{1}{R_2}\right)e_2 = 0$$

$$(29)$$

The characteristic equation is simply the determinant of the coefficients in (29) with the operator D replaced by the complex number s.

$$s^2 C_1 C_2 + s\left(\frac{C_1 + C_2}{R_1} + \frac{C_1}{R_2} - \frac{\mu C_1}{R_1}\right) + \frac{1}{R_1 R_2} = 0$$

or

$$s^2 + s\left(\frac{1}{R_1 C_2} + \frac{1}{R_1 C_1} + \frac{1}{R_2 C_2} - \frac{\mu}{R_1 C_2}\right) + \frac{1}{R_1 R_2 C_1 C_2} = 0 \qquad (30)$$

From (20) we write

$$\omega_r^2 = \frac{1}{R_1 R_2 C_1 C_2} \quad \text{and} \quad Q = \frac{\sqrt{R_1 R_2 C_1 C_2}}{R_1 C_1 + R_2 C_2 + R_2 C_1 (1 - \mu)}$$

If we assume that R_1, R_2, C_1, C_2, and μ are positive then the circuit has no damping if the coefficient of s term in (30) is zero, that is,

$$\frac{\mu}{R_1 C_2} = \frac{1}{R_1 C_2} + \frac{1}{R_1 C_1} + \frac{1}{R_2 C_2} \qquad (31)$$

Then, the characteristic equation is

$$s^2 + \omega_r^2 = 0$$

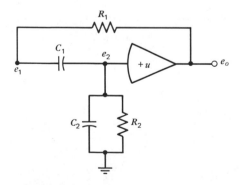

Figure 9.10 Oscillator.

so that

$$p_{1,2} = \pm j\omega_r \tag{32}$$

and the response

$$y(t) = A \cos(\omega_r t + \theta) \tag{33}$$

Note that with zero damping the circuit has a nondecaying sinusoidal response with zero input. The stored energy in the circuit is not dissipated. Such a circuit is called an oscillator. The parameters R_1, R_2, C_1, and C_2 adjust ω_r, the frequency of oscillation. The gain μ controls Q and can be adjusted to cause sustained oscillation ($\sigma = 0$) or make the circuit unstable ($\sigma > 0$).

In all of the above examples we have neglected to discuss the determination of the constants k_i, or the amplitude A and the phase θ in the complex root case. These constants are determined by the initial energy stored in the circuit and are discussed next.

9.2.2 INITIAL CONDITIONS

In the last section we saw that the form of the unforced solution was determined by the roots of the characteristic equation whose coefficients depended on the parameters (resistance, capacitance, inductance, and dependent source values) of the circuit. In a stable circuit the solution was a decaying exponential, or a damped sinusoid if the roots were complex. The magnitude of the response (and phase in the case of complex roots) is determined by the initial energy stored in the circuit. Let us show how to calculate these parameters of the solution.

Recall that the voltage across a capacitor determines the energy stored in it and that this voltage for $t \geq 0$ can be expressed as

$$v_c(t) = \left[v_{co} + \frac{1}{C} \int_0^t i_c(t)dt \right] u(t) \tag{34}$$

where v_{co} is the capacitor voltage at $t = 0$. From (34) we conclude that *as long as $i_c(t)$ is bounded, $v_c(t)$* (and hence the energy stored in the capacitor) is a continuous function of time. This implies that *the energy in a capacitor cannot change suddenly unless there is an impulse of current through the capacitor.* Similarly, for the inductor the current for $t \geq 0$ is

$$i_L(t) = \left[i_{Lo} + \frac{1}{L} \int_0^t v_L(t) \, dt \right] u(t) \tag{35}$$

so that in any time increment $i_L(t)$ is continuous *as long as $v_L(t)$ is bounded* (no voltage impulse across the inductor).

This observation allows us to *replace the capacitor with a voltage source and the inductor with a current source at any given time, and the value of these sources are determined by the voltage on the capacitor and the current in the inductor at that time.* Our circuit model now consists of independent sources, resistors, and possibly dependent sources, and we can solve for any of the other currents and voltages at that time by simply solving a set of algebraic equations. Also, *one must check that the current through the capacitor and the voltage across the inductor is bounded at that time (no unit impulse) so that the above model is valid.* The above capacitor and inductor properties are used to determine the constants k_i of the zero-input response. Several examples follow.

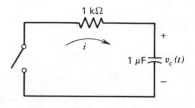

Figure 9.11   RC circuit.

circuit contains only one energy storage element and the circuit's response is determined by the first-order differential equation

$$R\frac{di}{dt}+\frac{1}{C}i=0 \tag{36}$$

which has the solution

$$i(t)=k\,e^{-t/RC} \tag{37}$$

In order to determine $i(t)$ for $t>0$ we need to calculate k. The constant k is computed from the initial condition $i(0^+)$, that is, the zero-plus initial condition. Note that $i(0)=0$ since the switch is open. In order to calculate $i(0^+)$ we assume that the energy stored in the capacitor does not change instantaneously when the switch is closed. Thus, the capacitor can be replaced by a 10 V voltage source as shown in Figure 9.12. Note that

$$i(0^+)=-10\text{ mA} \tag{38}$$

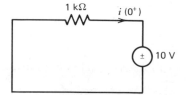

Figure 9.12   Computation of the 0^+ initial conditions.

Also observe that our model is correct since, although the current through the capacitor jumps from zero to -10 mA, it remains bounded so that the capacitor voltage does not change instantaneously when the switch is closed. We conclude that

$$i(t) = -10e^{-t/10^{-3}} \text{ mA} \qquad t > 0 \tag{39}$$

The time constant $\tau_1 = RC = 1$ ms.

As a second example let us return to the RLC example in Figure 9.5 and assume that $L = 1$ H, $C = 1$ F, and $R = 1/5\ \Omega$. Suppose that the voltage waveform across the capacitor is of interest and that $i_L(0) = i_{Lo}$ and $v_c(0) = v_{co}$. From (28) we write

$$v_c(t) = Ae^{-t/10} \cos(0.995t + \theta) \tag{40}$$

Next we model the circuit at $t = 0^+$ as shown in Figure 9.13. Note that the

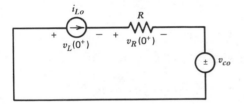

Figure 9.13   Computation of the 0^+ initial conditions.

voltage across the capacitor when $t = 0$ is

$$v_c(0^+) = v_{co} \tag{41}$$

which is assumed known and finite. Also,

$$i_c(0^+) = i_{Lo} \tag{42}$$

which is also bounded and known. An evaluation of (40) at $t = 0^+$ yields

$$v_c(0^+) = v_{co} = A \cos \theta \tag{43}$$

We need one more initial condition involving i_{Lo} in order to uniquely determine A and θ. Observe that

$$i_{Lo} = i(0^+) = \frac{C \, dv_c}{dt}\bigg|_{t=0^+} = C\left(-\frac{A}{10}\cos\theta - 0.995A\sin\theta\right) \tag{44}$$

Since $C = 1$ F (44) becomes

$$i_{Lo} = -\frac{A}{10}\cos\theta - 0.995A\sin\theta \tag{45}$$

Upon substitution of (43) into (45) we obtain

$$i_{Lo} = -\frac{1}{10} v_{co} - 0.995 v_{co} \tan \theta$$

$$\therefore \ \tan \theta = -\frac{i_{Lo} + 0.1 v_{co}}{0.995 v_{co}} \tag{46}$$

Suppose that $i_{Lo} = 1$ A and $v_{co} = 5$ V, then $\theta = -16.8°$ and $A = 5.22$ so that

$$v_c(t) = 5.22 e^{-t/10} \cos(0.995t - 16.8°) \tag{47}$$

Suppose that we also wished to obtain the voltage across the inductor, then we write

$$v_L(t) = \hat{A} e^{-t/10} \cos(0.995t + \hat{\theta}) \tag{40b}$$

From Figure 9.13 we obtain

$$v_L(0^+) = -(v_{co} + Ri_{Lo}) \tag{41b}$$

From (40b) and (41b) we write

$$-(Ri_{Lo} + v_{co}) = \hat{A} \cos \hat{\theta} \tag{43b}$$

We need one more equation to determine $\hat{A}$ and $\hat{\theta}$. Note that

$$i_L(t) = \frac{1}{L} \int v_L \, dt = \frac{1}{L} \int \hat{A} e^{-t/10} \cos(0.995t + \hat{\theta}) dt$$

From an integral table we obtain

$$i_L(t) = \hat{A} e^{-t/10} \frac{\left[\dfrac{-1}{10} \cos(0.995t + \hat{\theta}) + 0.995 \sin(0.995t + \hat{\theta}) \right]}{\left(\dfrac{1}{10}\right)^2 + (0.995)^2}$$

so that

$$i_L(0^+) = i_{Lo} \approx \hat{A} \left(\frac{-1}{10} \cos \hat{\theta} + 0.995 \sin \hat{\theta} \right) \tag{44b}$$

Next we solve (43b) and (44b) for $\hat{A}$ and $\hat{\theta}$. To obtain real numbers again suppose that $i_{Lo} = 1$ A and $v_{Co} = 5$ V, then from (43b) and (44b)

$$-\left[\left(\frac{1}{5}\right)(1) + 5 \right] = \hat{A} \cos \hat{\theta} \tag{45b}$$

$$1 = -0.1(\hat{A}) \cos \hat{\theta} + 0.995(\hat{A}) \sin \hat{\theta} \tag{46b}$$

The solution of which yields $\hat{\theta} \approx -5.3° \hat{A} \approx -5.22$ so that

$$v_L(t) = -5.22 e^{-t/10} \cos(0.995t - 5.3°) \tag{47b}$$

Of course an alternate approach to obtain (47b) would be to compute $i_L(t)$

and then $v_L(t) = L \, di_L/dt$. This approach involves the addition of sine and cosine terms of the same frequency that can be accomplished using the phasor approach in Chapter Two.

Let us conclude this section with one final example. Again consider the circuit in Figures 9.1 and 9.2. If the input $v_s = 0$ as shown in Figure 9.14,

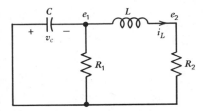

Figure 9.14   Circuit example.

then the node equations are

$$DCe_1 + \frac{e_1}{R} + \frac{1}{DL}(e_1 - e_2) = 0 \tag{48}$$

$$\frac{1}{DL}(e_2 - e_1) + \frac{e_2}{R_2} = 0 \tag{49}$$

and on the elimination of e_1 we obtain

$$\frac{d^2 e_2}{dt^2} + \left(\frac{R_2}{L} \, | \, \frac{1}{R_1 C}\right)\frac{de_2}{dt} + \frac{1}{LC}\left(1 + \frac{R_2}{R_1}\right)e_2 = 0 \tag{50}$$

If $R_2 = R_1 = 1\,\Omega$, $L = 1\,$H, and $C = 1\,$F, the characteristic equation

$$s^2 + 2s + 2 = 0 \tag{51}$$

has roots at $p_i = -1 \pm j$, $i = 1, 2$, so that

$$e_2(t) = Ae^{-t}\cos(t + \theta) \tag{52}$$

To obtain A and θ let us suppose that it is given that $v_C(0) = 10\,$V and $i_L(0) = 0.5\,$A. This time let us demonstrate how (48) and (49) can be used to find $e_2(0^+)$ and $de_2/dt(0^+)$, which are needed to calculate A and θ.

In (49) $1/L \int (e_2 - e_1)dt = -i_L(t)$ so that if we evaluate (49) at $t = 0^+$, then

$$-i_L(0^+) + \frac{e_2(0^+)}{R_2} = 0 \tag{53}$$

and we obtain one initial condition

$$e_2(0^+) = R_2 i_L(0^+) = 0.5\,\text{V} \tag{54}$$

Let us differentiate (49) to obtain de_2/dt at $t = 0^+$, which will serve as our second initial condition.

$$\frac{1}{L}[e_2(0^+)] - e_1(0^+)] + \frac{1}{R_2}\frac{de_2(0^+)}{dt} = 0 \tag{55}$$

so that

$$\frac{de_2(0^+)}{dt} = \frac{R_2}{L}[e_1(0^+) - e_2(0^+)] \tag{56}$$

An examination of the circuit reveals that

$$e_1(t) = -v_c(t) \tag{57}$$

Therefore, given (54) and (57), (56) yields

$$\frac{de_2(0^+)}{dt} = (-10 \text{ V} - 0.5 \text{ V}) = -10.5 \text{ V} \tag{58}$$

Evaluating (52) at $t = 0^+$ we write

$$0.5 = A \cos \theta \tag{59}$$

and taking the derivative of (52) and evaluating it at $t = 0^+$ we obtain

$$-10.5 = -A \cos \theta - A \sin \theta \tag{60}$$

The solution of (59) and (60) is $\theta = 87°$ and $A = 10.01$. These values are substituted into (52) to obtain the zero input solution.

The above calculations for the constants A and θ in the solution can become quite tedious, and in the case of higher-order systems the calculations could require several days or longer. Thus, computer-aided analysis is valuable even in cases where methods exist for determining the exact solution. Fortunately, many times one does not need to solve for these constants, but only the natural frequencies of the circuit are of interest since they indicate whether or not the circuit is stable, how fast the stored energy is dissipated if the circuit is stable, and the general waveshape of the response.

9.3 RESPONSE TO CONSTANT INPUTS

We have discussed the determination of circuit responses due to initial energy storage, but we have assumed zero input. In the nonzero input (forced) case the differential equation is determined as discussed in the first section of this chapter and has the form

$$\frac{d^n y}{dt^n} + b_{n-1}\frac{d^{n-1}y}{dt^{n-1}} + \cdots + b_1\frac{dy}{dt} + b_0 y = a_m\frac{d^m x}{dt} + \cdots + a_1\frac{dx}{dt} + a_0 x \tag{61}$$

where the forcing function $x(t)$ is a known function of time consisting of the independent sources in the circuit and derivatives of these independent sources with respect to time. Since (61) is linear, one method of solution is to let

$$y(t) = y_t(t) + y_{ss}(t) \tag{62}$$

where

$$\frac{d^n y_{ss}}{dt} + \cdots + b_1 \frac{d y_{ss}}{dt} + b_o y_{ss} = a_m \frac{d x^m}{dt} + \cdots + a_1 \frac{dx}{dt} + a_o x \tag{63}$$

and

$$\frac{d^n y_t}{dt^n} + \cdots + b_1 \frac{d y_t}{dt} + b_o y_t = 0 \tag{64}$$

This method works well when $x(t)$ has a simple mathematical description, for example, when $x(t)$ is of the form $\mathbf{X} e^{st}$ where $\mathbf{X}$ denotes the complex amplitude and s the complex input frequency. This includes the constant $(s = 0)$ and sinusoidal $(s = j\omega)$ inputs. From our previous discussion if the natural frequencies are distinct, then

$$y_t(t) = \sum_{i=1}^{n} k_i e^{p_i t} \tag{65}$$

Since $\lim_{t \to \infty} y_t(t) = 0$ if $\text{Re}\{p_i\} < 0$ for $i = 1, 2, \ldots, n$ the solution $y_{ss}(t)$ is usually referred to as the *steady state response*, and the solution $y_t(t)$ is called the *transient response*. In mathematics sometimes $y_t(t)$ is called the complementary function and $y_{ss}(t)$ the particular function.

In the case of constant inputs, $x(t) = X$, it is very easy to find $y_{ss}(t)$—it is simply a constant. Substitute the assumed solution $y_{ss}(t) = Y$ where Y is a constant, into (61), which becomes

$$b_o Y = a_o X$$

or

$$Y = \frac{a_o}{b_o} X \tag{66}$$

This result has an interesting circuit interpretation. To find the steady-state response to constant inputs, assume a constant capacitor voltage in the steady state so that $i_{css} = C d V_{css}/dt = 0$ since the capacitor voltage is constant, *which means that the capacitor can be replaced by an open circuit*. Similarly, *each inductor can be replaced by a short circuit* [$v_{Lss} = L(dI_{Lss}/dt) = 0$ since the inductor current is constant].

As an example consider the RC circuit in Figure 9.15a. The circuit equation is

$$\frac{d v_c}{dt} + \frac{1}{RC} v_c = \frac{1}{RC} V u(t) \tag{67}$$

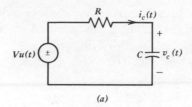

(a)

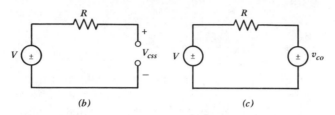

(b)   (c)

Figure 9.15 RC circuit with a step input. (a) Circuit. (b) Steady-state model. (c) 0^+ initial condition model.

The characteristic equation is

$$\left(s + \frac{1}{RC}\right) = 0 \tag{68}$$

so that the solution for the capacitor voltage has the form

$$v_c(t) = ke^{-t/RC} + V_{css} \tag{69}$$

To determine the steady-state response substitute the constant V_{css} into (67) or solve for V_{css} using the steady-state circuit model in Figure 9.15b. The result is that $V_{css} = V$ so that

$$v_c(t) = ke^{-t/RC} + V \tag{70}$$

To find k we need to examine the 0^+ conditions by means of Figure 9.15c. Note that $v_c(0^+) = v_{co}$ because $i_c(0^-) = -(v_{co}/R)$ and $i_c(0^+) = (V - v_{co})/R$, that is, the current through C is bounded so that $v_c(t)$ is continuous at $t = 0$. Thus,

$$k = v_c(0^+) - V \tag{71}$$

and

$$v_c(t) = [v_c(0^+) - V]e^{-t/RC} + V \tag{72}$$

Equation 72 is sketched in Figure 9.16 for $v_c(0^+) = 0$. In approximately $4RC$ seconds the response is 0.98 V. Essentially it has reached the steady state.

Let us conclude this section with a series RLC circuit example. See

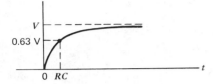

Figure 9.16   RC circuit step response.

Figure 9.17a. Recall (19) that the characteristic equation for this circuit is

$$s^2 + \frac{R}{L}s + \frac{1}{LC} = 0 \tag{73}$$

If we assume that $R = 1/5\ \Omega$ and $L = 1\ \text{H}$, and $C = 1\ \text{F}$, then from (27)

$$p_i = -0.1 \pm j0.995 \tag{74}$$

The solution has the form

$$v_c(t) = Ae^{-t/10}\cos(0.995t + \theta) + V_{css} \qquad t > 0 \tag{75}$$

To determine the steady-state response V_{css} we use the circuit model in Figure 9.17b, which yields

$$V_{css} = V_B \tag{76}$$

Next we utilize Figure 9.17c to determine the constants A and θ in (75).

$$v_c(0^+) = v_{co} = A\cos\theta + V_B \tag{77}$$

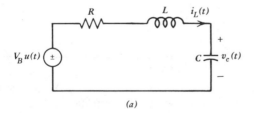

(a)

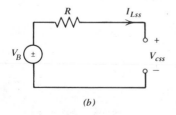

(b)

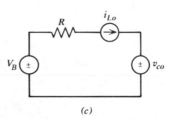

(c)

Figure 9.17   Series RLC circuit with a step input. (a) Circuit. (b) Steady-state model. (c) 0^+ initial condition model.

also,

$$C \frac{dv_c}{dt}\bigg|_{t=0^+} = i_{Lo} = C\left(-\frac{A}{10}\cos\theta - 0.995\,A\sin\theta\right) \tag{78}$$

From (77)

$$A = \frac{v_{co} - V_B}{\cos\theta} \tag{79}$$

and from (78) with $C = 1$ F and on the elimination of A we write

$$i_{Lo} = \frac{(V_B - v_{co})}{10} + 0.995(V_B - v_{co})\tan\theta \tag{80}$$

From (80) we obtain the phase angle θ and from (79) we obtain the amplitude of the solution.

Whenever we determine the forced circuit response with zero initial energy stored in the capacitors and inductors we call this response the zero-state response. For example if the *input is a step function* at $t = 0$ and *the initial state* at $t = 0$ *is zero*, then the response is called the zero-state step response or sometimes simply the *step response*.

9.4 TRANSIENT ANALYSIS WITH DEPENDENT STATES

Occasionally our circuit models are simplified to the point that the energy stored in the capacitors and inductors is not a continuous function of time, that is, sometimes a discontinuity in the input voltage or one of its derivatives, or the closing or opening of a switch can cause the voltage across a capacitor or the current through an inductor to change instantaneously. A good example of this situation is when a voltage source with a discontinuity appears in a loop with capacitors as shown in Figure 9.18a. If we assume that the capacitor voltages are continuous to develop the 0^+ model in Figure 9.18b, note that KVL is not satisfied when $V_B \neq v_{c1}(0^-) + v_{c2}(0^-)$. Thus, we have detected an inconsistency in our model. Actually, at $t = 0$ an impulse of current flows around the loop placing an equal amount of charge on each capacitor in the loop. Thus, we write

$$C_1[v_{c1}(0^+) - v_{c1}(0^-)] = \Delta Q \tag{81}$$

$$C_2[v_{c2}(0^+) - v_{c2}(0^-)] = \Delta Q \tag{82}$$

and KVL is satisfied at $t = 0^+$, that is,

$$v_{c1}(0^+) + v_{c2}(0^+) = V_B \tag{83}$$

Thus, we solve the above equations for $v_{c1}(0^+)$ and $v_{c2}(0^+)$ and continue our analysis.

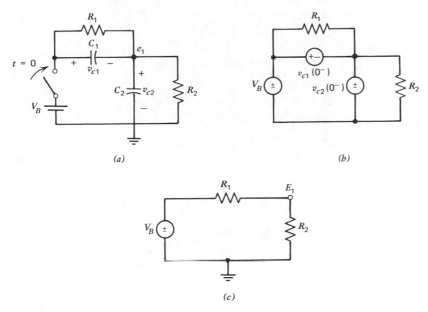

Figure 9.18   Capacitor-voltage source loop. (a) Circuit. (b) 0^+ model. (c) Steady-state circuit model.

The node equation for the circuit when $t > 0$ is

$$(C_2 + C_1)\frac{de_1}{dt} + \left(\frac{1}{R_1} + \frac{1}{R_2}\right)e_1 = \frac{V_B}{R_1} \tag{84}$$

The solution to (84) is

$$e_1(t) = ke^{-t/\tau} + E_1 \tag{85}$$

The time constant $\tau = (C_1 + C_2)R_1R_2/(R_1 + R_2)$ and E_1 is the steady-state node voltage. To solve for the steady-state node voltage E_1, substitute E_1 into (84) or use the steady-state circuit model in Figure 9.19c. In either case

$$E_1 = \frac{R_2}{R_1 + R_2} V_B \tag{86}$$

Next we determine the constant k from the 0^+ initial conditions

$$e_1(0^+) = v_{c2}(0^+) = k + \frac{R_2}{R_1 + R_2} V_B \tag{87}$$

To obtain $e_1(t)$ solve for k and substitute the solution of (87) and (86) into (85).

For example, suppose that $C_1 = 3$ F, $C_2 = 6$ F, $R_1 = 3\,\Omega$, $R_2 = 6\,\Omega$,

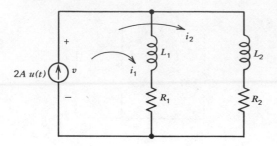

Figure 9.19 Inductor cutset example.

$v_{c1}(0^-) = v_{c2}(0^-) = 0$, and $V_B = 10$ V. From (81) and (82) we obtain

$$3 \text{ F } v_{c1}(0^+) = 6 \text{ F } v_{c2}(0^+)$$

and from (83)

$$v_{c1}(0^+) + v_{c2}(0^+) = 10 \text{ V}$$

Thus, $v_{c1}(0^+) = 20/3$ V and $v_{c2}(0^+) = 10/3$ V. From (86) or Figure 9.19c we see that the steady-state node voltage

$$E_1 = \frac{6}{(3+6)} 10 \text{ V} = \frac{20}{3} \text{ V}$$

The time constant $\tau = 18$ s so that

$$e_1(t) = \frac{-10}{3} e^{-t/18} + \frac{20}{3}$$

As our final example, let us suppose that rather than a loop of capacitors, we have a cutset of inductor branches and current sources, that is, if we cut all the inductor and current source branches the circuit will be in two separate pieces and there will be no branch connecting them (see Chapter Four). Such a circuit is shown in Figure 9.19. In this case a sudden change in one branch current will result in a sudden change in the other branch currents since KCL must be satisfied for the cutset branches. Now in Chapter Eight we saw that a discontinuity in inductor current causes an infinite voltage to be induced across the inductor. This is why one often observes sparks when a switch is opened that interrupts the current flow in a circuit.

Consider the circuit in Figure 9.19. The two loop equations involving the inductor currents are

$$L_1 \frac{di_1}{dt} + R_1 i_1 = v$$

and

$$L_2 \frac{di_2}{dt} + R_2 i_2 = v$$

but for $t > 0$

$$i_1 + i_2 = 2 \text{ A}$$

so that upon the elimination of i_2 and v in the above equations we obtain

$$(L_1 + L_2) \frac{di_1}{dt} + (R_1 + R_2) i_1 = R_2(2 \text{ A})$$

so that the solution is of the form

$$i_1(t) = k_1 e^{-t/\tau} + \left(\frac{R_2}{R_1 + R_2} \right) 2 \text{ A}$$

where $\tau = (L_1 + L_2)/(R_1 + R_2)$ is the circuit time constant, k_1 is determined by the 0^+ initial conditions discussed below, and $[R_2/(R_1 + R_2)]2$ A is the steady state current.

Let $i_1(0^-)$ and $i_2(0^-)$ represent arbitrary initial inductor currents. When the current source is switched on the currents will change to $i_1(0^+)$ and $i_2(0^+)$ where

$$i_1(0^+) + i_2(0^+) = 2 \text{ A}$$

This sudden change in current yields the infinite inductor voltages (see Chapter Eight)

$$v_1(0) = L_1[i_1(0^+) - i_1(0^-)]\delta(t)$$

and

$$v_2(0) = L_2[i_2(0^+) - i_1(0^-)]\delta(t)$$

Now KVL for the mesh of inductors requires that

$$L_1[i_1(0') - i_1(0^-)] = L_2[i_2(0^+) - i_1(0^-)]$$

because the resistor voltage drops do not contain a $\delta(t)$ term since the currents are finite. This equation implies that *the change in flux in the inductors is identical in this circuit when the current is switched on at the source.*

9.5 SWITCHED CIRCUITS

Next let us consider the analysis of some switched circuits. For example, the circuit in Figure 9.20 has a step input of 10 V at $t = 0$, then at $t = 1$ s a 20-kΩ resistor is connected in parallel with the capacitor. Sketch the response $v_c(t)$. First we ignore the fact that the switch is going to close at $t = 1$ s since the circuit certainly is not aware of this possibility. We see that the RC time constant is $(20 \text{ k}\Omega)(50 \text{ }\mu\text{F}) = 1$ s. Also, the steady-state

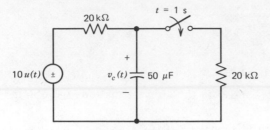

Figure 9.20   Switched RC circuit.

voltage across the capacitor is 10 V. Thus

$$v_c(t) = k\,e^{-t} + 10 \qquad t > 0 \tag{88}$$

If we assume that $v_c(0^+) = 0$ then

$$v_c(t) = 10(1 - e^{-t}) \qquad t > 0 \tag{89}$$

Now at $t = 1$ s,

$$v_c(1) = 10(1 - e^{-1}) \approx 6.3 \text{ V} \tag{90}$$

that is, at the moment of switching action we need to determine the energy stored in the capacitors and inductors. Now let us analyze the circuit in Figure 9.21 for $t > 1$ s.

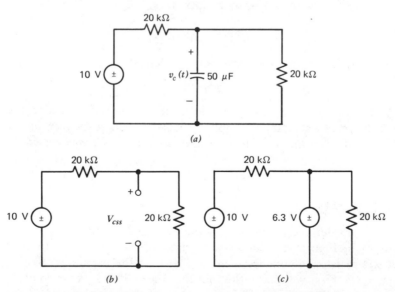

Figure 9.21   Circuit models for $t > 1$ s. (a) Circuit. (b) Steady state. (c) Initial conditions $t = 1$ s.

The differential equation is

$$DC v_c + \frac{v_c}{20 \text{ k}\Omega} + \frac{v_c - 10 \text{ V}}{20 \text{ k}\Omega} = 0 \qquad (91)$$

The characteristic equation is

$$s + \frac{1}{(10 \text{ k}\Omega)(C)} = 0 \qquad (92)$$

so that the time constant has changed to $(10 \text{ k}\Omega)(50 \times 10^{-6} \text{ F}) = 0.5 \text{ s}$ as a result of closing the switch. Note that the capacitance sees an equivalent Thévenin resistance of $10 \text{ k}\Omega$. The time constant could have easily been determined from this observation so that (91) is not needed. From Figure 9.21b the steady state response is 5 V. Therefore,

$$v_c(t) = k e^{-2t} + 5 \text{ V} \qquad (93)$$

To evaluate k we need the initial state of the capacitor at $t = 1$ s which is 6.3 V. Therefore,

$$6.3 = k e^{-2} + 5 \text{ V} \qquad (94)$$

and

$$k = 1.3 e^2 \qquad (95)$$

so that

$$v_c(t) = 1.3 e^{-2(t-1)} + 5 \text{ V} \qquad t \geqslant 1 \qquad (96)$$

The capacitor voltage is sketched in Figure 9.22 for $t \geqslant 0$. Note that in Figure 9.20, just before the switch is closed, the capacitor current is $(10 \text{ V} - 6.3 \text{ V})/20 \text{ k}\Omega = 0.185 \text{ mA}$ since the capacitor voltage is 6.3 V. However, just after the switch is closed the capacitor current is $0.185 \text{ mA} - 6.3 \text{ V}/20 \text{ k}\Omega = -0.13 \text{ mA}$. One should expect a negative capacitor current since the capacitor voltage is decreasing after the switch is closed. The capacitor current is discontinuous on switching, but it is bounded, so that the capacitor voltage is continuous as assumed.

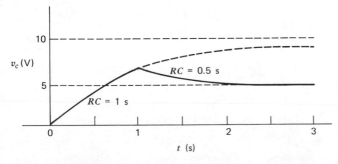

Figure 9.22   Response of switched RC circuit.

To illustrate a more dramatic change in time constants consider the simplified model of the automobile ignition system in Figure 9.23. Assume that the inductance of the ignition coil is 0.6 H and that the 10-kΩ resistor

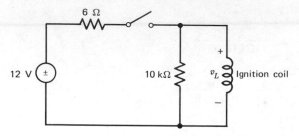

Figure 9.23   Automobile ignition (simplified).

represents the approximate resistance of the spark-plug gap. The equivalent resistance seen by the coil is 10 kΩ∥6 Ω or essentially 6 Ω. Therefore, the time constant is $L/R = 0.1$ s. The steady-state current in the inductor from Figure 9.24 is 2 A and the steady-state voltage across the inductor is 0 V. From Figure 9.25 the initial inductor voltage is 12 V

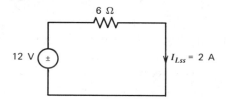

Figure 9.24   Steady-state response.

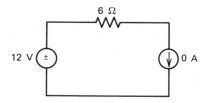

Figure 9.25 0^+ initial conditions.

assuming an initial state of zero for the inductor. Thus,

$$v_L(t) = 12\,e^{-10t} \qquad t > 0 \qquad\qquad (97)$$

and

$$i_L(t) = 2(1 - e^{-10t}) \qquad \text{for } t > 0 \qquad\qquad (98)$$

Now suppose that the switch is opened at $t = 0.1$ s, then the inductor current at the moment of switching is

$$i_L(0.1) = 2(1 - e^{-1}) \approx 1.26 \text{ A} \tag{99}$$

To find the voltage across the inductor and spark gap for $t > 0.1$ we analyze the circuit in Figure 9.26. Obviously $V_{Lss} = 0$, but $v_L(0.1)$ is determined from Figure 9.27. Note that we again assume that the energy in the

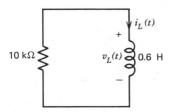

Figure 9.26   Ignition circuit $t > 0.1$ s.

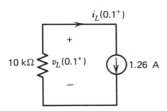

Figure 9.27 $t = 0.1^+$ model.

inductor does not change instantaneously so that $i_L(0.1^+) = 1.26$ A and

$$v_L(0.1^+) = -(1.26 \text{ A})(10 \text{ k}\Omega) = -12.6 \text{ kV} \tag{100}$$

Since this voltage is bounded our assumption that the inductor current is continuous when the switch is opened is valid. The time constant is $0.60/10 \text{ k}\Omega = 60 \ \mu\text{s}$. The solution is

$$v_L(t) = -12,600 \ e^{-16,667(t-0.1)} \qquad t > 0.1 \tag{101}$$

the voltage across the ignition coil is sketched in Figure 9.28 on a distorted scale due to the wide difference in time constants for the two switch positions and the difference in voltage levels. Note that a large voltage can be generated across an inductor by releasing its stored energy in a very short time interval.

In the next chapter we will study the response of linear time-invariant circuits to complex exponential inputs. Special cases of the exponential input are the constant (exponent = 0) and the sinusoid (sum of $e^{j\omega t}$ and $e^{-j\omega t}$

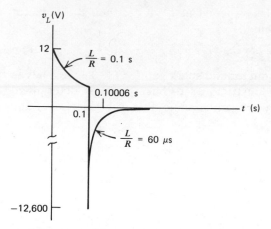

Figure 9.28   Voltage across the coil.

terms). The study of complex exponential inputs leads to the concept of the complex frequency circuit model and the complex frequency network function, both of which are extremely valuable in the analysis and design of circuits, because the step response (constant input with zero initial states) and the sinusoidal response of a circuit are frequently very important in engineering analysis and design.

PROBLEMS

1. For the circuit in Figure P9.1 write the differential equations that describe the relation between the input $v_s(t)$ and (a) the response $v_o(t)$, (b) the response $v_c(t)$, (c) the response $i_L(t)$, and (d) the response $v_L(t)$.

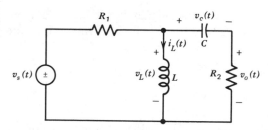

Figure P9.1

2. Find the differential equation that describes the relation between the input $i_s(t)$ and the response $v_o(t)$ in Figure P9.2. Sketch the loci of the roots of the characteristic equation for $L = 1\,\text{H}$, $C = 1\,\text{F}$, and $0 \leqslant R \leqslant \infty$.

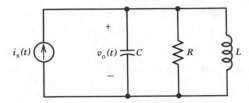

Figure P9.2

3. Repeat Problem 1(a) only for the circuit in Figure P9.3 with the input $v_s(t)$. If $R_1 = R_2 = R$ and $C_1 = C_2 = C$, can the roots of the characteristic equation be complex for some value of R and C?

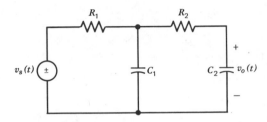

Figure P9.3

4. For each of the differential equations below find the roots of the characteristic equation and sketch the waveshape of the response for $t \geq 0$.

(a) $10^{-3}\dfrac{dy}{dt} + 10^3 y = 0$

(b) $\dfrac{dy}{dt} + 0.1y = 0$

(c) $\dfrac{d^2y}{dt^2} + 5\dfrac{dy}{dt} + 4y = 0$

(d) $\dfrac{d^2y}{dt^2} + 12\dfrac{dy}{dt} + 36y = 0$

(e) $\dfrac{d^2y}{dt^2} + 2 \times 10^6 \dfrac{dy}{dt} + 1.01 \times 10^8 y = 0$

5. In the series RLC circuit example, Figure 9.5, in which $L = 1\,\mathrm{H}$, $C = 1\,\mathrm{F}$, and $R = 1/5\,\Omega$, find the voltage $v_c(t)$ and the current $i_L(t)$ for $t \geq 0$ when $v_c(0) = 0$, and $i_L(0) = 1\,\mathrm{A}$. Sketch both responses.

6. In the series RLC circuit example, Figure 9.5, verify (47b) by solving for $i_L(t)$ and then using the relation $v_L(t) = L(di_L/dt)$.

7. Design the circuit in Figure 9.10 to oscillate at the frequency 100 kHz. Let $R_1 = R_2 = 1\,\mathrm{k}\Omega$ and assume $C_1 = C_2$. What values of C_1 and μ are needed?

8. In Figure 9.14 let $R_1 = 1\,\Omega$, $R_2 = 2\,\Omega$, $L = 2\,H$, $C = 1/2\,F$, $v_s(t) = 0$, $i_L(0) = 0$, and $v_c(0) = 10\,\mathrm{V}$. Find $e_2(t)$ for $t \geq 0$.

9. In Figure P9.9 find and sketch the responses $v_L(t)$ and $i_L(t)$ for $t \geqslant 0$ when

   (a) $v_s = 0$ and $i_L(0) = 5$ mA
   (b) $v_s(t) = 10u(t)$ and $i_L(0) = 5$ mA
   (c) $v_s(t) = u(t)$ and $i_L(0) = 0$ (step response)

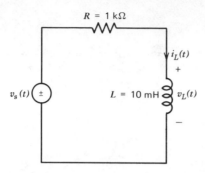

Figure P9.9

10. Repeat Problem 9 for $R = 10$ kΩ.
11. Find and sketch $v_o(t)$ in Figure P9.2 for $t \geqslant 0$ when $L = 1$ H, $C = 1$ F, $R = 0.1$ Ω, and

   (a) $i_s(t) = 0$, $v_c(0) = v_{co}$, and $i_L(0) = i_{Lo}$
   (b) $i_s(t) = 10u(t)$ mA, $v_c(0) = v_{co}$, and $i_L(0) = i_{Lo}$

12. Repeat Problem 11 for $R = 0.5$ Ω.
13. Repeat Problem 11 for $R = 1$ Ω.
14. In Figure P9.3 if $R_1 = R_2 = 1$ Ω and $C_1 = C_2 = 1$ F, find the step response $v_o(t)$, that is, the response when $v_s(t) = u(t)$ and $v_{c1}(0) = v_{c2}(0) = 0$.
15. Repeat Problem 14 when $R_1 = R_2 = 1$ kΩ and $C_1 = C_2 = 1$ μF.
16. The input-output relation for the circuit in Figure P9.16 is

$$\left[D^2 + \left(\frac{1}{R_2 C} + \frac{R_1}{L} \right) D + \frac{1}{LC} \left(1 + \frac{R_1}{R_2} \right) \right] v_o(t) = \frac{1}{LC} v_s(t)$$

If $R_1 = R_2 = 1$ Ω, $L = 1$ H, $C = 1$ F, and $v_s(t) = 10u(t)$

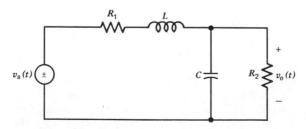

Figure P9.16

(a) Find the natural frequencies of the circuit.

(b) Find the steady state response.

(c) Assume that the circuit is in the zero state and find the complete response for $t \geqslant 0$.

17. In Figure P9.17 assume that the amplifier is ideal with gain $-A$ and show that

$$\lim_{A \to \infty} e_o(t) = -\frac{1}{RC} \int e_i(t) \, dt$$

that is, the circuit is an integrator.

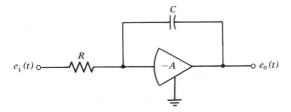

Figure P9.17

18. Find the voltage $v_o(t)$ in Figure P9.18 for $t \geqslant 0$. Assume that the capacitors are in the zero state before the switch is closed.

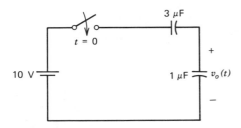

Figure P9.18

19. Find the voltage $v_{c2}(t)$ in Figure P9.19 for $t \geqslant 0$. Assume that $v_{c1}(0) = 0$ and $v_{c2}(0) = 10$ V.

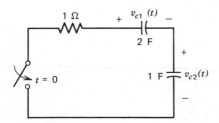

Figure P9.19

20. In Figure 9.18 let $R_1 = R_2 = 1\,\Omega$, $C_1 = 0.5$ F, $C_2 = 0.1$ F, and $V = 10$ V. Find $e_1(t)$ for $t \geqslant 0$ if $v_{c1}(0) = v_{c2}(0) = 0$. Repeat for $v_{c1}(0) = 5$ V and $v_{c2}(0) = 3$ V.

21. Find the step response $v_o(t)$ of the circuit in Figure P9.21.

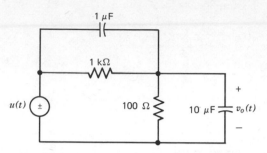

Figure P9.21

22. In Figure P9.22 find $v_c(t)$ and $i_c(t)$ for $t \geqslant 0$. Assume that $v_c(0) = 0$.

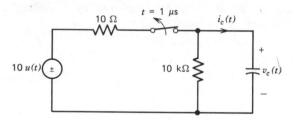

Figure P9.22

23. The circuit in Figure P9.23 is in the steady state when the switch is opened at $t = 0$. Find $v_c(t)$ and $i_c(t)$ for $t \geqslant 0$.

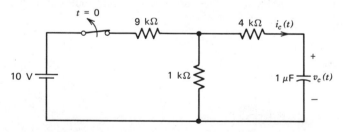

Figure P9.23

24. Find and sketch $i_L(t)$ and $v_L(t)$ in Figure P9.24 for $t \geqslant 0$. Assume that $i_L(0) = 0$.

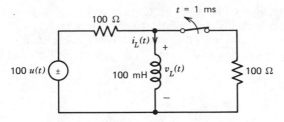

Figure P9.24

25. In Figure 9.19 let $i_1(0^-) = i_2(0^-) = 0$, $L_1 = 1\,\text{mH}$, $L_2 = 10\,\text{mH}$, $R_1 = 100\,\Omega$, and $R_2 = 50\,\Omega$. Find $i_1(t)$ and $i_2(t)$ for $t \geq 0$.

26. In Figure P9.26 find the current in each of the inductors and the voltage $v_o(t)$ for $t \geq 0$. Assume that initially all inductor currents are zero when the switch is closed at $t = 0$. Let $L_1 = 1\,\text{H}$, $L_2 = 2/3\,\text{H}$, and $L_3 = 1/3\,\text{H}$.

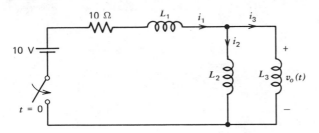

Figure P9.26

27. Repeat Problem 26 with $i_{L2}(0) = 1\,\text{A}$ and $i_{L3}(0) = -1\,\text{A}$.

28. A $1\text{-}\mu\text{F}$ capacitor is placed across the switch in the ignition circuit example as shown in Figure P9.28. Assume that the circuit is in the steady state before the switch is opened. Find and sketch $v_L(t)$ for $t \geq 0$.

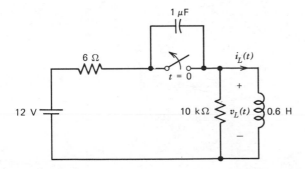

Figure P9.28

29. In Figure 9.17 assume that $R = 1/5 \, \Omega$, $L = 1 \, H$, $C = 1 \, F$, and $V_B = 1$. Find the initial states $v_c(0)$ and $i_L(0)$ such that the transient response is zero.

30. In Figure 9.19 what initial conditions $i_1(0^-)$ and $i_2(0^-)$ will result in the co-efficient k of the transient response equal to zero? Your answer will depend on R_1, R_2, L_1, and L_2.

Chapter Ten
Network Functions and Frequency Response

If we apply an input of the form $\mathbf{X} \, e^{st}$ to a linear time-invariant[1] system, then the steady state response has the form $\mathbf{Y} \, e^{st}$, and there exists a relationship between the input amplitude $\mathbf{X}$ and the steady-state response amplitude $\mathbf{Y}$ called a network function. This network function depends on the complex frequency $s = \sigma + j\omega$, but is independent of time. Although this input function is generally not physically realizable since s, $\mathbf{X}$, and $\mathbf{Y}$ are complex variables, nevertheless it is an extremely useful analysis and design tool because, for circuits that can be described by ordinary linear constant differential-algebraic equations, the network function is a ratio of two polynomials in s and the roots of these polynomials can give us (with a little training) significant insight into the steady-state sinusoidal and transient responses of the circuit. This chapter shows us how to determine the network function that relates a single output to a single input and how to use a network function in analysis and design. Since the circuit is linear, the principle of superposition can be applied in the case of more than one input.

---

[1] Implies constant circuit parameter values only. The independent sources and the circuit responses may be time dependent.

10.1 DRIVING-POINT AND TRANSFER FUNCTIONS

Consider the illustration in Figure 10.1. Let us assume that the elements in this circuit are lumped (no distributed elements such as transmission lines) so that by the process of elimination of variables we arrive at an ordinary differential equation relating the output signal $y(t)$ to the input signal $x(t)$. The output signal is some current or voltage in the circuit. If more than one input signal exists we can use superposition, since the

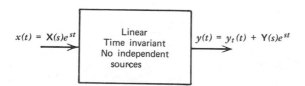

Figure 10.1   Response to a complex exponential input signal.

circuit is linear, and find differential equations for the relationship between the output and each of the inputs. In this single-input single-output example let us assume that the differential equation has the form

$$\frac{d^n y}{dt^{n-1}} + b_{n-1}\frac{d^{n-1}y}{dt^{n-1}} + \cdots + b_1\frac{dy}{dt} + b_o y = a_m\frac{d^m x}{dt^m} + \cdots + a_1\frac{dx}{dt} + a_0 x \qquad (1)$$

Now if the input has the form $x(t) = \mathbf{X}\,e^{st}$, then the steady-state response has the form $y_{ss}(t) = \mathbf{Y}\,e^{st}$ where $\mathbf{X}$ and $\mathbf{Y}$ denote the complex amplitudes of the input and output signals, respectively, and s is referred to as the complex frequency of the signals. In general, the input amplitude $\mathbf{X}$ and the response amplitude $\mathbf{Y}$ are frequency dependent so that we write $x(t) = \mathbf{X}(s)\,e^{st}$ and $y_{ss}(t) = \mathbf{Y}(s)\,e^{st}$. The complete response of the circuit consists of the transient response plus the steady-state response, that is,

$$y(t) = y_t(t) + \mathbf{Y}(s)\,e^{st} \qquad (2)$$

Upon substitution of $x(t)$ and $y(t)$ into (1) we obtain

$$\frac{d^n y_t}{dt^n} + b_{n-1}\frac{d^{n-1}y_t}{dt^{n-1}} + \cdots + \frac{b_1 dt_t}{dt} + b_o y_t + (s^n + b_{n-1}s^{n-1} + \cdots + b_1 s$$

$$+ b_o)\mathbf{Y}(s)\,e^{st} = (a_m s^m + \cdots + a_1 s + a_o)\mathbf{X}(s)\,e^{st} \qquad (3)$$

The complex amplitude of the steady state response, $\mathbf{Y}(s)$, is calculated from the equation

$$(s^n + b_{n-1}s^{n-1} + \cdots + b_1 s + b_o)\mathbf{Y}(s) = (a_m s^m + \cdots + a_1 s + a_o)\mathbf{X}(s) \qquad (4)$$

so that, as in Chapter Nine, the transient response is determined from the equation

$$\frac{d^n y_t}{dt^n} + b_{n-1}\frac{d^{n-1} y_t}{dt^{n-1}} + \cdots + b_1\frac{dy_t}{dt} + b_o y_t = 0 \tag{5}$$

The network function for the given input-output signals is defined as the ratio of the complex response amplitude to the complex input amplitude, that is, from (4),

$$\frac{Y(s)}{X(s)} = H(s) = \frac{a_m s^m + a_{m-1} s^{m-1} + \cdots + a_1 s + a_o}{s^n + b_{n-1} s^{n-1} + \cdots + b_1 s + b_o} \tag{6}$$

Note that (6) is a function of the complex frequency s of the input waveform and the coefficients a_i and b_j, which are determined by the resistor, inductor, capacitor, and dependent source values in the circuit. Remember that the derivation of (6) assumed a lumped, linear, time-invariant circuit with only one independent source whose voltage or current is $x(t) = X(s)\, e^{st}$. Sometimes the network function is written in factored form.

$$\frac{Y(s)}{X(s)} = H(s) = \frac{a_m (s - z_1)(s - z_2) \cdots (s - z_m)}{(s - p_1)(s - p_2) \cdots (s - p_n)} \tag{7}$$

where the roots of the numerator polynomial are called the *zeros* of the network function since, if the complex frequency $s = z_i$ for $i = 1, 2, \ldots$ or m, then $H(s)$ is zero. The roots of the denominator polynomial are called the *poles* of the network function because $H(s)$ is infinite whenever $s = p_i$, $i = 1, 2, \ldots$ or n.

The network function can be a driving-point immittance as shown in Figure 10.2. The term *immittance* means either an *impedance or an admittance*. The impedance function is the ratio of the complex voltage amplitude to the complex current amplitude at a terminal pair. It is a generalization of the concept of resistance and describes the relationship between the complex voltage and current amplitudes at a terminal pair in

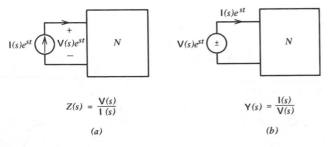

$$Z(s) = \frac{V(s)}{I(s)} \qquad\qquad Y(s) = \frac{I(s)}{V(s)}$$

(a)   (b)

Figure 10.2 Driving point immittances. (*a*) DP impedance. (*b*) DP admittance.

circuits that contain linear constant elements including inductors and capacitors and that are energized by independent sources of the form e^{st}. The admittance function is the reciprocal of the impedance function. The network function can also be a transfer immittance or transfer ratio as shown in Figure 10.3. A transfer ratio is the ratio of the complex amplitude of the output voltage to the complex amplitude of the input voltage, or the ratio of the complex amplitude of the output current to the complex amplitude of the input current, and is sometimes called a *gain function* ($|H(s)| > 1$) or *loss* function ($|H(s)| < 1$). In both of these figures the network N contains only lumped, linear, and constant elements and no independent sources.

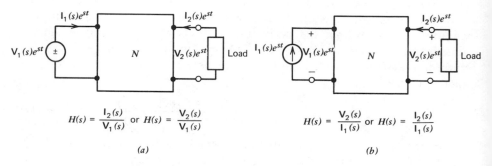

Figure 10.3 Transfer functions.

10.2 THE RELATION BETWEEN THE NETWORK FUNCTION AND THE CIRCUIT RESPONSE

The network function yields a wealth of information about the response of the circuit. First note that given the network function (6) one can easily find the differential equation describing the input-output relation. Simply write (6) in the form of (4) and replace s^k by the differential operator $D^k = d^k/dt^k$ and replace $\mathbf{Y}(s)$ and $\mathbf{X}(s)$ by $y(t)$ and $x(t)$, respectively. In the reverse direction we note that

$$a_k \frac{d^k}{dt^k}(\mathbf{X}(s) \, e^{st}) = a_k s^k (\mathbf{X}(s) \, e^{st}) \tag{8}$$

so that, from the differential equation (1), the network function (6) is easily formed. *Remember that x(t) is the input and y(t) is the response so that the coefficients in the numerator of (6) are the coefficients of the derivatives of the input in (1), and the coefficients in the denominator of (6) are the coefficients of the derivatives of the response in (1).*

10.2.1 TRANSIENT RESPONSE

Next observe that the characteristic equation whose roots determine the natural frequencies of the transient response in (5) has the form

$$s^n + b_{n-1}s^{n-1} + \cdots + b_1 s + b_o = 0 \tag{9}$$

The above polynomial is simply the denominator of the network function, that is, *the poles of the network function are the natural frequencies of the transient response* so that, assuming distinct poles,

$$y_t(t) = \sum_{i=1}^{n} k_i e^{p_i t} \tag{10}$$

The form of (10) in the case of identical poles is the same as the multiple root case already discussed in Chapter Nine.

Since the poles of the network function determines the waveform of the transient response, we can obtain a good picture of the transient response from a plot of the poles of the network function in the complex plane. For example, suppose that the network function of a circuit has three poles, a real pole $p_1 = -\sigma_1$ and a complex pair of poles $p_{2,3} = -\sigma_2 \pm j\omega_2$. The location of these poles in the complex plane is illustrated in Figure 10.4. Note that the poles lie to the left of the imaginary axis (left-half plane or

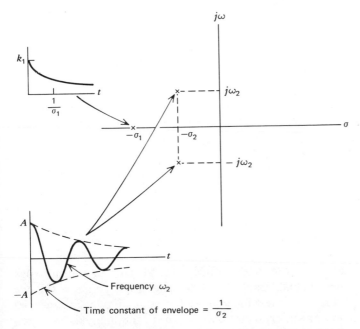

Figure 10.4   Relationship between poles and the transient response.

lhp) since they have negative real values. Poles in the lhp yield waveforms that decay exponentially. The time constant of the exponential decay is the reciprocal of the real part of the pole. In the case of complex poles the transient response oscillates with a frequency equal to the imaginary part of the pole. In our example

$$y_t(t) = k_1 e^{-\sigma_1 t} - A e^{-\sigma_2 t} \cos(\omega_2 t + \beta)$$

These waveforms are illustrated in Figure 10.4. Next we show how the network function can be used to determine the steady state response to a sinusoidal input.

10.2.2 STEADY-STATE RESPONSE

If the input signal is $\mathbf{X} e^{s_o t}$, then the steady-state response is

$$y_{ss}(t) = \mathbf{Y} e^{s_o t} \tag{11}$$

where the complex amplitude of the response

$$\mathbf{Y} = H(s_o)\mathbf{X} \tag{12}$$

that is, we evaluate the network function at the complex input frequency s_o. The complex function $H(s_o)$ tells us the relation of the magnitude and phase of $\mathbf{Y}$ to the input magnitude and phase $\mathbf{X}$. The complete response is

$$y(t) = \sum_{i=1}^{n} k_i e^{p_i t} + H(s_o)\mathbf{X} e^{s_o t} \tag{13}$$

We need n initial conditions to determine the constants k_i of the transient response.

In the real world we don't have complex frequency generators, instead we have sinusoidal generators of the form $X \cos(\omega_o t + \theta)$ and the steady-state response is of the form $Y \cos(\omega_o t + \phi)$ as shown in Figure 10.5.

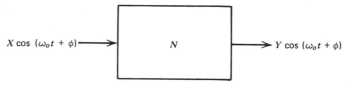

Figure 10.5 Sinusoidal steady-state response.

Nevertheless, the complex frequency concept that led us to the network function is an extremely valuable analysis and design tool. To show this we use Euler's identity to write the sinusoid as the sum of two exponential signals, then by means of superposition we separate the analysis into two separate parts as shown in Figure 10.6.

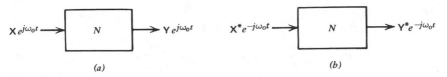

Figure 10.6 Analysis by superposition. (a) Positive frequency response. (b) Negative frequency response.

In Figure 10.6a the complex number

$$\mathbf{X} = X\,e^{j\theta} \qquad (14)$$

represents the amplitude and phase of the input signal. The frequency of our input signal is $s = j\omega_o$ so that from (12) the amplitude and phase of the response is

$$\mathbf{Y} = H(j\omega_o)\mathbf{X} \qquad (15)$$

In Figure 10.6b the input amplitude and phase is

$$\mathbf{X}^* = X\,e^{-j\theta}$$

and the input signal frequency is $s = -j\omega_o$. However, the coefficients of the network function (6) are real so that

$$H(-j\omega_o) = H^*(j\omega_o) \qquad (16)$$

which means that the amplitude and phase of the response in Figure 10.6b is simply the complex conjugate of that in Figure 10.6a.

$$\mathbf{Y}^* = H^*(j\omega_o)\mathbf{X}^* \qquad (17)$$

Since we have assumed a linear circuit, the complete steady-state response is simply one-half the sum[2] of both responses, that is,

$$y_{ss}(t) = \frac{1}{2}[H(j\omega_o)\mathbf{X}\,e^{j\omega_o t} + H^*(j\omega_o)\mathbf{X}^*\,e^{-j\omega_o t}] \qquad (18)$$

Since the negative frequency response is simply the complex conjugate of the positive frequency response,

$$y_{ss}(t) = \mathrm{Re}[H(j\omega_o)\mathbf{X}\,e^{j\omega_o t}] \qquad (19)$$

we need only analyze the circuit in Figure 10.6a and apply (19). Now let us express the complex network function in polar form.

$$H(j\omega_o) = |H(j\omega_o)|\,e^{j\alpha} \qquad (20)$$

then (19) becomes

$$y_{ss}(t) = Y\,\cos(\omega_o t + \phi) \qquad (21)$$

[2] Recall that $\cos(\omega t + \theta) = 1/2[e^{j(\omega t + \theta)} + e^{-j(\omega t + \theta)}]$.

where

$$Y = |H(j\omega_o)|X \qquad (22)$$

and

$$\phi = \alpha + \theta \qquad (23)$$

In simple language, *in sinusoidal steady-state analysis the amplitude of the response due to a sinusoidal input of frequency ω_o is the product of the magnitude of the network function evaluated at $s = j\omega_o$ with the amplitude of the input signal, and the phase of the response relative to the phase of the input signal is determined by the angle of the network function evaluated at $s = j\omega_o$.*

As an example, for a given input-output pair of signals let the network function

$$H(s) = \frac{10^3}{s + 10^3} \qquad (24)$$

and suppose that the input signal

$$x(t) = 10 \cos 2\pi \cdot 10^2 t \qquad (25)$$

The complex amplitude of the input signal is

$$\mathbf{X} = 10 \, e^{j0°} \qquad (26)$$

because the amplitude of the input is 10 and its phase is zero degrees. The complex frequency

$$s = j2\pi \cdot 10^2 \qquad (27)$$

since the input frequency is 10^2 Hz. The amplitude and phase of the steady-state response

$$\mathbf{Y} = H(j2\pi \cdot 10^2)\mathbf{X} \qquad (28)$$

but

$$H(j2\pi \cdot 10^2) = \frac{10^3}{j2\pi \cdot 10^2 + 10^3} = 0.85 \, e^{-j32°} \qquad (29)$$

From (26), (28), and (29) the amplitude and phase of the response is

$$\mathbf{Y} = 8.5 \, e^{-j32°} \qquad (30)$$

so that

$$y_{ss}(t) = 8.5 \cos(2\pi \cdot 10^2 t - 32°) \qquad (31)$$

In the next section a simple circuit model is developed from which the network function can easily be calculated. Several examples are given to illustrate the power of the network function in analysis and design.

10.3 THE COMPLEX FREQUENCY CIRCUIT MODEL

The network function for a given input-output pair of signals can easily be determined from the differential equation that describes the relationship between these signals. However, usually one begins with the circuit. Therefore, in this section we show how to determine the network function directly from a circuit model without the need to first calculate the differential equation.

Let us begin with the series RLC circuit example in Figure 10.7. The equations for this circuit are

$$v_s(t) = v_r(t) + v_l(t) + v_c(t) \tag{32}$$

where

$$v_r(t) = Ri(t) \tag{33}$$

$$v_l(t) = L\frac{di}{dt} \tag{34}$$

and

$$v_c(t) = \frac{1}{C}\int i\,dt \tag{35}$$

Suppose that $v_s(t) = \mathbf{V}_s(s)e^{st}$, then the steady-state responses have the form $v_r(t) = \mathbf{V}_r(s)e^{st}$, $v_l(t) = \mathbf{V}_s(s)e^{st}$, $v_c(t) = \mathbf{V}_c(s)e^{st}$, and $i(t) = \mathbf{I}(s)e^{st}$. Upon substitution of these steady-state responses into (32) to (35) we obtain the following relations among the complex amplitudes.

$$\mathbf{V}_s(s) = \mathbf{V}_r(s) + \mathbf{V}_l(s) + \mathbf{V}_c(s) \tag{36}$$

$$\mathbf{V}_r(s) = R\mathbf{I}(s) \tag{37}$$

$$\mathbf{V}_l(s) = sL\mathbf{I}(s) \tag{38}$$

and

$$\mathbf{V}_c(s) = \frac{1}{sC}\mathbf{I}(s) \tag{39}$$

where (36) is a Kirchhoff voltage law equation that the complex amplitudes must satisfy, and (37) to (39) are the branch constraints of the elements in

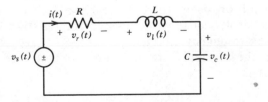

Figure 10.7   Series RLC circuit.

terms of their complex voltage and current amplitudes. We say that the resistor has an impedance

$$Z_r(s) = \frac{\mathbf{V}_r(s)}{\mathbf{I}(s)} = R \qquad (40)$$

Similarly, the impedance of the inductor is

$$Z_l(s) = \frac{\mathbf{V}_l(s)}{\mathbf{I}(s)} = sL \qquad (41)$$

and the impedance of the capacitor is

$$Z_c(s) = \frac{\mathbf{V}_c(s)}{\mathbf{I}(s)} = \frac{1}{sC} \qquad (42)$$

The complex frequency circuit model is derived by replacing every voltage and current by its complex amplitude and every branch constraint is replaced by its complex frequency constraint (see Table 10.1). In our

Table 10.1 Complex Frequency Branch Constraints

Element	Branch Constraint in Real World	Branch Constraint in Complex Frequency Domain
Linear constant resistor	$x = Ri$	$\mathbf{V}(s) = R\mathbf{I}(s)$
Linear constant inductor	$v = L\,di/dt$	$\mathbf{V}(s) = sL\mathbf{I}(s)$
Linear constant capacitor	$v = \dfrac{1}{C}\displaystyle\int i\,dt$	$\mathbf{V}(s) = \dfrac{1}{sC}\mathbf{I}(s)$
Linear dependent sources	$v_x = uv_y$ $i_x = g_m v_y$ $i_x = \beta i_y$ $v_x = \gamma i_y$	$\mathbf{V}_x(s) = u\mathbf{V}_y(s)$ $\mathbf{I}_x(s) = g_m\mathbf{V}_y(s)$ $\mathbf{I}_x(s) = \beta\mathbf{I}_y(s)$ $\mathbf{V}_x(s) = \gamma\mathbf{I}_y(s)$

example this leads to the circuit in Figure 10.8. Note that the equations for the complex frequency circuit model are algebraic. This circuit is simply a complex resistive circuit with complex resistors (impedances) sL, $1/sC$, and R. As a result, all of the useful analytical tools developed for resistive circuits carry over to the complex frequency circuit.

For example, in Figure 10.8 the elements are in series so that the impedance seen by the generator is

$$Z(s) = \frac{\mathbf{V}_s(s)}{\mathbf{I}(s)} = R + sL + \frac{1}{sC} = \frac{s^2 LC + sRC + 1}{sC} \qquad (43)$$

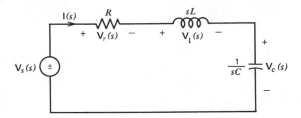

Figure 10.8 Complex frequency circuit model.

that is, it is simply the sum of the individual impedances. The admittance seen by the generator is

$$Y(s) = \frac{1}{Z(s)} = \frac{sC}{s^2LC + sRC + 1}$$ (44)

From (43) the differential equation relating the current to the input voltage is

$$LC\frac{d^2i}{dt^2} + RC\frac{di}{dt} + i = C\frac{dv_s}{dt}$$ (45)

The transfer function

$$H_c(s) = \frac{V_c(s)}{V_s(s)} = \frac{1/sC}{R + sL + \dfrac{1}{sC}} = \frac{1}{s^2LC + sRC + 1}$$ (46)

can be obtained from the voltage divider rule. The differential equation for this input-output function is

$$LC\frac{d^2v_c}{dt^2} + RC\frac{dv_c}{dt} + v_c = v_s(t)$$ (47)

In a similar manner the transfer function

$$H_r(s) = \frac{V_r(s)}{V_s(s)} = \frac{R}{sL + R + \dfrac{1}{sC}} = \frac{sRC}{s^2LC + sRC + 1}$$ (48)

Thus, the network function can be obtained from the solution of the algebraic equations for the complex frequency model, and, as in the above examples, circuit analysis methods which were developed for resistive circuits can be employed.

It is interesting to observe the asymptotic behavior of the network function as $s \to 0$, (zero frequency) and $s \to \infty$ (infinite frequency). For example, from (45) and (46)

$$\lim_{s \to 0} H_c(s) = 1$$ (49)

$$\lim_{s\to\infty} H_c(s) = \frac{1}{s^2 LC}\bigg|_{s\to\infty} \tag{50}$$

$$\lim_{s\to 0} Y(s) = sC\big|_{s\to 0} \tag{51}$$

and

$$\lim_{s\to\infty} Y(s) = \frac{sC}{s^2 LC}\bigg|_{s\to\infty} = \frac{1}{sL}\bigg|_{s\to\infty} \tag{52}$$

As $s = 0$ the impedance of the inductor is 0 and the impedance of the capacitor is infinite. Equation 49 says that the input voltage and output voltage are equal and (51) says that the capacitor is the dominant element as $s \to 0$. Examining Figure 10.8 reveals that all of these observations are in agreement with one another. Also, as $s \to \infty$ the impedance of the inductor becomes infinite and the impedance of the capacitor is zero. By examining Figure 10.8 we see that this observation is also in agreement with (50), which says that the voltage decrease is proportional to s^2, and from Figure 10.8 we see that the voltage goes to zero not only because the impedance of the capacitor is zero, but also because the series impedance of the inductor is infinite. Finally, as $s \to \infty$, (52) says that the inductor is the dominant element. After the network function has been calculated it is good to make these observations to see if an obvious error has been made.

10.4 APPLICATION OF THE NETWORK FUNCTION IN ANALYSIS

In this section a number of numerical examples will be given to indicate the utility of the network function in analysis.

Example 1

Consider the series RC circuit in Figure 10.9. The complex frequency circuit model is shown in Figure 10.10. Suppose that we want the capacitor voltage response to a sinusoidal input applied at $t = 0$, that is, suppose that

$$v_s(t) = (10 \cos 10^3 t) u(t) \tag{53}$$

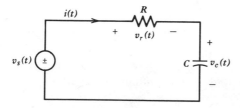

Figure 10.9   RC circuit example.

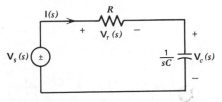

Figure 10.10 Complex frequency model.

The network function for the specified input-output pair is

$$H_c(s) = \frac{V_c(s)}{V_s(s)} = \frac{1/sC}{\frac{1}{sC} + R} = \frac{1}{sRC + 1} \tag{54}$$

From (53) we see that the input signal has an amplitude of 10 V at an angle of 0° and the frequency of the input signal is 10^3 rad/s so that $s = j10^3$. Thus, the amplitude and phase of the steady-state response

$$V_c = H(j10^3)10e^{j0°} \text{ V}$$

or upon substitution of (54) into the above equation

$$V_c = \frac{1}{(j10^3)RC + 1}(10 \text{ V}) \tag{55}$$

The form of the transient response is determined by the poles of the network function. Equation 54 has only one pole at $s = -1/RC$ so that the transient response is

$$v_{ct}(t) = ke^{-[(1/RC)t]} \tag{56}$$

Let us assume that $R = 1 \text{ k}\Omega$, $C = 1 \mu F$ and $v_c(0) = 0 \text{ V}$. Then, $RC = 10^{-3}$ so that from (55) the amplitude and phase of the steady-state response is

$$V_c = \frac{10}{j+1} = \frac{10}{\sqrt{2}}e^{-j45°} \tag{57}$$

The complete response is

$$v_c(t) = ke^{-10^3 t} + \text{Re}\left(\frac{10}{\sqrt{2}}e^{-j45°}e^{j10^3 t}\right)$$

or

$$v_c(t) = ke^{-10^3 t} + \frac{10}{\sqrt{2}}\cos(10^3 t - 45°) \tag{58}$$

In this circuit $v_c(0^+) = v_c(0^-) = 0$ so that

$$0 = k + \frac{10}{\sqrt{2}}\cos(-45°)$$

or

$$k = -5 \tag{59}$$

Thus, (58) becomes

$$v_c(t) = \underbrace{-5e^{-10^3t}}_{\text{transient}} + \underbrace{\frac{10}{\sqrt{2}}\cos(10^3t - 45°)}_{\text{steady state}}, \quad t \geq 0 \tag{60}$$

Note that the time constant of the transient solution is 1 ms so that in approximately 4 ms the circuit has reached the steady state solution. If one is not interested in the amplitude of the transient response, then there is no need to use the initial conditions to solve for k.

Example 2

As a second example, suppose that $R = 2\,\Omega$, $L = 1\,H$, and $C = 1/2\,F$ in the series *RLC* circuit in Figure 10.7. Find the capacitor voltage response to a step input $v_s(t) = 5u(t)$. Assume that the circuit is in the zero state at $t = 0$.

From (46) the transfer function

$$H_c(s) = \frac{2}{s^2 + 2s + 2} = \frac{2}{(s + 1 + j)(s + 1 - j)} \tag{61}$$

Since the input is constant for $t > 0$, this corresponds to an input with frequency $s = 0$ and amplitude $\mathbf{X} = 5\,V$. Thus, the steady-state response is

$$\mathbf{V}_c = H_c(s)\bigg|_{s=0} \cdot 5\,V = 5\,V \tag{62}$$

because $H_c(0) = 1$.

Since the two poles of (61) are complex conjugate pairs with a real part equal to -1 and an imaginary part equal to $\pm j$, then the transient response has the form

$$v_{ct}(t) = ke^{-t}\cos(t + \phi) \tag{63}$$

The complete response is

$$v_c(t) = ke^{-t}\cos(t + \phi) + 5\,V \tag{64}$$

From the given initial conditions $v_c(0) = 0$ and $i(0) = C[dv_c(0)/dt] = 0$, we find that

$$\phi = 0° \quad \text{and} \quad k = -5$$

so that

$$v_c(t) = -5e^{-t}\cos t + 5, \quad t \geq 0 \tag{65}$$

Now in this example suppose that the input is

$$v_s(t) = 5\cos\sqrt{2}\,t \tag{66}$$

that is, it has frequency $\sqrt{2}$ rads/s so that $s = j\sqrt{2}$. Furthermore, suppose that we are *only* interested in calculating the steady-state response. The amplitude and phase of the input signal is

$$\mathbf{V}_s = 5e^{j0°} \tag{67}$$

so that

$$\mathbf{V}_c = H_c(s)\big|_{s=j\sqrt{2}} \cdot (5e^{j0°})$$

$$\mathbf{V}_c = \frac{2}{(j\sqrt{2})^2 + 2(j\sqrt{2}) + 2} \cdot (5e^{j0°})$$

$$\mathbf{V}_c = \left(\frac{-j}{\sqrt{2}}\right)(5e^{j0°}) = \frac{5}{\sqrt{2}}e^{-j90°} \tag{68}$$

Thus, the amplitude of the response is $5/\sqrt{2}$ and its phase with respect to the phase of the input sinusoid is $-90°$. In the time domain, the steady-state response is

$$v_c(t) = \frac{5}{\sqrt{2}}\cos(\sqrt{2}\,t - 90°) \tag{69}$$

Note that the value of $\mathbf{V}_c$ obtained above from an evaluation of the transfer function at $s = j\sqrt{2}$ is equivalent to analyzing the complex frequency circuit model in Figure 10.8 with s replaced by $j\sqrt{2}$ as shown in Figure 10.11.

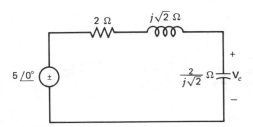

Figure 10.11   Complex frequency model for $s = j\sqrt{2}$.

Example 3

For our third example let us analyze a circuit that is frequently used to integrate signals. The circuit consists of a negative gain amplifier with a capacitor in a feedback loop around the amplifier as shown in Figure 10.12a. The complex frequency model for this circuit is shown in Figure 10.12b where it is assumed that the amplifier has infinite input resistance and zero output resistance. An analysis of this circuit yields

$$\left(\frac{1}{R} + sC\right)\mathbf{V}_i(s) - sC\mathbf{V}_o(s) = \mathbf{V}_s(s)/R$$

and

$$\mathbf{V}_o(s) = -A\mathbf{V}_i(s)$$

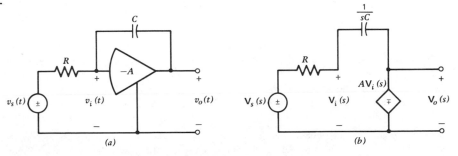

Figure 10.12   Integrator circuit. (a) Circuit. (b) Complex frequency model.

Eliminating $V_i(s)$ we obtain

$$H(s) = \frac{V_o(s)}{V_s(s)} = \frac{-1}{sRC\left(1 + \frac{1}{A}\right) + \frac{1}{A}} \tag{70}$$

Note that in the limit as the amplifier gain become large

$$\lim_{A \to \infty} H(s) = \frac{-1}{sRC} \tag{71}$$

which means that in the time domain

$$RC\frac{dv_o}{dt} = -v_s(t)$$

or

$$v_o(t) = v_o(0) - \frac{1}{RC}\int_0^t v_s(\tau)d\tau \tag{72}$$

Thus, for large A the output voltage is approximately equal to the integral of the input voltage. This circuit is a good sweep circuit since a constant input voltage yields a ramp output voltage.

Example 4

As our final example consider the active[3] RC circuit in Figure 10.13. From the complex frequency model the node equations are

$$\begin{pmatrix} G_1 + G_3 + sC_1 & -sC_1 + AG_3 \\ -sC_1 & G_2 + sC_1 + sC_2 \end{pmatrix}\begin{bmatrix} E_1(s) \\ E_2(s) \end{bmatrix} = \begin{bmatrix} G_1V_s(s) \\ 0 \end{bmatrix} \tag{73}$$

We obtain

$$V_o(s) = AE_2(s) = \frac{A\begin{vmatrix} G_1 + G_3 + sC_1 & G_1V_s(s) \\ -sC_1 & 0 \end{vmatrix}}{(G_1 + G_3 + sC_1)(sC_1 + sC_2 + G_2) - sC_1(sC_1 + AG_3)}$$

or

$$\frac{V_o(s)}{V_s(s)} = \frac{sC_1G_1A}{s^2C_1C_2 + s[(C_1 + C_2)(G_1 + G_2) + C_1G_2 - AC_1G_3] + G_2(G_1 + G_3)} \tag{74}$$

At this point let us make an interesting observation. Return to the RLC circuit in Figure 10.8 and, from (48),

$$\frac{V_r}{V_s} = H_r(s) = \frac{RCs}{LCs^2 + sRC + 1} \tag{75}$$

The polynomial in the numerator of (75) has the same form as the corresponding polynomial in (74), and the polynomials in the denominators of these two network functions are also of the same form. Thus, if we choose the element values in these two circuits such that the corresponding coefficients of the polynomials are identical, then these two circuits will have the same input-output behavior in both the

---

[3] Strictly speaking the adjective active implies that the small-signal circuit can deliver more energy to a load than is supplied by the input sources. Although frequently any circuit containing an amplifier is called active.

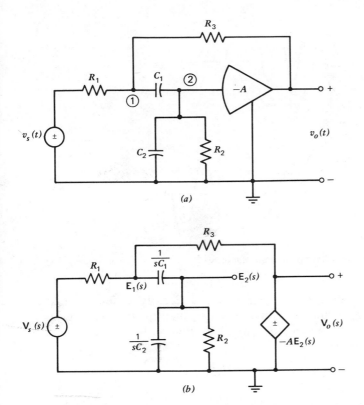

Figure 10.13   Active RC circuit. (a) Circuit. (b) Complex frequency model.

frequency domain and the time domain. In other words, the active RC circuit could replace the RLC circuit. This is desirable in many filter applications in which the active RC circuit is smaller and less costly than the RLC circuit. The amplifier is needed for situations in which the natural frequencies of the RLC circuits are complex numbers, because a passive RC circuit can only have real natural frequencies.[4]

10.5 STABILITY

A linear time-invariant lumped circuit is said to be stable if its natural frequencies have a negative real part, or in other words the poles lie in the lhp. If the poles lie on the jω axis or in the rhp the circuit is unstable. Figure 10.14 illustrates the transient response for jω axis and rhp poles. Circuits that have resistors, inductors, and capacitors that are positive cannot have poles in the rhp. In fact, one cannot even realize jω axis poles since ideal[5]

[4] M. E. Van Valkenburg, *Introduction to Modern Network Synthesis*, Chapter 6, Wiley, 1960.
[5] Ideal in the sense that they have no resistive loss.

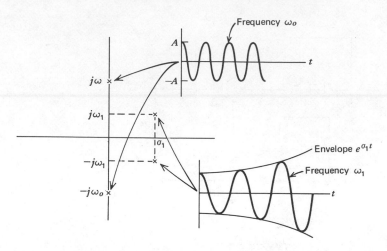

Figure 10.14 Oscillator and unstable circuit responses.

inductors and capacitors do not exist in practice. In order to make an oscillator one must include an active element such as an amplifier.

Let us return to the active RC circuit in Figure 10.13 whose transfer function (74) can be written as

$$H(s) = \frac{s\dfrac{G_1 A}{C_2}}{s^2 + s\dfrac{[(C_1 + C_2)(G_1 + G_2) + C_1 G_2 - A C_1 G_3]}{C_1 C_2} + \dfrac{G_2(G_1 + G_3)}{C_1 C_2}} \tag{76}$$

Let $C_1 = C_2 = 1$ F and $G_1 = G_2 = G_3 = 1$ mho, then

$$H(s) = \frac{sA}{s^2 + s(5 - A) + 2} \tag{77}$$

In order to make this circuit into an oscillator ($j\omega$ axis poles) choose the gain A of the amplifier such that

$$A = 5 \tag{78}$$

then

$$H(s) = \frac{5s}{s^2 + 2} \tag{79}$$

and the frequency of oscillation is $\sqrt{2}$ rads/s, that is, the poles are located at $\pm j\sqrt{2}$ and from (79) assuming zero input (short circuit)

$$\frac{d^2 v_o}{dt} + 2v_o = 0 \tag{80}$$

so that

$$v_o(t) = A \sin(\sqrt{2}\, t + \phi) \tag{81}$$

where A and ϕ are determined by the initial conditions in theory. However, in practice one cannot place the poles of $H(s)$ precisely on the $j\omega$ axis. The gain of the amplifier is adjusted so that the poles are actually slightly in the rhp so that the small-signal circuit is unstable. As the unstable oscillation grows, the gain of the amplifier actually decreases at large signals due to the nonlinear effects of the electronic devices. The circuit quickly reaches a steady-state oscillation whose frequency is approximately given by the poles of the small-signal network function and whose amplitude is actually determined by the nonlinear characteristics of the amplifier.

10.6 POLE-ZERO LOCI

In the manufacturing process it is impossible to realize the exact design values for a component. Also, the values of components change with the environment such as temperature, humidity, and so on. Thus, it is important for a designer to study the effects of the variation of the component values on the response of the circuit. One approach is to observe how the poles and zeros change with respect to changes in the component values. In order to keep the problem tractable typically the variation of the roots of the network function are plotted in the complex plane with respect to the variation in only one parameter value at a time. To further simplify the presentation we will restrict ourselves to the loci of only second order network functions.

Let us return to the RC amplifier circuit in Example 4 whose transfer function is given by (77). The denominator polynomial is

$$D(s) = s^2 + (5 - A)s + 2 \tag{82}$$

Let us examine how the roots of $D(s)$ change with the amplifier gain A. The roots are

$$s_{1,2} = -\frac{(5-A)}{2} \pm \sqrt{\left(\frac{5-A}{2}\right)^2 - 2} \tag{83}$$

When $A = 0$, the roots are -4.56 and -0.44. Also, when $A = 5$, clearly $s_{1,2} = \pm j/\sqrt{2}$. Thus, as A increases from 0 to 5 the roots move from the negative real axis to the imaginary axis. In order to determine the locus of this motion we need additional points between the gains 0 and 5 or we need a geometrical formula that describes the trajectory of the roots as A varies. Observe that when the roots are complex then

$$s_{1,2} = -\left(\frac{5-A}{2}\right) \pm j\sqrt{2 - \frac{(5-A)^2}{2}} \tag{84}$$

and the magnitude of this vector is

$$|s_{1,2}| = \sqrt{2} \tag{85}$$

that is, the locus is a constant distance from the origin as shown in Figure 10.15. For complex roots the locus is a circle of radius $\sqrt{2}$. As A increases from 0 the roots move along the negative real axis toward the point $-\sqrt{2}$ where they split and become complex. At $A = 5$ the roots are on the $j\omega$ axis, the circuit is an oscillator. As A increases beyond 5 the roots move in the rhp and back toward the real axis where they meet at $+\sqrt{2}$ and split with one moving along the positive real axis toward the origin and the other moving along the positive real axis away from the origin. As $A \to +\infty$ the roots of (82) are zero and infinity.

From this root locus we conclude that if the design calls for poles near the $j\omega$ axis, that is, A less than but approximately equal to 5, then we should worry about the tolerances on A because if A is equal to or greater than 5 our circuit will become unstable and oscillate.

As a final example consider the polynomial

$$(s + \sigma_1)(s + \sigma_2) + K \tag{86}$$

where σ_1, σ_2, and K are greater than zero. Let us plot the locus of the roots as a function of K. For $K = 0$ the roots lie on the negative real axis at $-\sigma_1$ and $-\sigma_2$. In general, the roots for any value of K are

$$s_{1,2} = \frac{-(\sigma_1 + \sigma_2)}{2} \pm \sqrt{\left(\frac{\sigma_1 + \sigma_2}{2}\right)^2 - K} \tag{87}$$

As K increases the roots move together and meet at the point $-(\sigma_1 + \sigma_2)/2$ where $K = [(\sigma_1 + \sigma_2)/2]^2$. As K continues to increase the roots become complex, but their real part is constant at $-(\sigma_1 + \sigma_2)/2$. This locus is shown in Figure 10.16. As K increases the imaginary part of the roots increases but the circuit does not become unstable.

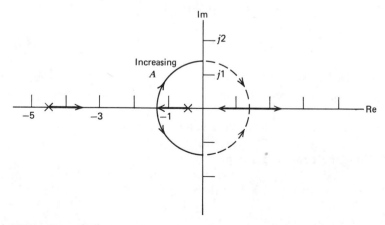

Figure 10.15 Root locus of RC active circuit.

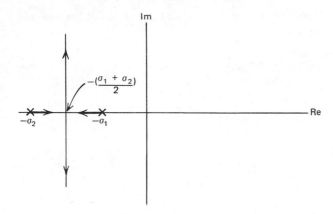

Figure 10.16   Root locus example.

10.7 MULTIPLE INPUTS

If a circuit is linear and contains more than one independent source, then the superposition principle can be applied in order to obtain the response. For example, consider the circuit in Figure 10.17*a*. One independent source has an amplitude V_1, frequency ω_1, and phase θ_1, while the amplitude of the other source is V_2, its frequency is ω_2, and its phase is θ_2. The complex

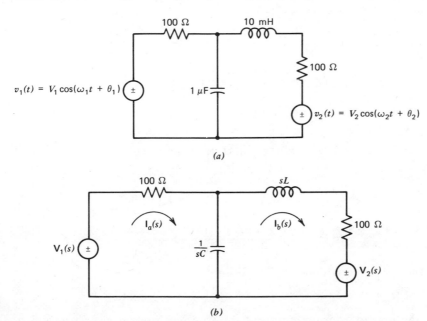

Figure 10.17   Multiple input example. (*a*) Circuit. (*b*) Complex frequency model.

frequency model is illustrated in Figure 10.17b. Let us solve for the current $\mathbf{I}_a(s)$. The mesh equations are

$$\mathbf{V}_1(s) = \left(100 + \frac{1}{sC}\right)\mathbf{I}_a(s) - \frac{1}{sC}\mathbf{I}_b(s) \tag{88}$$

and

$$-\mathbf{V}_2(s) = -\frac{1}{sC}\mathbf{I}_a(s) + \left(\frac{1}{sC} + sL + 100\right)\mathbf{I}_b(s)$$

Therefore,

$$\mathbf{I}_a(s) = \frac{\begin{vmatrix} \mathbf{V}_1(s) & -\dfrac{1}{sC} \\[2ex] -\mathbf{V}_2(s) & \dfrac{1}{sC} + sL + 100 \end{vmatrix}}{\begin{vmatrix} 100 + \dfrac{1}{sC} & -\dfrac{1}{sC} \\[2ex] -\dfrac{1}{sC} & \dfrac{1}{sC} + sL + 100 \end{vmatrix}} \tag{89}$$

which yields

$$\mathbf{I}_a(s) = \frac{\left(\dfrac{1}{sC} + sL + 100\right)\mathbf{V}_1(s) - \dfrac{1}{sC}\mathbf{V}_2(s)}{(100)\left(\dfrac{1}{sC} + sL + 100\right) + \dfrac{1}{sC}(sL + 100)}$$

or

$$\mathbf{I}_a(s) = \frac{(s^2LC + 100Cs + 1)\mathbf{V}_1(s) - \mathbf{V}_2(s)}{s^2LC(100) + s(10^4C + L) + 200} \tag{90}$$

From (90) we see that the response depends linearly on the inputs $\mathbf{V}_1(s)$ and $\mathbf{V}_2(s)$. We write

$$\mathbf{I}_a(s) = H_1(s) \cdot \mathbf{V}_1(s) + H_2(s) \cdot \mathbf{V}_2(s) \tag{91}$$

where

$$H_1(s) = \frac{s^2LC + 100Cs + 1}{s^2LC(100) + s(10^4C + L) + 200}$$

and

$$H_2(s) = \frac{-1}{s^2LC(100) + s(10^4C + L) + 200}$$

From the above equations

$$i_a(t) = k_1 e^{p_1 t} + k_2 e^{p_2 t} + I_{a1}\cos(\omega_1 t + \phi_{a1}) + I_{a2}\cos(\omega_2 t + \phi_{a2}) \tag{92}$$

where the first two terms represent the transient response, and p_1 and p_2 are the natural frequencies of the circuit (poles of the transfer functions). It is assumed that $p_1 \neq p_2$, if not, then the transient response has the form

$k_1 e^{p_1 t} + k_2 t e^{p_1 t}$. Finally note that p_1 and p_2 are roots of the characteristic polynomial, which is simply the circuit determinant, that is, the denominator of (89). These natural frequencies are dependent only on the structure and element values of the zero input circuit shown in Figure 10.18. Thus, both $H_1(s)$ and $H_2(s)$ have the same denominator.

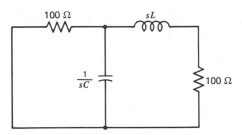

Figure 10.18   Zero input circuit.

I_{a1} is the amplitude and ϕ_{a1} is the phase of the sinusoidal steady-state response due to the input $v_1(t)$, $(v_2(t) = 0)$, while I_{a2} is the amplitude and ϕ_{a2} is the phase of the sinusoidal steady-state response due to the input $v_2(t)$, $(v_1(t) = 0)$. Below we compute the sinusoidal steady-state response for some given input voltages.

Suppose that $V_1 = 10$ V, $\theta_1 = 0°$, and $\omega_1 = 2 \times 10^4$ rads/s; also let $V_2 = 20$ V, $\theta_2 = 0°$, and $\omega_2 = 10^4$ rads/s. Recall that $L = 10$ mH and $C = 1$ uF. From (91) the amplitude and phase of the steady-state response due to only $v_1(t)$ is

$$\mathbf{I}_{a1} = H_1(j2 \times 10^4) \cdot 10 \, e^{j0°}$$

$$\mathbf{I}_{a1} = \frac{(j2 \times 10^4)^2(10^{-2})(10^{-6}) + 100(10^{-6})(j2 \times 10^4) + 1}{(j2 \times 10^4)^2(10^{-2})(10^{-6})(100) + (j2 \times 10^4)(10^4 \cdot 10^{-6} + 10^{-2}) + 200} \cdot 10 \, e^{j0°}$$

which yields

$$\mathbf{I}_{a1} = \frac{-3 + j2}{-200 + j400} \cdot 10 \, e^{j0°} = 80.6 \, e^{j29.7} \text{ mA} \tag{93}$$

This result could also have been obtained from an analysis of the circuit in Figure 10.19a.

The amplitude and phase of the current $\mathbf{I}_{a2}$ is

$$\mathbf{I}_{a2} = H_2(j10^4) \cdot 20 \, e^{j0°}$$

or

$$\mathbf{I}_{a2} = \frac{-20}{(j10^4)(10^{-2})(10^{-6})(100) + (j10^4)(10^4 \cdot 10^{-6} + 10^{-2}) + 200}$$

$$\mathbf{I}_{a2} = \frac{0.2}{1 + j2} = 89.4 \, e^{-j63.4°} \text{ mA} \tag{94}$$

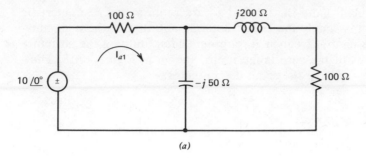

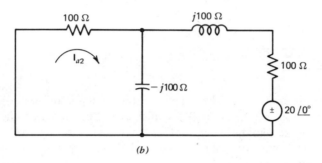

Figure 10.19 Analysis by superposition; (a) $\omega = 2 \times 10^4$, $V_2 = 0$. (b) $\omega = 10^4$, $V_1 = 0$.

This result could have been obtained from an analysis of the circuit in Figure 10.19b. From (93) and (94) the sinusoidal steady-state response is

$$i_a(t) = 80.6 \cos(2 \times 10^4 t + 29.7°) + 89.4 \cos(10^4 t - 63.4°) \text{ mA} \qquad (95)$$

If both sources have the *same frequency*, then the phasors $\mathbf{I}_{a1}$ and $\mathbf{I}_{a2}$ can be combined into one phasor. For example, if $\omega_2 = 2 \times 10^4$ also, then using the fact that $H_2(s)$ has the same denominator as $H_1(s)$ and the frequencies are the same we obtain

$$\mathbf{I}_{a2} = \frac{-1}{-200 + j400} \cdot 20 \, e^{j0°} \qquad (96)$$

From (93) and (96)

$$\mathbf{I}_{a1} + \mathbf{I}_{a2} = \frac{-30 + j20 - 20}{-200 + j400} = 120.4 \, e^{j41.6°} \text{ mA}$$

so that

$$i_a(t) = 120.4 \cos(2 \times 10^4 t + 41.6°) \text{ mA} \qquad (97)$$

The constants k_1 and k_2 in (92) can be computed from the knowledge of $i_a(0^+)$ and $(di_a/dt)(0^+)$. For example, suppose that both sources are switched on simultaneously at $t = 0$. Also, suppose that the capacitor voltage and

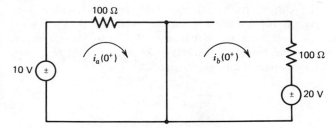

Figure 10.20 Zero-plus initial condition circuit model.

inductor current are zero at $t = 0$. The $t = 0^+$ circuit is illustrated in Figure 10.20. Note that

$$i_a(0^+) = 0.1 \text{ A} \tag{98}$$

The derivative of this current at $t = 0^+$ can be computed from the first equation in (88). Let us multiply this equation by s and replace s with the differential operator, then

$$\frac{dv_1}{dt} = 100 \frac{di_a}{dt} + \frac{1}{C} i_a(t) - \frac{1}{C} i_b(t). \tag{99}$$

The evaluation of (99) at $t = 0^+$ yields

$$\frac{di_a}{dt}(0^+) = \frac{-i_a(0^+)}{100\,C} = -10^3 \tag{100}$$

because $(dv_1/dt)|_{t=0} = 0$ and $i_b(0^+) = 0$.

The poles p_1 and p_2 are roots of the equation

$$s^2 LC(100) + s(10^4 C + L) + 200 = 0$$

which for $L = 10 \text{ mH}$ and $C = 1 \text{ } \mu\text{F}$ becomes

$$s^2 + 2 \times 10^4 \, s + 2 \times 10^8 = 0$$

so that

$$p_{1,2} = -10^4 \pm j10^4 \tag{101}$$

Given (98), (100), (101), and the sinusoidal steady-state response one can compute the complex amplitudes k_1 and k_2 of the transient response. The time constant of the transient decay is 10^{-4} s or $100 \text{ } \mu\text{s}$.

PROBLEMS

1. Listed below are the transfer functions of four linear time-invariant circuits that have a single independent source. The first transfer function is a transfer ratio (voltage input-voltage response), the second transfer function is a transfer

admittance (voltage input-current response), the third transfer function is a transfer impedance (current input-voltage response), and the final transfer function is a transfer ratio (current input-current response). From these transfer functions, sketch the transient response waveform for each circuit. Since no initial conditions are given assume arbitrary constants.

(a) $\dfrac{V_0(s)}{V_s(s)} = \dfrac{10^4}{s + 10^3}$

(b) $\dfrac{I_0(s)}{V_s(s)} = \dfrac{0.1\, s}{s + 10^3}$

(c) $\dfrac{V_0(s)}{I_s(s)} = \dfrac{10^6}{s^2 + 10^2 s + 10^6}$

(d) $\dfrac{I_0(s)}{I_s(s)} = \dfrac{10^3 s}{s^2 + 2 \times 10^3 s + 10^6}$

2. Find the steady-state responses of the circuits in Problem 1 for each of the following inputs: (a) $0.1 \cos 10^3 t$, (b) $0.1 \cos 2\pi \times 10^3 t$, and (c) 0.1 (constant input).

3. From the transfer functions given in Problem 1, write the differential equation that describes the input-output relation for each circuit.

4. The input-output relation for a linear time-invariant circuit with a single input is

$$\frac{d^2 v_o}{dt^2} + 15 \frac{dv_o}{dt} + 36 v_o = \frac{d^2 v_s}{dt}$$

(a) Find the transfer function $V_o(s)/V_s(s)$
(b) Find the steady-state response if $v_s = 5$ V.
(c) Find the transient response for zero input if $v_o(0^+) = 0$, and $(dv_o/dt)(0^+) = 10$.
(d) Find the steady-state response if $v_s(t) = 5 \cos 2\pi t$.

5. Let $x(t) = X \sin(\omega_o t + \theta)$ be the input signal to a linear time-invariant circuit. Show that the amplitude and phase of the sinusoidal response is

$$Y = H(j\omega_o) \cdot X$$

where $H(s)$ is the transfer function and $X = X e^{j\theta}$. Also, show that the steady-state response

$$y(t) = \text{Im}[H(j\omega_o) \cdot X e^{j\omega_o t}]$$

6. For the series RL circuit in Figure P10.6 draw the complex frequency circuit model and find the transfer ratio $V_L(s)/V_s(s)$. Investigate the asymptotic behavior of the circuit as $s \to 0$ and $s \to \infty$. If $R = 100\,\Omega$ and $L = 0.1$ H find the response to a input (a) $v_s(t) = 10u(t)$ and (b) $v_s(t) = (10 \cos 2\pi \times 150t)u(t)$. Assume that the circuit is in the zero state when the input is applied.

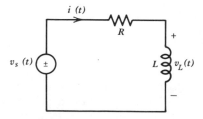

Figure P10.6

7. Repeat Problem 6 only let the current $i(t)$ be the response.
- 8. Repeat Problem 6 for the parallel RL circuit in Figure P10.8 only now the input is the current source $i_s(t)$ and assume that the response is $v(t)$.

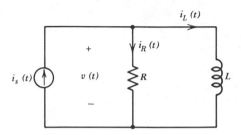

Figure P10.8

9. Repeat Problem 8 only for the response $i_R(t)$.
10. Repeat Problem 8 only for the response $i_L(t)$.
11. For the parallel RLC circuit in Figure P10.11 draw the complex frequency circuit model and find the driving-point impedance $V(s)/I_s(s)$. Investigate the asymptotic behavior of the circuit as $s \to 0$ and $s \to \infty$. If $R = 1\,\Omega$, $L = 1\,H$, and $C = 1\,F$ find the steady state response for $i_s(t) = 2\cos t$. How long does it take for the transient response of this circuit to decay to approximately 2% of its initial value?

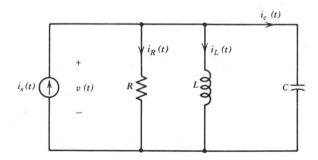

Figure P10.11

12. Repeat Problem 11 for the transfer function $I_L(s)/I_s(s)$.
- 13. Repeat Problem 11 for the transfer function $I_C(s)/I_s(s)$.
14. For the circuit in Figure P10.14 draw the complex frequency circuit model and find the driving-point impedance $V(s)/I(s)$ and the transfer ratio $V_o(s)/V(s)$. Investigate the asymptotic behavior of these two transfer functions as $s \to 0$ and $s \to \infty$. Find the natural frequencies of this circuit for $R_1 = R_2 = 1\,\Omega$, $L = 1\,H$ and $C = 1\,F$ for (a) a voltage source input, (b) a current source input.

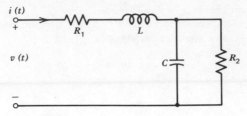

Figure P10.14

15. Repeat Problem 14 for the circuit in Figure P10.15.

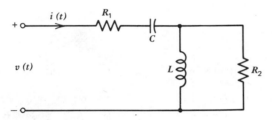

Figure P10.15

16. Figure P10.16 represents the small-signal model of a transistor circuit. Draw the complex frequency model and find the transfer ratio $V_2(s)/V_s(s)$. If the input $v_s(t) = 10^{-2}u(t)$ V find the steady-state response. How many seconds are required for the output voltage to reach approximately 98% of its final value?

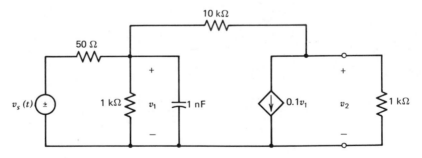

Figure P10.16

17. Repeat Problem P10.16 with the 10-kΩ feedback resistor reduced to 1 kΩ.
18. The ideal response of the integrator circuit in Figure 10.12 to a step input $Vu(t)$ is $v_o(t) = (-V/RC)t$ for $v_o(0) = 0$. Find the RC product such that if the input is $-0.01u(t)$ then the output is a ramp with $v_o(1\ \mu s) = 10$ V. If the gain $A = 10^3$, what is the error between the actual response obtained from the transfer

function (70) and the ideal response (72) at $t = 1\,\mu s$? What is the error if $A = 10^5$?

19. For the given transfer function plot the locus of the poles for $0 \leqslant A < \infty$. For what range of values of A is the circuit unstable?

$$H(s) = \frac{1}{(s+1)(s+3) - As}$$

20. Plot the locus of the poles of the transfer functions below and indicate the range of values of A for which the circuit is unstable.

(a) $\dfrac{1}{(s+1)(s+3) + As^2}$ (b) $\dfrac{1}{(s+1)(s+3) + A(s^2+1)}$

(c) $\dfrac{1}{(s+1)(s+3) - A}$ (d) $\dfrac{1}{(s+1)(s+3) - A(s+2)}$

21. In Figure P10.21 use superposition to find the steady-state response $v_R(t)$ when $\omega_1 = \omega_2 = 1$ rad/s.

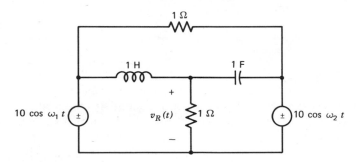

Figure P10.21

22. Repeat Problem 21 for $\omega_1 = 1$ rad/s and $\omega_2 = 5$ rad/s.

Chapter Eleven
AC Analysis

Frequently circuits are designed to have a specific sinusoidal steady-state performance. The transient response is often of secondary importance and after the design is completed the transient response may be checked simply to insure that the circuit is not unstable. In this chapter we examine some of the concepts useful for the sinusoidal steady-state (ac) analysis and design of circuits. We restrict our discussion to circuits whose branch constraints are linear and constant; we say that the circuit is linear and time invariant.

11.1 SINUSOIDAL ANALYSIS WITH PHASORS

In the previous chapter it was shown that when the input to a linear time-invariant circuit is sinusoidal, then the steady-state response waveforms are sinusoidal and have the same frequency as the input waveform. In the case of more than one input signal one can use the superposition principle. Sinusoidal steady-state analysis is conveniently carried out by examining the response of the circuit to a complex exponential input of the form e^{st}. With this complex exponential input all of the steady-state responses also have the form e^{st}. Thus, the factor e^{st} is common to all terms in the circuit equations and can be factored from the equations. What remains is a set of linear algebraic equations that relate the amplitude and phase of the steady-state responses to the amplitude and phase of the input signal. Corresponding to this set of equation is a circuit model that we call the complex frequency circuit model. It is formed from the actual circuit by replacing each element by its complex frequency constraint (see Table 10.1). Since we have simply removed the factor e^{st} from the steady state equations, it follows that the complex voltages must obey Kirchhoff's voltage law and the complex currents must obey Kirchhoff's current law. Thus, all of the analytical tools developed in Chapters Three to Five for resistive circuits readily extend to the complex frequency circuit model. This is the beauty of the complex exponential input signal. Several examples follow.

Consider the parallel RLC circuit in Figure 11.1. The complex frequency or phasor circuit model is shown in Figure 11.2. The complex voltages and currents V_1, I_1, I_C, I_L, and I_R are called phasors. Recall from Chapter 2 that a phasor is a complex number where magnitude denotes the amplitude of a sinusoidal waveform and its angle denotes the phase of the waveform.

Since Kirchhoff's laws also apply to the complex frequency circuit, the current I_1 is the vector sum

$$\mathbf{I}_1 = \mathbf{I}_L + \mathbf{I}_R + \mathbf{I}_C \tag{1}$$

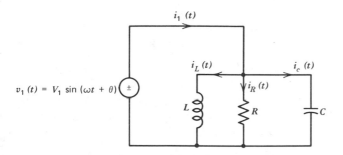

Figure 11.1 Parallel RLC circuit with a sinusoidal input.

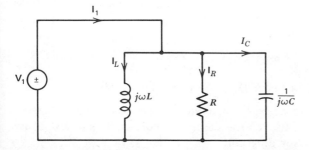

Figure 11.2 Phasor model of the parallel RLC circuit.

Since $I_R = V_1/R$, $I_C = j\omega CV_1$, and $I_L = V_1/j\omega L$, equation 1 becomes

$$I_1 = \frac{V_1}{j\omega L} + \frac{V_1}{R} + j\omega CV_1 \tag{2}$$

A vector diagram of these phasors, such as the one shown in Figure 11.3, is called a *phasor diagram*. The phasor V_1 is the phasor voltage across each

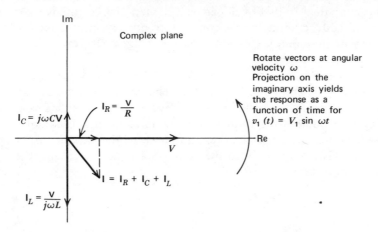

Figure 11.3   Phasor diagram.

of the parallel elements. In the diagram it is assumed that the phase θ of the input voltage is $0°$. Since the resistor is real its current is in phase with the voltage across it, whereas *the inductor current lags behind its voltage by 90°, and the capacitor current leads its voltage by 90°*. This result also can be obtained from the branch relations in the time domain.

$$i_L(t) = \frac{1}{L} \int V_1 \sin(\omega t + \theta) \, dt$$

$$= -\frac{V_1}{\omega L} \cos(\omega t + \theta) \equiv \frac{V_1}{\omega L} \sin(\omega t + \theta - 90°) \tag{3}$$

and

$$i_c(t) = C \frac{d}{dt} (V_1 \sin(\omega t + \theta))$$

$$= \omega CV_1 \cos(\omega t + \theta) \equiv \omega CV_1 \sin(\omega t + \theta + 90°) \tag{4}$$

Note that the inductor current lags the inductor voltage $V_1 \sin(\omega t + \theta)$ by $90°$ and the capacitor current leads the capacitor voltage $V_1 \sin(\omega t + \theta)$ by $90°$ as indicated in the phasor diagram.

The circuit in Figure 11.2 has a single input V_1, hence all of the response phasors can be expressed as a function of $\mathbf{V}_1$. For example, from (2)

$$\mathbf{I}_1 = \left[\frac{1}{R} + j\left(\omega C - \frac{1}{\omega L}\right)\right]\mathbf{V}_1 \tag{5}$$

Note that the second and third terms tend to cancel each other. Again this is clear from the phasor diagram. Note that in Figure 11.3 the currents $\mathbf{I}_L$ and $\mathbf{I}_C$ are 180° out of phase. There is an input frequency such that these currents cancel, that is,

$$\mathbf{I}_L - \mathbf{I}_C = \left(-\frac{1}{\omega L} + j\omega C\right)\mathbf{V}_1 = 0 \tag{6}$$

which is equivalent to

$$\omega C - \frac{1}{\omega L} = 0 \tag{7}$$

The frequency at which cancellation occurs, the solution of (7), is called the resonant frequency,

$$\omega_r = \frac{1}{\sqrt{LC}} \tag{8}$$

of the parallel RLC circuit. In Figure 11.3 $|I_L| > |I_C|$, which means that $\omega < \omega_r$ for this circuit so that the source current phasor $\mathbf{I}_1$ lags the source voltage phasor $\mathbf{V}_1$. We say that the circuit is inductive for frequencies $\omega < \omega_r$. For frequencies $\omega > \omega_r$, the phasor $\mathbf{I}_1$, leads $\mathbf{V}_1$ and we say that the circuit is capacitive for $\omega > \omega_r$.

We can write (5) as

$$\mathbf{I}_1 = Y(j\omega)\mathbf{V}_1 \tag{9}$$

In polar form

$$Y(j\omega) = |Y(j\omega)|e^{j\alpha}$$

where

$$|Y(j\omega)| = \left[\left(\frac{1}{R}\right)^2 + \left(\omega C - \frac{1}{\omega L}\right)^2\right]^{1/2} \tag{10}$$

and

$$\tan \alpha = R\left(\omega C - \frac{1}{\omega L}\right) \tag{11}$$

Thus the magnitude of the source current is equal to the product of the magnitude of the admittance seen by the source times the magnitude of the input voltage. The phase angle of the current $i_1(t)$ is the phase angle of the admittance plus the phase angle of the input voltage, that is, since the input is a sine function

$$i_1(t) = \text{Im}\left\{\left[\left(\frac{1}{R}\right)^2 + \left(\omega C - \frac{1}{\omega L}\right)^2\right]^{1/2} e^{j\alpha} \cdot V_1 e^{j\omega t}\right\} \tag{12}$$

or

$$i_1(t) = \left[\left(\frac{1}{R} \right)^2 + \left(\omega C - \frac{1}{\omega L} \right)^2 \right]^{1/2} V_1 \sin(\omega t + \theta + \alpha) \tag{13}$$

If the input was $\cos(\omega t + \theta)$, then we would take the real part of (12). In other words, in Figure 11.3 rotate all the phasors by θ degrees, the phase angle of the input voltage, then spin the phasors in the counterclockwise direction with the angular velocity ω of the input frequency. The projection of the vectors on the real axis is the solution for a $\cos(\omega t + \theta)$ input and the projection on the imaginary axis is the solution for a $\sin(\omega t + \theta)$ input. Next let us put some numerical values to this example.

In the parallel RLC circuit, Figure 11.1, suppose that $R = 100 \, \Omega$, $C = 10 \, \mu F$, $L = 100 \, mH$. From (8) the resonant frequency is

$$\omega_r = \frac{1}{\sqrt{10^{-5} \times 10^{-1}}} = 10^3 \, \text{rads/s} \tag{14}$$

Now if the input voltage is $10 \sin 2\pi 60t$, that is, a 10-V peak and a frequency of 60 Hz, then (5) becomes

$$\mathbf{I}_1 = \left[\frac{1}{100} + j \left(2\pi 60 \times 10^{-5} - \frac{1}{2\pi \times 60 \times 10^{-1}} \right) \right] 10$$

$$\mathbf{I}_1 = [0.1 - j0.23] = 0.251 \, e^{-j66.5°} \, \text{A} \tag{15}$$

Therefore,

$$i_1(t) = 0.251 \sin(2\pi 60t - 66.5°) \, \text{A} \tag{16}$$

The source current lags behind its voltage by 66.5° so that the circuit is inductive as expected since $\omega = 2\pi \cdot 60 = 377 \, \text{rads/s} < \omega_r$.

In the above problem if we let $\omega = 2 \times 10^3 \, \text{rads/s}$, then

$$\mathbf{I}_1 = \left[\frac{1}{100} + j \left(2 \times 10^3 \times 10^{-5} - \frac{1}{2 \times 10^3 \times 10^{-1}} \right) \right] 10$$

$$\mathbf{I}_1 = 0.1 + j.15 = 0.18 \, e^{j56.3°} \, \text{A} \tag{17}$$

so that

$$i_1(t) = 0.18 \sin(2 \times 10^3 t + 56.3°) \tag{18}$$

Now the circuit is capacitive since the phase of the current $i_1(t)$ leads the input voltage phase by 56.3°.

11.2 IMPEDANCE AND ADMITTANCE PROPERTIES

In the phasor circuit model the voltage and current at a terminal pair are related by the equation

$$\mathbf{V} = Z(j\omega)\mathbf{I} \tag{19}$$

where the constant of proportionality Z is called the *impedance* seen at that terminal pair and it has units of ohms. The impedance is a complex number dependent on the input frequency. We write

$$Z(j\omega) = R(j\omega) + jX(j\omega) \tag{20}$$

where the real part R is resistive and the imaginary part X is called the *reactance*. If $X > 0$ we say the impedance is inductive, and if $X < 0$ we say the impedance is capacitive.

Also, we could write

$$\mathbf{I} = Y(j\omega)\mathbf{V} \tag{21}$$

where Y is called the *admittance* and has units of mhos.

$$Y(j\omega) = \frac{1}{Z(j\omega)} = G(j\omega) + jB(j\omega) \tag{22}$$

The real part of Y is the conductance and the imaginary part of Y is called *susceptance*. From (20) and (22) note that

$$G + jB = \frac{1}{R + jX} = \frac{R - jX}{R^2 + X^2} \tag{23}$$

Thus

$$G = \frac{R}{R^2 + X^2} \tag{24}$$

and

$$B = \frac{-X}{R^2 + X^2} \tag{25}$$

Often we will write Z instead of $Z(j\omega)$, R instead of $R(j\omega)$, X instead of $X(j\omega)$, and similarly for the admittance function. However, always remember that in the phasor model these parameters are, in general, frequency dependent. From (24) and (25) we conclude that

$$\text{Re}[Y] \neq \frac{1}{\text{Re}[Z]} \tag{26}$$

and

$$\text{Im}[Y] \neq \frac{1}{\text{Im}[Z]} \tag{27}$$

unless $R = 0$, $G = 0$, $X = 0$, or $B = 0$.

11.3 PHASOR CIRCUIT EQUIVALENTS

Recall that for sinusoidal waveforms ($s = j\omega$) the impedance of the inductor is $j\omega L$ and its admittance is $1/j\omega L$ or $-j/\omega L$, while the impedance of the capacitor is $1/j\omega C$ or $-j/\omega C$ and its admittance is $j\omega C$. In the phasor circuit the equivalent impedance of a series connection is the sum of the impedances of the individual series elements, and in the case of parallel elements the admittance of the parallel combination is the sum of the admittances of each parallel element. These results are easily proved by the application of Kirchhoff's laws to the circuits in Figures 11.4 and 11.5.

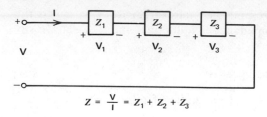

$$Z = \frac{V}{I} = Z_1 + Z_2 + Z_3$$

Figure 11.4   Series circuit.

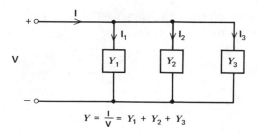

$$Y = \frac{I}{V} = Y_1 + Y_2 + Y_3$$

Figure 11.5   Shunt circuit.

As an example consider the circuit in Figure 11.6. The admittance

$$Y(j\omega) = \frac{1}{R} + \frac{1}{j\omega L} + j\omega C \qquad (28)$$

If $R = 100\,\Omega$, $C = 10^{-5}\,$F, and $L = 10^{-1}\,$H, let us compute the admittance and impedance for $\omega = 377$, 10^3, and $2 \times 10^3\,$rads/s. We obtain

$$Y(j377) = 0.010 + \frac{1}{j377 \times 10^{-1}} + j377 \times 10^{-5}$$

$$= 0.010 - j0.023 \text{ mho} \qquad (29)$$

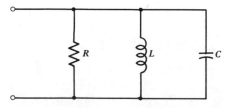

Figure 11.6   Parallel RLC circuit.

and

$$Z(j377) = \frac{1}{0.010 - j0.023} = \frac{0.010 + j0.023}{(0.010)^2 + (0.023)^2}$$

$$= 15.9 + j36.6 \, \Omega \tag{30}$$

Thus, (30) implies that at $\omega = 377 \, \text{rads/s}$ the parallel RLC circuit is equivalent to a series RL with $R = 15.9 \, \Omega$ and $L = 36.6/377 = 97 \, \text{mH}$ as shown in Figure 11.7a. The circuit is inductive for $\omega = 377 \, \text{rads/s}$.

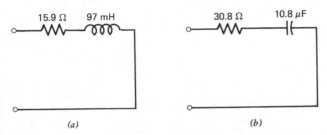

(a) (b)

Figure 11.7 Equivalent circuits. (a) $\omega = 377 \, \text{rad/s}$. (b) $\omega = 2 \times 10^3 \, \text{rad/s}$.

If $\omega = 10^3 \, \text{rads/s}$, then

$$Y(j10^3) = 0.01 \, \text{mho} \tag{31}$$

and

$$Z(j10^3) = 100 \, \Omega \tag{32}$$

The inductor and capacitor susceptance cancel and the equivalent circuit is simply the resistor R. Note that $10^3 \, \text{rads/s}$ is the resonant frequency.

Finally, when $\omega = 2 \times 10^3 \, \text{rads/s}$ we obtain

$$Y(j2 \times 10^3) = 0.010 + j0.015 \, \text{mho} \tag{33}$$

and

$$Z(j2 \times 10^3) = \frac{1}{0.010 + j0.015} \, 30.8 - j46.2 \, \Omega \tag{34}$$

The parallel RLC circuit is equivalent to the series RC circuit in Figure 11.7b when $\omega = 2 \times 10^3 \, \text{rads/s}$ because the circuit is capacitive at this frequency. Note that $R = 30.8 \, \Omega$ and $46.2 \, \Omega = 1/[(2 \times 10^3 C)]$, so that $C = 10.8 \, \mu\text{F}$ in Figure 11.7b.

As a second example consider the circuit in Figure 11.8. Let us compute the driving-point function $\mathbf{I}_1/\mathbf{V}$ and the transfer function $\mathbf{V}_2/\mathbf{V}$. First let us use the mesh method. From Figure 11.8b

$$\left(R_1 + \frac{1}{j\omega C}\right)\mathbf{I}_a - \frac{1}{j\omega C}\mathbf{I}_b = \mathbf{V} \tag{35}$$

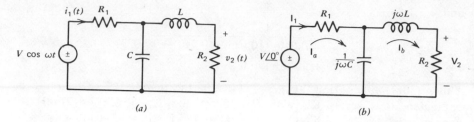

Figure 11.8   RLC circuit with a single sinusoidal input. (*a*) Circuit. (*b*) Phasor model.

$$-\frac{1}{j\omega C}\mathbf{I}_a + \left(R_2 + j\omega L + \frac{1}{j\omega C}\right)\mathbf{I}_b = 0 \tag{36}$$

so that

$$\mathbf{I}_1 = \mathbf{I}_a = \frac{\begin{vmatrix} \mathbf{V} & -\dfrac{1}{j\omega C} \\[2mm] 0 & R_2 + j\omega L + \dfrac{1}{j\omega C} \end{vmatrix}}{\left(R_1 + \dfrac{1}{j\omega C}\right)\left(R_2 + j\omega L + \dfrac{1}{j\omega C}\right) - \left(\dfrac{1}{j\omega C}\right)^2}$$

or

$$\mathbf{I}_1 = \frac{\mathbf{V}\left(R_2 + j\omega L + \dfrac{1}{j\omega C}\right)}{R_1\left(R_2 + j\omega L + \dfrac{1}{j\omega C}\right) + \dfrac{1}{j\omega C}(R_2 + j\omega L)} \tag{37}$$

The DP admittance seen by the sinusoidal generator is

$$Y_g(j\omega) = \frac{\mathbf{I}_1}{\mathbf{V}} = \frac{R_2 + j\omega L + \dfrac{1}{j\omega C}}{R_1\left(R_2 + j\omega L + \dfrac{1}{j\omega C}\right) + \dfrac{1}{j\omega C}(R_2 + j\omega L)} \tag{38}$$

and the impedance is

$$Z_g(j\omega) = \frac{\mathbf{V}}{\mathbf{I}_1} = R_1 + \frac{\dfrac{1}{j\omega C}(R_2 + j\omega L)}{R_2 + j\omega L + \dfrac{1}{j\omega C}} \tag{39}$$

In (38) and (39), $\mathbf{I}_1$, Y_g, and Z_g are functions of the frequency ω of the input as well as the circuit parameters R_1, R_2, L, and C.

Equations 38 and 39 could have easily been determined by means of series-parallel combinations. In Figure 11.8*b* we note that R_2 is in series with L, and this combination is in parallel with C. If impedances Z_1 and Z_2 are in parallel, then the impedance of the parallel combination is $Z_1 Z_2/(Z_1 +$

Z_2). Using this result, and the fact that R_1 is in series with this parallel combination, we write

$$Z_g(j\omega) = R_1 + \cfrac{\dfrac{1}{j\omega C}(R_2 + j\omega L)}{R_2 + j\omega L + \dfrac{1}{j\omega C}} \tag{40}$$

Note that (40) is in agreement with (39).

Next we compute $\mathbf{V}_2$ from (35) and (36).

$$\mathbf{V}_2 = \mathbf{I}_b R_2 = \cfrac{R_2 \begin{vmatrix} R_1 + \dfrac{1}{j\omega C} & \mathbf{V} \\[2mm] -\dfrac{1}{j\omega C} & 0 \end{vmatrix}}{R_1\left(R_2 + j\omega L + \dfrac{1}{j\omega C}\right) + \dfrac{1}{j\omega C}(R_2 + j\omega L)}$$

or

$$\mathbf{V}_2 = \cfrac{+\dfrac{R_2}{j\omega C}\mathbf{V}}{R_1\left(R_2 + j\omega L + \dfrac{1}{j\omega C}\right) + \dfrac{1}{j\omega C}(R_2 + j\omega L)} \tag{41}$$

The transfer function is

$$H(j\omega) = \frac{\mathbf{V}_2}{\mathbf{V}} = \cfrac{+R_2}{R_1 j\omega C\left(R_2 + j\omega L + \dfrac{1}{j\omega C}\right) + (R_2 + j\omega L)} \tag{42}$$

where $\mathbf{V}_2$ depends on the frequency ω of the input in addition to the circuit parameters R_1, R_2, L, and C.

Let us suppose that $R_1 = R_2 = 100\,\Omega$, $L = 10\,\text{mH}$, $C = 1\,\mu\text{F}$, and $\omega = 2 \times 10^4\,\text{rads/sec}$, then from (40)

$$Z_g(j2 \times 10^4) = 100 + \frac{(100 + j2 \times 10^4 \times 10^{-2})}{1 - 4 \times 10^8 \times 10^{-2} \times 10^{-6} + j100 \times 2 \times 10^4 \times 10^{-6}}$$

$$= 100 + \frac{100 + j200}{-3 + j2} \tag{43}$$

If we multiply the numerator and denominator of the last term by $-3 - j2$ we obtain

$$Z_g(j2 \times 10^4) = 100 + \frac{100 - j800}{13}$$

$$= 107.7 - j61.5 = 124\, e^{-j29.7°}\,\Omega \tag{44}$$

The resistance is 107.7 Ω and the reactance is -61.5Ω. Since the reactance is negative, the DP impedance is capacitive at the frequency 2×10^4 rads/s. The admittance is

$$Y_g(j2 \times 10^4) = \frac{1}{124e^{-j29.7°}} = 8.06e^{j29.7} \text{ mmho}$$

$$= 7 + j4 \text{ mmho} \tag{45}$$

so that the conductance is 0.007 mho and the suspectance is 0.004 mho. From (38) if $\mathbf{V} = 10\underline{/0°}$ V, then

$$I_1(j2 \times 10^4) = Y_g(j2 \times 10^4) \cdot \mathbf{V} = 80.6e^{j29.7} \text{ mA} \tag{46}$$

so that

$$i_1(t) = 80.6 \cos(2 \times 10^4 t + 29.7°) \text{ mA} \tag{47}$$

Note that the current leads the voltage since the DP impedance is capacitive.

Similarly, from (42) we obtain

$$H(j2 \times 10^4)$$

$$= \frac{100}{100(1 - 4 \times 10^8 \times 10^{-2} \times 10^{-6} + j2 \times 10^4 \times 10^{-6} \times 10^2) + 100 + j2 \times 10^4 \times 10^{-2}}$$

$$= \frac{1}{-2 + j4}$$

$$= -0.1 - j0.2 = 0.224e^{j243.4°} \tag{48}$$

Again, if $\mathbf{V} = 10\underline{/0°}$, then

$$\mathbf{V}_2 = 2.24e^{j243.4°} \tag{49}$$

and

$$v_2(t) = 2.24 \cos(2 \times 10^4 t + 243.4°) \tag{50}$$

Sinusoidal analysis can also aid in model simplification. Recall that the admittance of the capacitor is $j\omega C$, therefore

$$|\mathbf{I}_C| = \omega C |\mathbf{V}_C| \tag{51}$$

From (51) we conclude that when $\omega = 0$, $\mathbf{I}_C = 0$. *The capacitor is an open circuit for dc.* Also, when $\omega \to \infty$ we obtain $\mathbf{V}_C = 0$ for finite values of $|\mathbf{I}_C|$. Thus, *at sufficiently high frequencies the capacitor behaves as a short circuit.* A similar examination of the inductor reveals that the *inductor is a short circuit for dc signals, and an open circuit at sufficiently high frequencies.*

To better illustrate this point consider the circuit in Figure 11.9a. Using phasor circuit model in Figure 11.9b the magnitude of the voltage drops

$$|\mathbf{V}_R| = (1 \text{ k}\Omega)(10^{-3} \text{ A}) = 1 \text{ V} \tag{52}$$

and

$$|\mathbf{V}_L| = |j\omega L|(10^{-3} \text{ A}) = (10^3)(10^{-3})(10^{-3} \text{ A}) = 10^{-3} \text{ V} \tag{53}$$

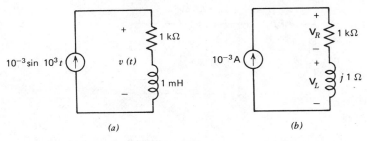

Figure 11.9   Sinusoidal analysis of series RL circuit. (*a*) Circuit. (*b*) Phasor model.

In this series circuit the impedance of the resistor is much greater than the impedance of the inductor $(R \gg |j\omega L|)$ so that in the *series connection shown* the inductor can be neglected without seriously affecting the analysis.

In Figure 11.10 note that

$$|\mathbf{I}_R| = \frac{1 \text{ V}}{10^3 \text{ }\Omega} = 10^{-3} \text{ A} \tag{54}$$

and

$$|\mathbf{I}_C| = \omega C |\mathbf{V}_C| = (10^3)(10^{-9})(1) = 10^{-6} \text{ A} \tag{55}$$

This time the current in the shunt capacitor could be neglected since the admittance of the resistor is much greater than the admittance of the capacitor $(C = 1/R \gg |j\omega C| \text{ or } R \ll 1/\omega C)$.

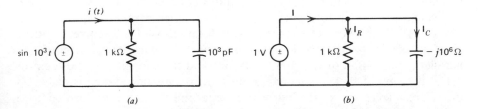

Figure 11.10   Sinusoidal analysis of shunt RC circuit. (*a*) Circuit. (*b*) Phasor model.

We conclude that frequently the analysis of a circuit can be greatly simplified if elements that have a negligible effect on the response are eliminated. Of course, if the model is insufficiently accurate, then the analytical results will not agree with physical observations. Good engineering judgment is required.

11.3.1 THÉVENIN AND NORTON EQUIVALENT CIRCUITS

We conclude this section with an example of Thévenin and Norton equivalent circuits. Let us compute the equivalent DP circuit seen by the

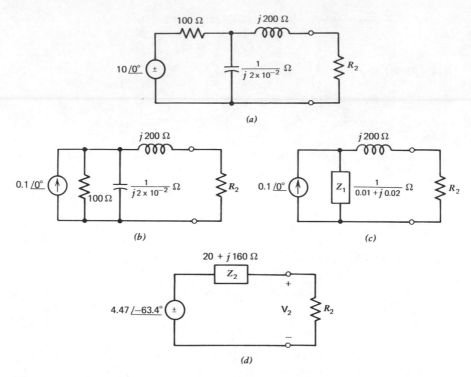

Figure 11.11   Reduction to the Thévenin equivalent.

resistor R_2 at the output port in Figure 11.11a, that is, the Thévenin equivalent circuit at this port. Let us use series-parallel reductions. In Figure 11.11b the voltage source is transformed to a 0.1-A current source at $0°$ phase in parallel with the 100-Ω resistor. The impedance of the parallel combination of this 100-Ω resistor and the capacitor is the reciprocal of the sum of their respective admittances. This impedance is called Z_1 in Figure 11.11c. Next the current source and Z_1 are transformed to a voltage source and the series combination of Z_1 and the inductor in Figure 11.11d is called Z_2. Note that the impedance at the output port is inductive. From Figure 11.11d we see that the equivalent impedance seen by the resistor R_2 is $Z_2 = 20 + j160 \, \Omega$, and the magnitude and phase of the open circuit voltage is $4.47e^{-j63.4°}$. From the voltage divider transfer function we can compute the output voltage for $R_2 = 100 \, \Omega$.

$$V_2 = \frac{R_2}{R_2 + Z_2} \, 4.47\underline{/-63.4°}$$

$$= \frac{100}{120 + j160} \, 4.47\underline{/-63.4°} = 2.24\underline{/-116.6°} \tag{56}$$

The circuit in Figure 11.11 is the same as the circuit in Figure 11.8 and the parameter values are the same as the second example in this section. Note further that (56) yields the same answer as our mesh analysis (49).

Finally, as a check, let us compute the Thévenin equivalent circuit in Figure 11.11 by the method of the elimination of variables. In Figure 11.12 the mesh equations are

$$(100 - j50)\mathbf{I}_a + j50\mathbf{I}_b = 10 \tag{57}$$

$$j50\mathbf{I}_a + j150\mathbf{I}_b = -\mathbf{V}_o \tag{58}$$

We are only interested in the relation between the current $\mathbf{I}_o = -\mathbf{I}_b$ and the voltage $\mathbf{V}_o$ at the output port. Thus, solve (57) and (58) for $\mathbf{I}_b$.

$$\mathbf{I}_b = \frac{\begin{vmatrix} 100 - j50 & 10 \\ j50 & -\mathbf{V}_o \end{vmatrix}}{(100 - j50)(j150) - (j50)^2} \tag{59}$$

Therefore,

$$\mathbf{I}_o = -\mathbf{I}_b = \frac{\mathbf{V}_o(100 - j50) + j500}{10,000 + j15,000} \tag{60}$$

and on simplification (60) becomes

$$\mathbf{V}_o = (20 + j160)\mathbf{I}_o + 2 - j4 \tag{61}$$

or

$$\mathbf{V}_o = (20 + j160)\mathbf{I}_o + 4.47\underline{/-63.4°} \tag{62}$$

From (62) we see that the impedance of the equivalent circuit is $20 + j160\ \Omega$ in series with a voltage source of magnitude 4.47 V and a phase of $-63.4°$.

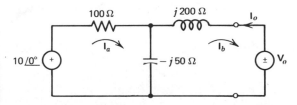

Figure 11.12   Circuit model for computing the Thévenin equivalent circuit.

11.4 AVERAGE AND COMPLEX POWER

Figure 11.13 illustrates two circuits connected by a pair of terminals. It is assumed that the voltage and current associated with this pair of terminals is sinusoidal with the same frequency but not necessarily the same phase. The power is

$$p(t) = VI\cos(\omega t + \beta)\cos(\omega t + \alpha) \tag{63}$$

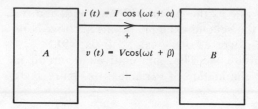

Figure 11.13   Calculation of power.

If $p(t) > 0$ in any finite time interval, then according to our standard reference system circuit B is absorbing energy in that interval and circuit A is supplying energy, and vice versa if $p(t) < 0$. Now since $p(t)$ is periodic usually the average power is of more interest. From Chapter Three recall that

$$P_{avg} = \frac{1}{T} \int_0^T p(t)\,dt \tag{64}$$

where T is the period of the signals. Upon substitution of (63) into (64) we obtain

$$P_{avg} = \frac{1}{2} VI \cos(\beta - \alpha) \tag{65}$$

or

$$P_{avg} = V_{rms} I_{rms} \cos(\beta - \alpha) \tag{66}$$

The quantity $\cos(\beta - \alpha)$ is called the *power factor* (PF). A power factor of 1 means that the voltage and current are in phase and the circuit is resistive.

Let us now express the power in terms of the phasors $Ve^{j\beta}$ and $Ie^{j\alpha}$. We define a *complex power*

$$\mathbf{P} = \frac{1}{2} \mathbf{V}\,\mathbf{I}^* = \mathbf{V}_{rms}\,\mathbf{I}^*_{rms} \tag{67}$$

which can be written as

$$\mathbf{P} = \frac{1}{2} VI\, e^{j(\beta - \alpha)} \tag{68}$$

or

$$\mathbf{P} = \frac{1}{2} VI[\cos(\beta - \alpha) + j \sin(\beta - \alpha)] \tag{69}$$

which, in terms of rms values instead of peak values, becomes

$$\mathbf{P} = V_{rms} I_{rms}[\cos(\beta - \alpha) + j \sin(\beta - \alpha)] \tag{70}$$

Obviously the complex power is a complex number

$$\mathbf{P} = P_{avg} + jQ \tag{71}$$

and the magnitude of the complex power

$$|\mathbf{P}| = \frac{1}{2}|\mathbf{V}||\mathbf{I}| = [(P_{avg})^2 + Q^2]^{1/2} \tag{72}$$

is called the *apparent power* and has units of volt-amperes. The real part of
$\mathbf{P}$ is the average power dissipated ($P_{avg} > 0$) or supplied ($P_{avg} < 0$), that is,

$$P_{avg} = \frac{1}{2} VI \cos(\beta - \alpha) = V_{rms} I_{rms} \cos(\beta - \alpha) \text{ W} \tag{73a}$$

The imaginary part of (71) is called the *reactive power* and is measured in
units of vars (volt-amperes reactive):

$$Q = \frac{1}{2} VI \sin(\beta - \alpha) \dot{=} V_{rms} I_{rms} \sin(\beta - \alpha) \text{ vars} \tag{73b}$$

In power transmission systems the reactive power is monitored, particularly at industrial plants that have large inductive machines. If the
power factor is too low the electrical rate is increased. The reason for this
is demonstrated with the simplified example in Figure 11.14. First let us
assume that a customer has a load Z_L, which requires 22 kW at 220 V rms.
Let us further assume with the power factor of the load is unity so that $\mathbf{I}_L$
is in phase with $\mathbf{V}_L$ and $|\mathbf{I}_L| = 100$ A rms. Suppose that the resistance of the
power lines is 0.5 Ω, then the power company must generate 22 kW +
(0.5 Ω)(100 A)^2 = 27 kW to supply the customer with 22 kW.

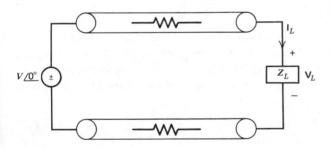

Figure 11.14 Transmission system.

Now let us consider the same example, but suppose that the current $\mathbf{I}_L$
lags the voltage $\mathbf{V}_L$ by 60° so that the power factor is 0.5. Thus

$$\mathbf{P}_L = (\mathbf{V}_L)(\mathbf{I}_L)^* = |\mathbf{V}_L||\mathbf{I}_L|e^{j60°}$$

The real power is

$$22 \text{ kW} = 220 \text{ V } |\mathbf{I}_L|_{rms} \cos 60° \tag{74}$$

so that

$$|\mathbf{I}_L|_{rms} = 200 \text{ A} \tag{75}$$

The reactive power

$$Q = (220 \text{ V})(200 \text{ A}) \sin 60° = 38.1 \text{ kvars}$$

and the apparent power

$$\sqrt{P^2 + Q^2} = |\mathbf{V}_L||\mathbf{I}_L| = 44 \text{ kVA}$$

The total real power dissipated by the load and the line is

$$22 \text{ kW} + (0.5 \text{ }\Omega)(200)^2 = 42 \text{ kW} \tag{76}$$

In the first case the customer is paying for 81.5% of the power generated by the power company, but in the second case the customer is paying for only 52.4% of the power generated by the power company. Thus, when a power company enters into an agreement with an industrial firm to supply it with electrical power, the agreement typically contains a clause that if the power factor is less than a certain value, typically 0.85 lagging, then the electrical rate is increased in a prescribed manner. If the industrial plant has inductive machinery it is common practice to shunt the power lines with banks of capacitors to increase the lagging power factor and reduce the electrical rate.

11.4.1 MULTIPLE FREQUENCY INPUTS

Let us return to Figure 11.13, but now suppose that this circuit contains two independent sources that are both sinusoidal, one with frequency ω_1 and the other with frequency ω_2. If the circuit is linear and time invariant, then the steady state current and voltage indicated in Figure 11.13 will have the form

$$i(t) = I_1 \cos(\omega_1 t + \alpha_1) + I_2 \cos(\omega_2 t + \alpha_2) \tag{77}$$

and

$$v(t) = V_1 \cos(\omega_1 t + \beta_1) + V_2 \cos(\omega_2 t + \beta_2) \tag{78}$$

In this case, the power is

$$
\begin{aligned}
p(t) = \ & V_1 I_1 \cos(\omega_1 t + \beta_1) \cos(\omega_1 t + \alpha_1) \\
& + V_2 I_2 \cos(\omega_2 t + \beta_2) \cos(\omega_2 t + \alpha_2) \\
& + V_1 I_2 \cos(\omega_1 t + \beta_1) \cos(\omega_2 t + \alpha_2) \\
& + V_2 I_1 \cos(\omega_2 t + \beta_2) \cos(\omega_1 t + \alpha_1)
\end{aligned}
\tag{79}
$$

or

$$p(t) = \frac{1}{2} V_1 I_1 [\cos(\beta_1 - \alpha_1) + \cos(2\omega_1 t + \beta_1 + \alpha_1)]$$

$$+ \frac{1}{2} V_2 I_2 [\cos(\beta_2 - \alpha_2) + \cos(2\omega_2 t + \beta_2 + \alpha_2)]$$

$$+ \frac{1}{2} V_1 I_2 \{\cos[(\omega_1 - \omega_2)t + \beta_1 - \alpha_2]$$

$$+ \cos[(\omega_1 + \omega_2)t + \beta_1 + \alpha_2]\} \tag{80}$$

$$+ \frac{1}{2} V_2 I_1 \{\cos[(\omega_2 - \omega_1)t + \beta_2 - \alpha_1]$$

$$+ \cos[(\omega_2 - \omega_1)t + \beta_2 + \alpha_1]\}$$

In order to compute the average value of $p(t)$ we need to know the period T of $p(t)$. Unfortunately $p(t)$ is periodic if and only if the frequencies f_1 and f_2 are rationally related, that is,

$$\frac{f_1}{f_2} = \frac{m}{n} \tag{81}$$

where m and n are positive integers. If the fraction m/n is irreducible, then the period

$$T = \frac{n}{f_2} = \frac{m}{f_1} \tag{82}$$

In other words, every T seconds both signals with frequency ω_1 and signals with frequency ω_2 simultaneously have returned to the value they had T seconds previously. However, if the frequencies are not rationally related, for example, $\omega_1 = 1$ and $\omega_2 = \pi$, then the waveforms are *nonperiodic*. To avoid this difficulty let us define

$$P_{avg} = \lim_{T \to \infty} \frac{1}{T} \int_0^T p(t)\, dt \tag{83}$$

Upon applications of (83) to (80) we obtain

$$P_{avg} = \frac{1}{2} [V_1 I_1 \cos(\beta_1 - \alpha_1) + V_2 I_2 \cos(\beta_2 - \alpha_2)] \tag{84}$$

provided that $\omega_2 \neq \omega_1$. If $\omega_1 = \omega_2$, then

$$P_{avg} = \frac{1}{2} [V_1 I_1 \cos(\beta_1 - \alpha_1) + V_2 I_2 \cos(\beta_2 - \alpha_2)$$

$$+ V_2 I_1 \cos(\beta_2 - \alpha_1) + V_1 I_2 \cos(\beta_1 - \alpha_2)] \tag{85}$$

In this case the phasors I_1 and I_2, and the phasors V_1 and V_2 should each be combined into a single current phasor and a single voltage phasor so that there is no need to use (85) when the frequencies are identical.

11.5 MAXIMUM POWER TRANSFER

In many applications we would like to have a sinusoidal source deliver the maximum possible power to a circuit in the sinusoidal steady state. Let us assume that the circuit can be modeled by linear constant elements. In Figure 11.15a we illustrate a sinusoidal source connected to a pair of

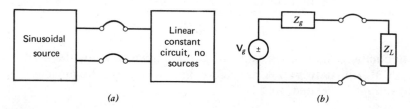

Figure 11.15   Maximum power transfer.

terminals of a circuit. The phasor model is given in Figure 11.15b. The impedances Z_g and Z_L denote the Thévenin equivalent impedances of the source and the circuit respectively at the frequency ω, the frequency of the sinusoidal source. In this circuit the voltage drop across the impedance Z_L is

$$\mathbf{V}_L = \frac{Z_L}{Z_L + Z_g} \mathbf{V}_g \tag{86}$$

and the current through the load is

$$\mathbf{I}_L = \frac{\mathbf{V}_g}{Z_L + Z_g} \tag{87}$$

Thus, the complex power delivered to the load is

$$\mathbf{P}_L = \frac{1}{2} \mathbf{V}_L \mathbf{I}_L^* = \frac{1}{2} \left(\frac{Z_L}{Z_L + Z_g} \mathbf{V}_g \right) \left(\frac{\mathbf{V}_g}{Z_L + Z_g} \right)^* \tag{88}$$

or

$$\mathbf{P}_L = \frac{1}{2} \frac{Z_L}{|Z_L + Z_g|^2} |\mathbf{V}_g|^2 \tag{89}$$

If we express the impedances in terms of their real and imaginary parts, that is, $Z_L = R_L + jX_L$ and $Z_g = R_g + jX_g$, then (89) becomes

$$\mathbf{P}_L = \frac{1}{2} \frac{Z_L}{(R_L + R_g)^2 + (X_L + X_g)^2} |\mathbf{V}_g|^2 \tag{90}$$

and the real power is

$$P_{avg} = \frac{1}{2} \frac{R_L}{(R_L + R_g)^2 + (X_L + X_g)^2} |V_g|^2 \tag{91}$$

11.5.1 MAXIMUM POWER THEOREM

Given a sinusoidal source with impedance Z_g the source delivers maximum real power to a circuit if the Thévenin equivalent impedance of the circuit $Z_L = Z_g^*$. It is assumed that $R_L > 0$ and $R_g > 0$.

To prove this theorem, first note that in order to maximize (91) we should design the circuit so that $X_L = -X_g$, then

$$P_{avg} = \frac{1}{2} \frac{R_L}{(R_L + R_g)^2} |V_g|^2 \tag{92}$$

To find the value of R_L which maximizes (92) we solve the equation

$$\frac{\partial P_{avg}}{\partial R_L} = \left[\frac{1}{(R_L + R_g)^2} - \frac{2R_L}{(R_L + R_g)^3} \right] \frac{|V_g|^2}{2} = 0 \tag{93}$$

The solution of (93) is

$$R_L + R_g - 2R_L = 0$$

which yields

$$R_g = R_L \tag{94}$$

It can be verified that (94) is a maxima of (92), so that when $Z_L = Z_g^*$, maximum power is delivered to the load, and

$$P_{max} = \frac{1}{2} |V_g|^2 \frac{1}{4R_g} \tag{95}$$

is called the maximum available power from the source. When $Z_L = Z_g^*$ we say that the load is *matched* to the source.

11.6 SINGLE FREQUENCY IMPEDANCE MATCHING

Frequently the load and source impedances are not matched. In such cases a lossless matching two-port can be inserted between the load and source as shown in Figure 11.16. Ideally the matching circuit does not dissipate any energy (lossless) but does change impedance levels to achieve a match. The transformer is a two-port device that can be used to match impedances and will be discussed in Chapter Thirteen. In this section we discuss some simple LC two-ports matching circuits that can be designed to *match a load to a source at a single input frequency*.

In single frequency matching the lossless two-port in Figure 11.16 must transform the impedance Z_L to Z_g^* at a given frequency. This implies that

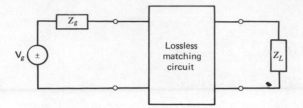

Figure 11.16   Impedance matching circuit.

we need two independent parameters. Consider the simple circuit in Figure 11.17. The DP admittance at terminals ⓐ–ⓑ is

$$Y_{ab} = jB + \frac{1}{R_2 + jX} = jB + \frac{R_2 - jX}{R_2^2 + X^2} \tag{96}$$

In order to obtain an impedance match we require that $Y_{ab} = 1/R_1$ so that in (96) we require that

$$B = \frac{X}{R_2^2 + X^2} \tag{97}$$

and

$$R_1 = \frac{R_2^2 + X^2}{R_2} = R_2 + \frac{X^2}{R_2} \tag{98}$$

Clearly from (98) we cannot obtain an impedance match unless $R_1 > R_2$.

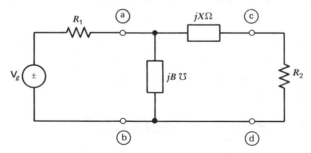

Figure 11.17   Single-frequency matching circuit, $R_1 > R_2$.

Next consider the matching circuit in Figure 11.18. The DP impedance at terminals ⓐ–ⓑ is

$$Z_{ab} = jX + \frac{1}{G_2 + jB} = jX + \frac{G_2 - jB}{G_2^2 + B^2} \tag{99}$$

Now to obtain a match we require that

$$X = \frac{B}{G_2^2 + B^2} \tag{100}$$

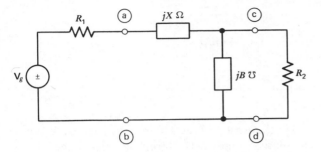

Figure 11.18   Single-frequency matching circuit, $R_2 > R_1$.

and

$$R_1 = \frac{G_2}{G_2^2 + B^2} = \frac{R_2}{1 + (BR_2)^2} \tag{101}$$

From (101) a match is only possible if $R_1 < R_2$. Let us apply these results to a practical problem.

Example

Let us assume that we have a 100-Ω antenna feeding a 100-Ω transmission line so that in the Thévenin equivalent circuit $R_1 = 100\ \Omega$. Suppose that our load $R_2 = 50\ \Omega$. Therefore, we must use the circuit in Figure 11.17. Furthermore, assume that we must match R_1 to R_2 at the frequency $\omega = 10^6$ rads/s. In (97) note that the susceptance B and the reactance X must have the same sign, so that we should choose the shunt arm to be a capacitor and the series arm to be an inductor, or vice versa. Let us choose the former as shown in Figure 11.19.

From (98) we obtain

$$100 = 50 + \frac{(\omega L)^2}{50}$$

Since $\omega = 10^6$, then

$$L = \frac{50}{10^6} = 50\ \mu\text{H}$$

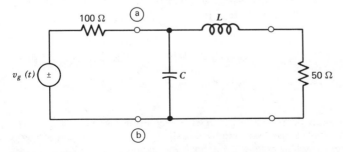

Figure 11.19   Single-frequency impedance matching example.

From (97) we obtain

$$10^6 C = \frac{10^6 L}{(50)^2 + (10^6 L)^2}$$

so that since $L = 50\ \mu H$,

$$C = \frac{50 \times 10^{-6}}{(50)^2 + (10^6 \times 50 \times 10^{-6})^2}$$

$$C = 0.01\ \mu F$$

Note that at terminals ⓐ–ⓑ in Figure 11.19 the DP impedance is now 100 Ω so that maximum power is delivered to the port ⓐ–ⓑ. However, since both L and C cannot dissipate any power, this means that the maximum available power from the sinusoidal source is delivered to the 50-Ω load in the sinusoidal steady state.

11.7 THREE-PHASE SYSTEMS

One of the most basic electrical systems in all countries is the electrical power system for the generation and transmission of electric energy to consumers. The heart of the electrical power system is the electric generator, which is driven by a steam turbine or a hydroturbine. In the case of steam turbines the primary source of energy is nuclear energy or the fossil fuels such as coal or oil. The hydroturbine is of course driven by water power. In the early stages of development most of the generators were dc, but the dc generators were quickly replaced by a three-phase (3-ϕ) alternating current synchronous generator. The 3-ϕ synchronous generator consists of three coils 120° out of phase and the voltages generated are almost sinusoidal with a frequency of 60 Hz in the United States, but 50 Hz in many other countries. The ac waveform is used because voltage and current levels can easily be changed by means of transformers which allow for very efficient transmission of power.[1] Also, the 3-ϕ generator is used rather than the single-phase generator because the 3-ϕ transmission is more efficient than single phase for a given amount of conductor. In this section we discuss some of the basic concepts for the analysis and operation of three-phase power systems.

The three coils of the synchronous generator are connected in a Y arrangement as shown in Figure 11.20, the three voltages

$$v_{an} = V_p \sin \omega t$$
$$v_{bn} = V_p \sin(\omega t - 120°) \tag{102}$$
$$v_{cn} = V_p \sin(\omega t - 240°)$$

Because of the corona effect (ionization of the surrounding air) in high voltage ac power transmission, there is serious discussion of returning to dc signals for the transmission of energy at extremely high voltages over long distances. Energy is transmitted at high voltages because the higher the voltage the smaller the current is for a given power, and a small current in the conductor means a small $i^2 R$ loss in the transmission line.

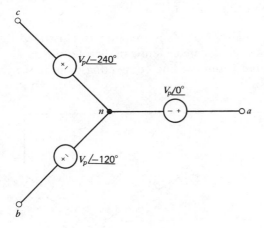

Figure 11.20 Three phase Y connection.

are called the *phase voltages* and we say that the *phase sequence is abc* and the reference voltage is v_{an} since its phase angle is zero degrees. The phase sequences associated with Figure 11.20 is also called a *positive phase sequence.* The voltages

$$v_{ab} = \sqrt{3}\, V_p\, \sin(\omega t + 30°)$$
$$v_{bc} = \sqrt{3}\, V_p\, \sin(\omega t - 90°)$$
$$v_{ca} = \sqrt{3}\, V_p\, \sin(\omega t - 210°)$$

(103)

are called the *line voltages* and are computed from the phasor diagram in Figure 11.21.

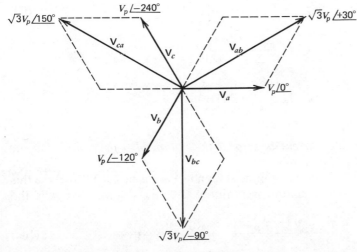

Figure 11.21 Phase voltages and line voltages.

11.7.1 THE Y-Y CONNECTION

In Figure 11.22 the *four-wire balanced Y-Y connection* is illustrated. The load impedances in each arm are assumed identical. The terminals n and N are called the *neutral terminals* and the line connecting them is the

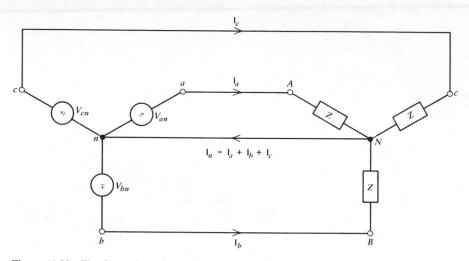

Figure 11.22 The four-wire balanced Y-Y connection.

neutral line. Note that if we assume a positive phase sequence, then

$$\mathbf{I}_a = \frac{V_p}{Z} = \frac{V_p}{|Z|\, e^{j\alpha}}$$

$$\mathbf{I}_b = \frac{V_p\, e^{-j120°}}{Z} = \frac{V_p}{|Z|}\, e^{-j(120°-\alpha)} \qquad (104)$$

$$\mathbf{I}_c = \frac{V_p\, e^{-j240°}}{Z} = \frac{V_p}{|Z|}\, e^{-j(240°-\alpha)}$$

so that

$$\mathbf{I}_a + \mathbf{I}_b + \mathbf{I}_c = \frac{V_p}{|Z|}\, e^{-j\alpha}(1 + e^{-j120°} + e^{-j240°})$$

$$= 0 \qquad (105)$$

Since the neutral line current is zero in the balanced Y-Y connection the neutral line can be eliminated.

The *Three-wire Y-Y* connection is shown in Figure 11.23. To analyze this circuit only one node equation is required. Let us assume that n is the datum node, then

$$\frac{\mathbf{E}_N - \mathbf{V}_{an}}{Z_a} + \frac{\mathbf{E}_N - \mathbf{V}_{bn}}{Z_b} + \frac{\mathbf{E}_N - \mathbf{V}_{cn}}{Z_c} = 0 \qquad (106)$$

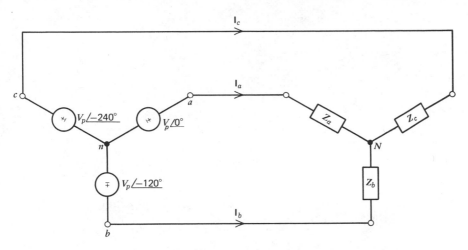

Figure 11.23   The three-wire unbalanced Y-Y connection.

where the phase voltages (also node voltages) $\mathbf{V}_{an}$, $\mathbf{V}_{bn}$, $\mathbf{V}_{cn}$ are known. We have assumed that the generator and line impedance is zero. If the load impedances are not identical then we say that the system is *unbalanced*.

Example

The unbalanced load in Figure 11.24 is supplied by a balanced, three-phase, three-wire source with *abc* phase sequence. The line-to-line voltage is 200 V rms. Find the amplitude and phase of the line currents using v_{ab} as the zero phase reference.

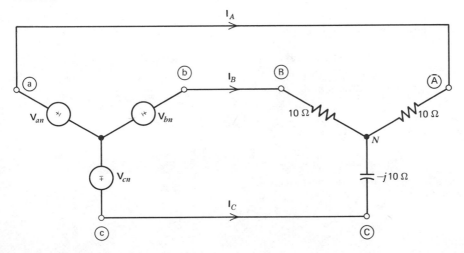

Figure 11.24   Unbalanced load example.

From (102) and (103) we obtain

$$\mathbf{V}_{an} = \frac{200}{\sqrt{3}} \underline{/-30°} \text{ V rms}$$

$$\mathbf{V}_{bn} = \frac{200}{\sqrt{3}} \underline{/-150°} \text{ V rms}$$

$$\mathbf{V}_{cn} = \frac{200}{\sqrt{3}} \underline{/90°} \text{ V rms}$$

Note that the phasor diagram has been rotated by $-30°$ since v_{ab} is the reference voltage. From (106)

$$\frac{\left(\mathbf{E}_N - \frac{200}{\sqrt{3}} e^{-j30°}\right)}{10} + \frac{\left(\mathbf{E}_N - \frac{200}{\sqrt{3}} e^{-j150°}\right)}{10} + \frac{\left(\mathbf{E}_N - \frac{200}{\sqrt{3}} e^{j90°}\right)}{-j10} = 0$$

or

$$(2+j)\mathbf{E}_N = \frac{200}{\sqrt{3}} (e^{-j30°} + e^{-j150°} + j e^{j90°})$$

Therefore, the node voltage

$$\mathbf{E}_N = \frac{200}{\sqrt{3}(2+j)} (-1-j) = 73.0 \, e^{-j161.6°} \text{ V rms}$$

Now we know all of the node voltages in the system and can proceed to calculate the line currents

$$\mathbf{I}_A = \frac{\mathbf{V}_{an} - \mathbf{E}_N}{10} = \left(\frac{200}{\sqrt{3}} e^{-j30°} - 73.0 \, e^{-j161.6°}\right)/10$$

The above expression reduces to

$$\mathbf{I}_A = 16.93 - j3.46 = 17.3 \underline{/-11.6°} \text{ A rms}$$

Next we compute the line current

$$\mathbf{I}_B = \frac{\mathbf{V}_{bn} - \mathbf{E}_N}{10} = \left(\frac{200}{\sqrt{3}} e^{-j150°} - 73.0 \, e^{-j161.6°}\right)/10$$

$$= -3.07 - j3.47 = 4.63 \underline{/228.5°} \text{ A rms}$$

Finally, we compute the current in line c.

$$\mathbf{I}_C = \frac{\mathbf{V}_{cn} - \mathbf{E}_N}{-j10} = \frac{20}{\sqrt{3}} e^{j180°} + 7.3 \, e^{-j251.6°}$$

or

$$\mathbf{I}_C = -13.85 + j6.93 = 15.49 \underline{/153.4°}$$

11.7.2 THE Y-Δ CONNECTION

Another way in which loads can be connected is the Y-Δ connection shown in Figure 11.25. Since we know the line voltages from (103), then

$$I_1 = \frac{V_{ab}}{Z_1}$$

$$I_2 = \frac{V_{bc}}{Z_2} \tag{107}$$

$$I_3 = \frac{V_{ca}}{Z_3}$$

and

$$I_a = I_1 - I_3$$
$$I_b = I_2 - I_1 \tag{108}$$
$$I_c = I_3 - I_2$$

The balanced case in which $Z_1 = Z_2 = Z_3$ is of particular interest. In this case I_1, I_2, and I_3 have the same magnitudes but lag each other respectively by 120°. In this case (108) becomes

$$I_a = \frac{\sqrt{3}V_p}{|Z|\,e^{j\alpha}}(e^{j30°} - e^{-j210°}) = \frac{3V_p}{|Z|}\,e^{-j\alpha}$$

$$I_b = \frac{\sqrt{3}V_p}{|Z|\,e^{j\alpha}}(e^{-j90°} - e^{j30°}) = \frac{3V_p}{|Z|}\,e^{-j(120°+\alpha)} \tag{109}$$

$$I_c = \frac{\sqrt{3}V_p}{|Z|\,e^{j\alpha}}(e^{-j210°} - e^{-j90°}) = \frac{3V_p}{|Z|}\,e^{-j(240°+\alpha)}$$

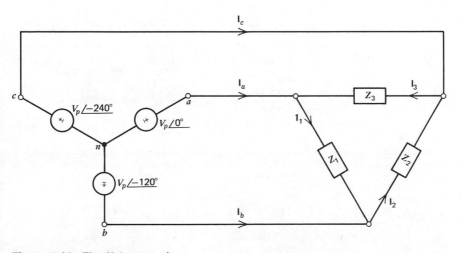

Figure 11.25   The Y-Δ connection.

Note that the line currents are 120° out of phase with each other in the balanced circuit.

Table 11.1 summarizes the results for both the balanced Y-Y system and the balanced Y-Δ system. The first two rows denote the source and line voltages that are *independent of whether or not the circuit is balanced*. However, the line and load current results assume a balanced load. The last row presents the startling result that in both balanced circuits the *instantaneous power* delivered by the synchronous generator is *constant* and so the average power is the instantaneous power. The quantities V_ℓ and I_ℓ denote the *magnitude* of the line voltage and line current, respectively.

Table 11.1 Balanced Systems

Quantity Measured	Y Load	Δ Load
Source voltage	$V_p \sin \omega t$ $V_p \sin(\omega t - 120°)$ $V_p \sin(\omega t - 240°)$	(same)
Line voltage	$\sqrt{3}\, V_p \sin(\omega t + 30°)$ $\sqrt{3}\, V_p \sin(\omega t - 90°)$ $\sqrt{3}\, V_p \sin(\omega t - 210°)$	(same)
Line current	$\dfrac{V_p}{\|Z\|} \sin(\omega t - \alpha)$ $\dfrac{V_p}{\|Z\|} \sin(\omega t - 120° - \alpha)$ $\dfrac{V_p}{\|Z\|} \sin(\omega t - 240° - \alpha)$	$\dfrac{3V_p}{\|Z\|} \sin(\omega t - \alpha)$ $\dfrac{3V_p}{\|Z\|} \sin(\omega t - 120° - \alpha)$ $\dfrac{3V_p}{\|Z\|} \sin(\omega t - 240° - \alpha)$
Load currents	Equal to line currents above	$\dfrac{\sqrt{3}\, V_p}{\|Z\|} \sin(\omega t + 30° - \alpha)$ $\dfrac{\sqrt{3}\, V_p}{\|Z\|} \sin(\omega t - 90° - \alpha)$ $\dfrac{\sqrt{3}\, V_p}{\|Z\|} \sin(\omega t - 210° - \alpha)$
Instantaneous power $p(t)$	$\dfrac{3V_p{}^2}{2\|Z\|} \cos \alpha$ $= \dfrac{3}{2} V_p I_\ell \cos \alpha$ $= \dfrac{\sqrt{3}}{2} V_\ell I_\ell \cos \alpha$	$\dfrac{9V_p{}^2}{2\|Z\|} \cos \alpha$ $= \dfrac{3}{2} V_p I_\ell \cos \alpha$ $= \dfrac{\sqrt{3}}{2} V_\ell I_\ell \cos \alpha$

Next we prove that the instantaneous power is constant in the balanced three-phase system.

The instantaneous power delivered by the generator in either balanced circuit is

$$p(t) = V_p I_\ell [\sin \omega t \sin(\omega t - \alpha)$$
$$+ \sin(\omega t - 120°) \sin(\omega t - 120° - \alpha) \tag{110}$$
$$+ \sin(\omega t - 240°) \sin(\omega t - 240° - \alpha)]$$

where $I_\ell = V_p/|Z|$ in the balanced Y-Y circuit, and $I_\ell = 3V_p/|Z|$ in the balanced Y-Δ circuit. Using the trigonometric identity $\sin x \sin y = 1/2[\cos(x - y) - \cos(x + y)]$, we can write (110) as

$$p(t) = \frac{1}{2} V_p I_\ell [\cos \alpha - \cos(2\omega t - \alpha) + \cos \alpha - \cos(2\omega t - 240° - \alpha)$$
$$+ \cos \alpha - \cos(2\omega t - 120° - \alpha)] \tag{111}$$

But

$$\cos(2\omega t - \alpha) + \cos(2\omega t - 120° - \alpha) + \cos(2\omega t - 240° - \alpha) \equiv 0 \tag{112}$$

since the sum of three sinusoids of the same frequency but 120° out of phase with each other is zero for all t [see (105)]. Thus, the total instantaneous power delivered by the three-phase generator to a balanced load is

$$p(t) = \frac{3}{2} V_p I_\ell \cos \alpha \tag{113}$$

If rms values are used for the voltage and current, then the 1/2 factor can be deleted from (113). Since the total instantaneous power delivered by the 3-ϕ generator is constant this means that the 3-ϕ generator has less vibration than the single-phase generator. Also since power can be delivered to three loads by means of only three wires instead of six, the transmission is more efficient since the conductor loss is less.

11.8 THE Y-Δ TRANSFORMATION

The Y load (sometimes referred to as the *tee* load) can always be transformed to an equivalent Δ load or vice versa. Consider the circuits in Figure 11.26. We wish to find the relationship between Z_a, Z_b, Z_c of the Y circuit and Z_1, Z_2, Z_3 of the Δ circuit such that $\mathbf{I}_a = \hat{\mathbf{I}}_a$, $\mathbf{I}_b = \hat{\mathbf{I}}_b$, $\mathbf{V}_{ab} = \hat{\mathbf{V}}_{ab}$ and $\mathbf{V}_{ca} = \hat{\mathbf{V}}_{ca}$. Note that we need not worry about $\mathbf{I}_c$ since $\mathbf{I}_c + \mathbf{I}_a + \mathbf{I}_b = 0$, also $\mathbf{V}_{cb} = \mathbf{V}_{ca} + \mathbf{V}_{cb}$. We use the substitution property to generate the circuits in Figure 11.27. It is assumed that the circuits are equivalent so that the line voltages in Figure 11.27a and 11.27b are equal. The mesh equations for the Y network are

$$\begin{bmatrix} \mathbf{V}_{ab} \\ \mathbf{V}_{ca} \end{bmatrix} = \begin{bmatrix} Z_a + Z_b & -Z_a \\ -Z_a & Z_a + Z_c \end{bmatrix} \begin{bmatrix} \mathbf{I}_a \\ \mathbf{I}_c \end{bmatrix} \tag{114}$$

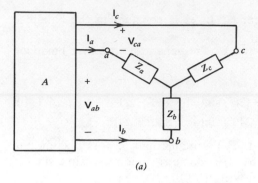

(a)

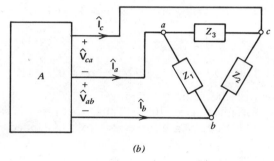

(b)

Figure 11.26 The Y and Δ connection. (a) Y. (b) Δ.

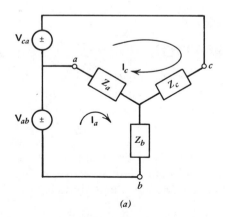

(a)

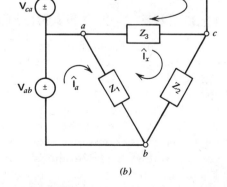

(b)

Figure 11.27 Application of substitution.

and the mesh equations for the Δ network are

$$\begin{bmatrix} \mathbf{V}_{ab} \\ \mathbf{V}_{ca} \\ 0 \end{bmatrix} = \begin{bmatrix} Z_1 & 0 & -Z_1 \\ 0 & Z_3 & -Z_3 \\ -Z_1 & -Z_3 & Z_\Sigma \end{bmatrix} \begin{bmatrix} \hat{\mathbf{I}}_a \\ \hat{\mathbf{I}}_c \\ \hat{\mathbf{I}}_x \end{bmatrix} \tag{115}$$

where $Z_\Sigma = Z_1 + Z_2 + Z_3$. In the first two equations of (115), we need to eliminate $\hat{\mathbf{I}}_x$ so that we can equate these two equations to (114) and establish the equivalency. Therefore, multiply the last equation of (115) by Z_1/Z_Σ and add it to the first equation; also multiply the last equation of (115) by Z_3/Z_Σ and add it to the second equation. Equation 115 becomes

$$\begin{bmatrix} \mathbf{V}_{ab} \\ \mathbf{V}_{ca} \\ 0 \end{bmatrix} = \begin{bmatrix} Z_1 - Z_1^2/Z_\Sigma & -Z_3 Z_1/Z_\Sigma & 0 \\ -Z_1 Z_3/Z_\Sigma & Z_3 - Z_3^2/Z_\Sigma & 0 \\ -Z_1 & -Z_3 & Z_\Sigma \end{bmatrix} \begin{bmatrix} \hat{\mathbf{I}}_a \\ \hat{\mathbf{I}}_c \\ \hat{\mathbf{I}}_x \end{bmatrix} \tag{116}$$

Now that we have eliminated the internal mesh current $\hat{\mathbf{I}}_x$ we see that Figure 11.26a and b are equivalent if the following equality is satisfied.

$$\begin{bmatrix} Z_a + Z_b & -Z_a \\ -Z_a & Z_a + Z_c \end{bmatrix} = \begin{bmatrix} Z_1 - Z_1^2/Z_\Sigma & -Z_3 Z_1/Z_\Sigma \\ -Z_1 Z_3/Z_\Sigma & Z_3 - Z_3^2/Z_\Sigma \end{bmatrix} \tag{117}$$

The above equation yields

$$Z_a = \frac{Z_1 Z_3}{Z_1 + Z_2 + Z_3} \tag{118}$$

$$Z_a + Z_b = Z_1 - Z_1^2/(Z_1 + Z_2 + Z_3)$$

but from (118) we obtain

$$Z_b = \frac{Z_1 Z_2}{Z_1 + Z_2 + Z_3} \tag{119}$$

and

$$Z_c = \frac{Z_2 Z_3}{Z_1 + Z_2 + Z_3} \tag{120}$$

Formulas 118 to 120 can be remembered by noting that the equivalent Y impedance that connects a given node is equal to the product of the two Δ impedances that connect to the corresponding node (e.g., Z_1, Z_3, are all connected to node (a)) divided by the sum of the impedances in the Δ. If we solve (118) to (120) for Z_1, Z_2, and Z_3 we obtain

$$Z_1 = \frac{Z_a Z_b + Z_b Z_c + Z_a Z_c}{Z_c} \tag{121}$$

$$Z_2 = \frac{Z_a Z_b + Z_b Z_c + Z_a Z_c}{Z_a} \tag{122}$$

$$Z_3 = \frac{Z_a Z_b + Z_b Z_c + Z_a Z_c}{Z_b} \tag{123}$$

The proof of these formulas is left as an exercise. This transformation finds application in analysis and design.

11.9 MULTITERMINAL POWER MEASUREMENTS

Consider the three terminal network in Figure 11.28. For example, this could be a three-phase system. Let us now determine what voltage and current measurements are necessary in order to calculate the instantaneous power being delivered to network B.

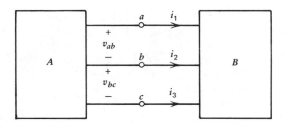

Figure 11.28　Power flow in multiterminal networks.

First, on the application of Kirchhoff's laws to these terminals, we find that

$$v_{ab} + v_{bc} + v_{ca} = 0 \tag{124}$$

and

$$i_1 + i_2 + i_3 = 0 \tag{125}$$

Note that there are two independent voltages in (124) and two independent currents in (125). By means of the substitution property we can separate networks A and B by means of two current sources whose values are any of the two currents in (125), or by means of two voltage sources whose values are any of the two voltages in (124). The six possible substitutions are shown in Figure 11.29.

From Figure 11.29 it is very easy to see that the total instantaneous power in network B is

$$p_T(t) = v_{ac} i_1 + v_{bc} i_2 \tag{126}$$

or

$$p_T(t) = v_{ab} i_1 + v_{cb} i_3 \tag{127}$$

or

$$p_T(t) = v_{ca} i_3 + v_{ba} i_2 \tag{128}$$

All three of these quantities are identical by virtue of (124) and (125). The

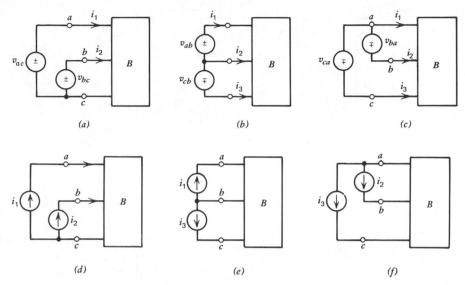

Figure 11.29 Substitution of independent variables with independent sources.

results in Section 11.4 can be used to compute the average value of the power when the signals are sinusoidal.

In the case when the signals are low frequency (e.g., 60 Hz), then wattmeters can be used to measure the power delivered to a circuit. The wattmeter consists of two magnetically coupled coils, a low resistance large and fixed current coil, and a small high-resistance movable voltage coil. Each coil is marked with a ± or dot at one terminal. The pointer of the

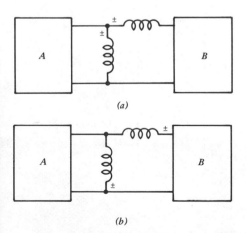

Figure 11.30 Wattmeter connection.

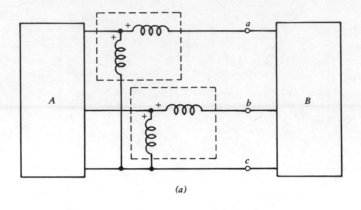

(a)

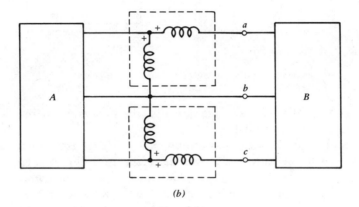

(b)

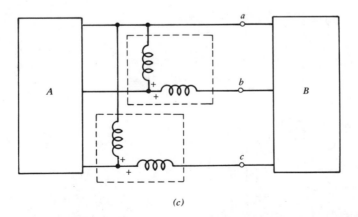

(c)

Figure 11.31 Two wattmeter power measurement in three terminal networks.

wattmeter is connected to the movable coil whose deflection is proportional to vi, the voltage v across the voltage coil and the current i through the current coil. Since the natural frequencies of the mechanical system are much lower than the frequency of the electrical signals, the pointer does not indicate the instantaneous value of vi, but rather the average value of the vi product.

In order to measure the average power delivered to network B in Figure 11.30 by network A either of the wattmeter connections shown is correct. If a negative deflection is obtained, then network B is supplying power to network A and one of the coils should be reversed in order to obtain a positive deflection.

Figure 11.31 indicates the three possible ways that two wattmeters could be connected in a three-wire system in order to measure the total average power supplied by network A to network B. The total power is the sum of both readings. In case a deflection is negative, one can reverse one of the coils of that meter to obtain a positive deflection, but this value must be substracted from the other reading.

In an n-terminal system $n-1$ wattmeters would be required since the Kirchhoff equations of the type (124) and (125) would have $n-1$ independent currents and $n-1$ independent voltages.

Example

Let us return to the unbalanced load example in Figure 11.24. Suppose that the total power delivered to this load is measured with two wattmeters as shown in Figure 11.31b. Let us compute (a) the power read by each wattmeter, (b) the total real power, (c) the total reactive power, and (d) the total apparent power.

For line A the complex power

$$\mathbf{P}_A = \mathbf{V}_{AB} \cdot \mathbf{I}_A^* = (200e^{j0°})(17.3e^{+j11.6°})$$
$$= 3460e^{+j11.6°} = 3389 + j696$$

Therefore, in line A the power read by the wattmeter is

$$3.389 \text{ kW}$$

and the reactive power is

$$696 \text{ vars}$$

In line C the complex power

$$\mathbf{P}_C = \mathbf{V}_{CB} \cdot \mathbf{I}_C^* = (200e^{j60°})(15.49e^{-j153.4°})$$
$$= 3098e^{-j93.4°} = -183.7 - j3092.5$$

Therefore, in line C the power read by the wattmeter is

$$-183.7 \text{ W}$$

and the reactive power is

$$-3092.5 \text{ vars}$$

In conclusion, the total real power

$$P_{avg} = (3.389 - 0.184)\,kW = 3.205\,kW$$

This result can be checked as follows. Note that in Figure 11.24 only the 10-Ω resistors can dissipate energy, hence

$$P_{avg} = (17.3\,A)^2 \cdot (10\,\Omega) + (4.63\,A)^2(10\,\Omega) = 3.207\,kW$$

Our answers disagree due to rounding errors, but the error is less than 0.1%.
The total reactive power is

$$(696 - 3093) = -2397\,vars$$

Again we check this answer by noting that the only reactive element is in line C and the current in line C is 15.49 A and the load in that branch is $-j10\,\Omega$ so that the complex power is strictly reactive.

$$(15.49)^2(-10) = -2399\,vars$$

Again the answers are in slight disagreement due to rounding erros.
Finally, the total apparent power is the magnitude of the total complex power, that is,

$$|\mathbf{P}_T| = [(3205)^2 + (2397)^2]^{1/2} = 4.002\,kVA$$

REFERENCES

1. Balabanian, N. *Fundamentals of Circuit Theory*, Allyn and Bacon, Boston, 1961, Chapters 4 and 5.
2. Hayt, W. H. Jr. and J. E. Kemmerly. *Engineering Circuit Analysis*, 2nd edition, McGraw-Hill, New York, 1971, Chapters 9–12.
3. Cruz, J. B. Jr. and M. E. Van Valkenburg. *Signals in Linear Circuits*, Houghton Mifflin, Boston, 1974, Chapters 9, 10, and 16.
4. Desoer, C. A. and E. S. Kuh. *Basic Circuit Theory*, Chapter 7, McGraw-Hill, New York, 1969.

PROBLEMS

1. For the series RC circuit in Figure P11.1, (a) draw the phasor circuit model, (b) compute the phasors $\mathbf{V}_R$, $\mathbf{V}_C$, $\mathbf{I}$ and draw a phasor diagram in which $\mathbf{V}_s = 10$ V is the reference vector, (c) compute the impedance $\mathbf{V}_s/\mathbf{I}$ seen by the source, and (d) compute the transfer ratios $\mathbf{V}_R/\mathbf{V}_s$ and $\mathbf{V}_C/\mathbf{V}_s$.

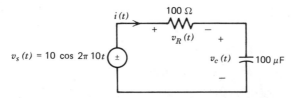

Figure P11.1

2. Repeat Problem 1 only let the input frequency be 100 Hz instead of 10 Hz.

3. For the series RL circuit in Figure P11.3, (a) draw the phasor circuit model, (b) compute the phasors V_R, V_L, and I and draw a phasor diagram in which $V_s = 10$ V is the reference vector, (c) compute the impedance V_s/I seen by the source, and (d) compute the transfer ratios V_R/V_s and V_L/V_s.

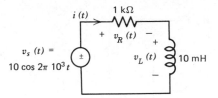

Figure P11.3

4. Repeat Problem 3 only let the input frequency be 5000 Hz instead of 1000 Hz.
5. For the parallel RLC circuit in Figure P11.5, (a) draw the phasor circuit model, (b) compute the phasors V, I_R, I_C, and I_L and draw the phasor diagram (let $I_s = 0.1$ A be the reference vector) for each of the following frequencies $\omega = 0.5$, 1.0, and 2.0 rad/s. For each of these frequencies compute the impedance V/I_s seen by the source and the transfer ratios I_R/I_s, I_C/I_s, and I_L/I_s.

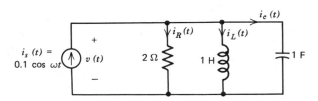

Figure P11.5

6. For the series RLC circuit in Figure P11.6, (a) draw the phasor circuit model, (b) compute the phasors I, V_R, V_L, and V_C and draw the phasor diagram for each of the following frequencies, $\omega = 10^5$, 10^6, and 10^7 rad/s. Let $V_s = 10$ V be the reference vector. For each of these frequencies compute the impedance V_s/I seen by the source and the transfer ratios V_R/V_s, V_L/V_s, and V_C/V_s.

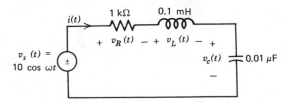

Figure P11.6

7. In Figure 11.8 calculate the amplitude and phase of the steady-state current $i_1(t)$ and voltage $v_2(t)$ for $R_1 = 1$ kΩ, $R_2 = 500$ Ω, $C = 2$ μF, $L = 10$ mH, and $f = 60$ Hz.

8. —Repeat Problem 7 for $f = 10^3$ Hz.
9. For the circuit in Figure P11.9 (a) draw the phasor circuit model, (b) compute $\mathbf{I}_L$, $\mathbf{I}_C$, and $\mathbf{I}_R$ and draw the phasor diagram, and (c) find the input impedance $Z(j\omega)$ at the frequency 3×10^4 rads/s.

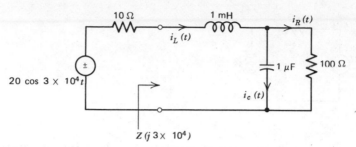

Figure P11.9

10. Find the Thévenin equivalent of the circuit in Figure P11.10 at the frequency $\omega = 10$ rads/s. First write the network equations and eliminate variables to find the equivalent circuit. Check your answer by using source transformations to reduce the circuit. Let $L = 1$ H and $C = 0.005$ F.

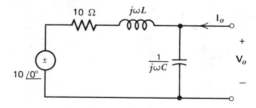

Figure P11.10

11. Repeat Problem 10 for $\omega = 1$ rad/s.
12. Repeat Problem 10 for $\omega = 20$ rads/s.
13. Use source splitting (Chapter Five) and source transformation to find the Thévenin equivalent of the circuit in Figure P11.13.

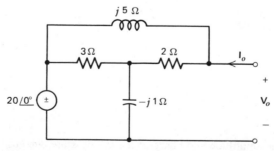

Figure P11.13

14. Repeat Problem 13 for the circuit in Figure P11.14.

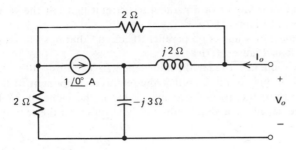

Figure P11.14

15. In Figure P11.15 show that

$$I_2 = \frac{Z_1}{Z_1 + Z_2} I \quad \text{and} \quad I_1 = \frac{Z_2}{Z_1 + Z_2} I$$

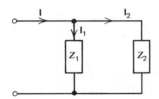

Figure P11.15

16. In the bridge circuit in Figure P11.16 show that the ammeter current I_m is zero when $Z_1 Z_x = Z_2 Z_3$.

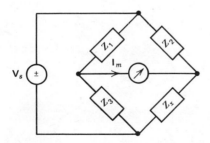

Figure P11.16

17. In the bridge circuit in Figure P11.16 let $Z_2 = Z_3 = 1 \text{ k}\Omega$. If the frequency of the source is 10^3 Hz, $Z_1 = 1/j\omega C$, and $I_m = 0$ when $C = 5 \mu F$ state whether the element in arm x is an inductor or capacitor and find its value.

18. In the bridge circuit in Figure P11.16 suppose that $Z_2 = Z_3 = R_a$, $Z_1 = G + j\omega C$ and $Z_x = r_x + j\omega L$. Under what conditions will the bridge be balanced?

19. Find the resistor and inductor values of a parallel RL circuit that has the same impedance as the circuit in Figure 11.7a.

20. Find the resistor and capacitor values of a parallel RC circuit that has the same impedance as the circuit in Figure 11.7b.

21. In Figure P11.21 at the frequency 10 kHz the impedance $Z_L(j2\pi 10^4) = 10 + j20$. Find the capacitance C in this circuit such that the reactive component of the impedance $Z = V/I$ is zero at the given frequency. What is the value of Z? If the shunt capacitor is replaced by a shunt inductor can you make the reactive component of Z zero?

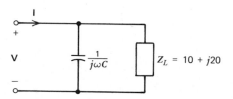

Figure P11.21

22. In Problem 21 connect the capacitor C in series with Z_L. What value of capacitance is needed to cancel the reactive component of Z_L? What is the value of the input impedance now?

23. In Figure P11.23 find the frequency range in hertz such that the magnitude of the voltage across the 1-μF coupling capacitor is less than 1% of the magnitude of the voltage across the network N. Assume that the network N has an impedance 1000 Ω for all ω. Repeat for $C = 100\ \mu$F.

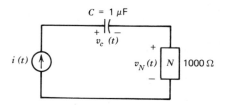

Figure P11.23

24. Repeat Problem 23 only assume that the impedance of the network N is (a) 10 Ω, and (b) 1 MΩ.

25. In Figure P11.25 the parasitic lead inductance is 1 nH. Assuming that the impedance of the network N is 1000 Ω for all ω, find the frequency range in Hertz such that $|V_L| < 0.01|V_N|$.

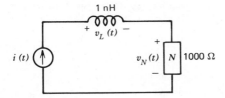

Figure P11.25

26. Repeat Problem 25 for (a) $Z_N = 100\,\Omega$, and (b) $Z_N = 1\,\mathrm{M\Omega}$.
27. In Figure P11.27 find the frequency range in hertz such that $|\mathbf{I}_C| < 0.01|\mathbf{I}_N|$. Assume that the impedance of the network N is $1000\,\Omega$ for all ω.

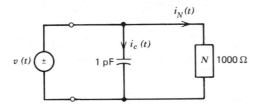

Figure P11.27

28. Repeat Problem 27 for $C = 10\,\mathrm{pF}$.
29. For the circuit N in Figure P11.29 find:
    (a) The complex power $\mathbf{P}$.
    (b) The average real power.
    (c) The average imaginary power.
    (d) The impedance and admittance of N at the frequency $\omega = 10^3$ rads/s. Is the circuit capacitive or inductive at this frequency?

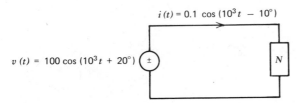

Figure P11.29

30. Repeat Problem 29 for $v(t) = 50 \sin(10^3 t + 10°)$ and $i(t) = 0.2 \sin(10^3 t + 55°)$.
31. In Problem 29 compute the average real power dissipated if

$$v(t) = 10 \cos \omega_1 t + 100 \sin(2\omega_1 t + 10°) + 50 \cos(3\omega_1 t + 20°) \text{ and}$$
$$i(t) = 0.4 \cos(\omega_1 t - 20°) + 0.1 \sin(2\omega_1 t + 40°) + 0.8 \cos(3\omega_1 t - 70°).$$

Also assume that the circuit is linear and time invariant and compute the impedance of the circuit N at the frequencies ω_1, $2\omega_1$, and $3\omega_1$.

32. A linear time-invariant circuit has two independent sources, one with frequency 15 kHz and the other with frequency 25 kHz. Find the common period T of the responses from (81) and (82).

33. For the circuit in Figure P11.33:
    (a) Draw the phasor circuit model.
    (b) Find the voltage across each element in the phasor model.
    (c) Compute the impedance seen by the source.
    (d) Compute the apparent, real, and reactive power supplied by the source.

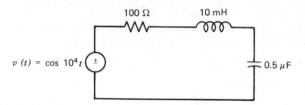

Figure P11.33

34. In Figure P11.34 find L and C such that the source delivers maximum power to the load at $\omega = 10^4$ rads/s. What is the power delivered to the load when (a) $\omega = 10^3$ rads/s, and (b) $\omega = 10^5$ rads/s? Let $\mathbf{V}_s = 100/\underline{0°}$.

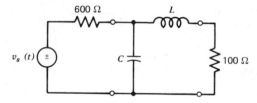

Figure P11.34

35. Repeat Problem 34 for the circuit in Figure P11.35.

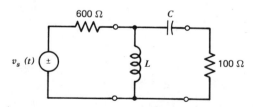

Figure P11.35

36. Repeat Problem 34 for the circuit in Figure P11.36.

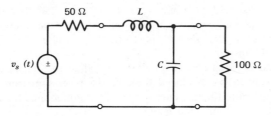

Figure P11.36

37. Repeat Problem 34 for the circuit in Figure P11.37.

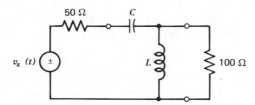

Figure P11.37

38. The normal residential electric system is the single phase three wire system shown in Figure P11.38. Lower power appliances are designed to operate on 120 V rms and are connected between terminals A-N or terminals B-N. In order that larger diameter wires are not required to reduce line losses, larger power appliances are designed to operate on 240 V rms and are connected between terminals A-B. In Figure P11.38 suppose that load 1 is a 60-W light bulb rated at 120 V rms, load 2 is a 100-W light bulb rated at 120 V rms, and load 3 is a 4800-W motor rated at 240 V rms and with a power factor of 0.8. Calculate Z_1, Z_2, and Z_3, respectively. If $r_a = r_b = 0.1\,\Omega$ and $r_n = 0.5\,\Omega$, compute $\mathbf{I}_a$, $\mathbf{I}_b$, $\mathbf{I}_c$, and $\mathbf{I}_a - \mathbf{I}_b$. Calculate the power delivered by each source, and the power dissipated by r_a, r_b, r_n, Z_1, Z_2, and Z_3.

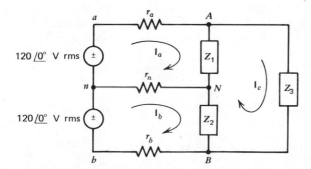

Figure P11.38

39. In Figure 11.23 let $Z_a = Z_b = Z_c = 3 - j4\,\Omega$ and $V_{an} = 150$ V rms. Calculate the line currents and the total real and reactive power delivered by the sources.

40. In Figure 11.25 assume that $Z_1 = Z_2 = Z_3 = 3 - j4\,\Omega$ and $V_{an} = 150$ V rms. Repeat the calculations in Problem 39.

41. A balanced three-phase load is rated at 10 kW, 500 V rms (line to line) and a power factor of 0.85. The source is a balanced three-phase, three-wire source with abc phase sequence. Compute the line currents and load impedance for (a) a Y connected load, and (b) a Δ connected load.

42. In Figure 11.22 let $Z = 10 + j10\,\Omega$ and $V_{an} = 240$ V rms. Suppose that line a–A is open so that $\mathbf{I}_a = 0$. Find the line currents $\mathbf{I}_b$, $\mathbf{I}_c$, and $\mathbf{I}_n$. Compute the real, reactive, and apparent power delivered by the sources.

43. Repeat Problem 42 for the circuit in Figure 11.23. Let $Z_b = Z_c = 10 + j10\,\Omega$.

44. In Figure 11.23 let $Z_a = 3 + j4\,\Omega$, $Z_b = 5\,\Omega$, $Z_c = 4 + j3\,\Omega$, and the line-to-line voltage is 173 V rms. Calculate the line currents and the voltages $\mathbf{V}_{an}$, $\mathbf{V}_{bn}$, and $\mathbf{V}_{cn}$. Draw a phasor diagram. Compute the real, reactive, and apparent power delivered by the sources.

45. In Figure 11.25, let $Z_1 = 3 + j4\,\Omega$, $Z_2 = 4 + j3\,\Omega$, $Z_c = 5\,\Omega$, and the line-to-line voltage is 100 V rms. Calculate the line and load currents. Draw a phasor diagram. Compute the real and reactive power delivered by the sources.

46. Repeat Problem 45 only assume that a line a is open so that $\mathbf{I}_a = 0$. Compute the load voltages also.

47. In a balanced Y–Y connection $Z_a = Z_b = Z_c = 10 + j10\,\Omega$. Place a Y network of capacitors in parallel with the Y connected load so that the line currents and source voltages are in phase. Convert the Y network of capacitors to an equivalent Δ network so that the capacitors are now connected from line to line. Assume that $f = 60$ Hz.

48. Convert the Y load in Problem 44 to an equivalent Δ load.

49. Convert the Δ load in Problem 45 to an equivalent Y load.

50. A balanced 2300 V rms (line to line), three-phase source of acb positive phase sequence supplies power to a balanced three-wire load. The complex line current $\mathbf{I}_a = 20 + j0$A with $\mathbf{V}_{ab}$ as the reference phasor. Find (a) the power factor, (b) total apparent power, (c) total real power, and (d) total reactive power.

51. In Figure P11.51 let $Z_a = 1\,\Omega$, $Z_b = -j1\,\Omega$, and $Z_c = j1\,\Omega$. Calculate the real average powers P_1 and P_2 indicated by each of the wattmeters in the connection shown. What is the total real power supplied by the sources?

52. In Figure P11.51 assume that the load is balanced and that $Z_a = Z_b = Z_c = 4 + j3\,\Omega$. Compute P_1 and P_2. Repeat with $Z_a = Z_b = Z_c = 4 - j3\,\Omega$.

53. In Figure P11.51 assume that the load is balanced and that $Z_a = Z_b = Z_c = |Z|\underline{/\alpha}$. Show that

$$P_1 + P_2 = \frac{3V^2}{|Z|}\cos\alpha$$

and

$$P_1 - P_2 = \frac{\sqrt{3}\,V^2}{|Z|}\sin\alpha$$

where V is the rms value of the sources, and P_1 and P_2 are the real average

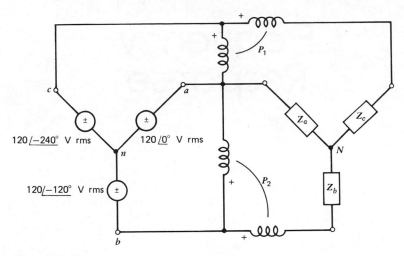

Figure P11.51

powers indicated by the wattmeters. Note that one can determine the power factor of a balanced load from the equation

$$\sqrt{3}\,\frac{P_1 - P_2}{P_1 + P_2} = \tan\alpha$$

where $PF = \cos\alpha$.

Chapter Twelve
Frequency Response

In the previous chapter we analyzed the steady-state sinusoidal behavior of circuits at a single input frequency. In this chapter we examine the sinusoidal steady-state behavior as a function of the input frequency ω. In Chapter Ten recall that the relationship between a single-input signal and a single-output signal in the complex frequency domain is characterized by the complex network function

$$H(s) = \frac{\mathbf{Y}(s)}{\mathbf{X}(s)} \tag{1}$$

where $\mathbf{Y}(s)$ is the complex amplitude of the output signal and $\mathbf{X}(s)$ is the complex amplitude of the input signal. For sinusoidal steady-state analysis $s = j\omega$ and we write

$$H(j\omega) = M(\omega)e^{j\phi(\omega)} \tag{2}$$

where $M(\omega)$ is the magnitude of the complex function (2) and $\phi(\omega)$ is its phase. Furthermore, the plots of $M(\omega)$ and $\phi(\omega)$ are called the magnitude and phase frequency response plots, respectively. Examining these plots quickly tells us how the magnitude and phase of the sinusoidal steady state response varies with the input frequency ω. For example, if the input is a cosine function with amplitude V, frequency ω_o and phase θ_o, then the sinusoidal steady-state response is

$$y(t) = M(\omega_o) \cdot V \cos[\omega_o t + \theta_o + \phi(\omega_o)] \tag{3}$$

Note that the amplitude of the sinusoidal steady-state response for an input frequency ω_o is $M(\omega_o) \cdot V$ and its phase is $\theta_o + \phi(\omega_o)$ where the input amplitude is V and the input phase is θ_o. If we assume that V and θ_o are constant with respect to ω, then the amplitude and phase variation of the output signal over the frequency range are determined by the frequency response plots of $M(\omega)$ and $\phi(\omega)$.

In the frequency response plots very often a logarithmic scale is used for the magnitude function $M(\omega)$ as well as for the frequency variable ω. The natural logarithm of (2) is

$$\ln[H(j\omega)] = \ln[M(\omega)] + j\phi(\omega) \tag{4}$$

The gain $\ln[M(\omega)]$ is measured in *nepers* and the angle function $\phi(\omega)$ is *radians*. The usual unit for the gain is the *decibel* (dB) which is defined as

$$\text{Gain in decibels} = 20 \log_{10} M(\omega)^{1} \tag{5}$$

Similarly, usually the angle $\phi(\omega)$ is expressed in *degrees* rather than radians. In the next section we examine the relationship between the location of the poles and zeros of the network function and the magnitude and phase of the sinusoidal steady state response.

12.1 THE RELATION BETWEEN THE FREQUENCY RESPONSE AND THE POLES AND ZEROS OF THE NETWORK FUNCTION

The network function of a lumped linear time-invariant circuit is a function of the complex frequency s and is of the form

$$\frac{Y(s)}{X(s)} = H(s) = K \frac{s^m + a_{m-1}s^{m-1} + \cdots + a_1 s + a_o}{s^n + b_{n-1}s^{n-1} + \cdots + b_1 s + b_o} \tag{6}$$

where the complex variable $s = \sigma + j\omega$. We can factor the numerator and the denominator polynomials of $H(s)$ and rewrite (6) as

$$H(s) = K \frac{(s - z_1)(s - z_2) \cdots (s - z_m)}{(s - p_1)(s - p_2) \cdots (s - p_n)} \tag{7}$$

where the m roots $z_i, i = 1, 2, \ldots, m$, of the numerator of (6) are called the *zeros* of the function since $H(z_i) = 0$, and the n roots $p_i, i = 1, 2, \ldots, n$, of the denominator of (6) are called the *poles* of the function since $H(p_i)$ is infinite. The poles and zeros can be complex, but complex roots must occur in conjugate pairs since the coefficients a_i and b_j of (6) are real.

In the sinusoidal steady state, if the input frequency is ω_o, then (7)

---

[1] Henceforth the symbol ln will denote the natural logarithm (base e) and the symbol log will denote the logarithm to the base 10.

becomes

$$H(s)|_{s=j\omega_o} = \frac{K(j\omega_o - z_1)(j\omega_o - z_2) \cdots (j\omega_o - z_m)}{(j\omega_o - p_1)(j\omega_o - p_2) \cdots (j\omega_o - p_n)} \tag{8}$$

Let us examine a typical factor of (8), which we write as $(j\omega_o - r)$ where r is a root of one of the polynomials in (6). In Figure 12.1 we represent the complex numbers $j\omega_o$ and r as vectors. The difference between them is also shown. It has length R and angle α and can be found by simply drawing a vector from point r (the value of the root) to point $j\omega_o$ (the sinusoidal input frequency) as shown in Figure 12.1.

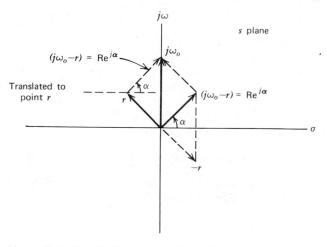

Figure 12.1   Graphic illustration of the difference between two complex numbers.

Assume that we have a network function

$$H(s) = \frac{(s - z_1)(s - z_2)}{(s - p_1)(s - p_2)} \tag{9}$$

then

$$H(j\omega_0) = \frac{Z_1 Z_2 \, e^{j(\alpha_1 + \alpha_2)}}{P_1 P_2 \, e^{j(\beta_1 + \beta_2)}} \tag{10}$$

or

$$H(j\omega_0) = \frac{Z_1 Z_2}{P_1 P_2} e^{j(\alpha_1 + \alpha_2 - \beta_1 - \beta_2)} \tag{11}$$

where the magnitudes Z_i and P_j and the angles α_i and β_j can be determined graphically from Figure 12.2. For example, P_1 is just the length of the vector from the pole p_1 to the point $j\omega_o$ where ω_o is the frequency of the input signal. The angle β_1 is simply the angle between the vector $(j\omega_o - p_1)$ and a horizontal line through the root. In sinusoidal analysis the maxima of $H(j\omega)$ occur when the frequency of the input signal is in the vicinity of a

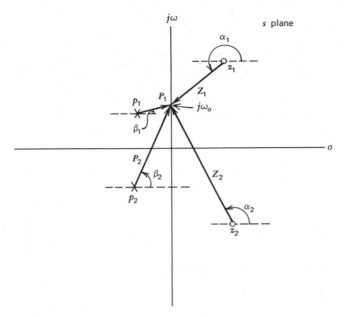

Figure 12.2   Graphical calculation of the ac response.

pole. For example, in Figure 12.2, P_1 is smallest when $\beta_1 = 0°$, that is, $\omega_o = Im[p_1]$ and $H(j\omega_o)$ is large because it is proportional to $1/P_1$.

Add a third dimension to Figure 12.2 and think of the poles as mountains infinite in height and the zeros as forming valleys at sea level ($H(z_i) = 0$). Assume that you are driving along the $j\omega$ axis. Obviously, as you move into the vicinity of a nearby mountain (pole), you must climb in elevation (the magnitude of the sinusoidal response increases). As you move toward a zero the elevation decreases (the magnitude of the sinusoidal response decreases). Suppose that you wanted to design a filter to pass certain frequencies but not others. The network function would have poles in the neighborhood of the band of frequencies that are desirable and zeros in the band of frequencies which are undesirable.

12.1.1 THE FIRST-ORDER NETWORK FUNCTION

Let us reconsider the series RC circuit in Figure 12.3 and its network functions

$$\frac{\mathbf{V}_c(s)}{\mathbf{V}_s(s)} = H_c(s) = \frac{\dfrac{1}{RC}}{s + \dfrac{1}{RC}} \tag{12}$$

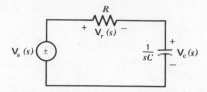

Figure 12.3   First-order network function example.

and

$$\frac{\mathbf{V}_r(s)}{\mathbf{V}_s(s)} = H_r(s) = \frac{s}{s + \dfrac{1}{RC}} \tag{13}$$

Both of these functions have a pole on the real axis at $s = -1/RC$. However, (12) has a zero at infinity since this function behaves like $1/RCs$ as s become large, but (13) has a zero at $s = 0$. The two-dimensional pole-zero plots for these two network functions are shown in Figure 12.4,

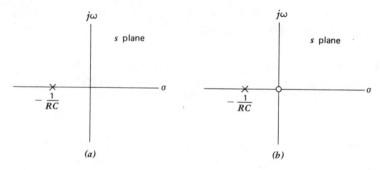

Figure 12.4   Pole-zero plots for RC circuit. (a) Capacitor voltage output. (b) Resistor voltage output.

and the three-dimensional plots are shown in Figure 12.5 where the third dimension represents the *magnitude* of $H(s)$.

In sinusoidal analysis we are only interested in the behavior of the network function along the $j\omega$ axis. To determine the magnitude and phase of the network function along the $j\omega$ axis we substitute $s = j\omega$ into (12).

$$H_c(j\omega) = \frac{1}{RCj\omega + 1} \tag{14}$$

and note that

$$M(\omega) = \frac{1}{[(RC\omega)^2 + 1]^{1/2}} \tag{15}$$

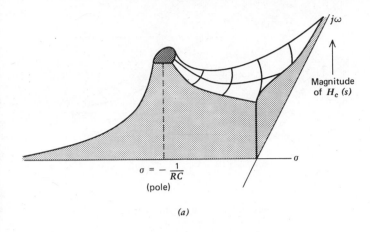

(a)

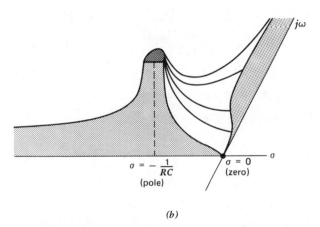

(b)

Figure 12.5   Three dimensional pole-zero plot. (a) Capacitor voltage response. (b) Resistor voltage response.

and

$$\phi(\omega) = -\tan^{-1}\frac{\omega RC}{1} \tag{16}$$

Functions 15 and 16 are plotted in Figure 12.6. In plotting first-order network functions, one keys on three frequencies, $\omega \to 0$, $\omega \to \infty$, and the *break point* or *break frequency* $\omega = 1/RC$ where

$$H_c\!\left(j\frac{1}{RC}\right) = \frac{1}{1+j1} = 0.707\ \underline{/-45°} \tag{17}$$

The break frequency is so called because it is the frequency at which the

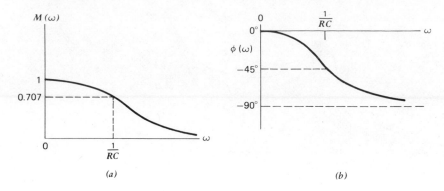

Figure 12.6   Frequency response of low-pass filter. (a) magnitude, (b) phase.

high frequency and low frequency asymptotes intersect. For example,

$$20 \log M(\omega) = 20 \log \left[\frac{1}{1 + (\omega RC)^2} \right]^{1/2}$$

$$= -10 \log[1 + (\omega RC)^2] \, dB \qquad (18)$$

Remember that $\log 1/A = -\log A$. Note that as $\omega \to 0$

$$20 \log M(\omega) \approx 20 \log 1 = 0 \, dB \qquad (19)$$

This is the low-frequency asymptote. As $\omega \to \infty$

$$20 \log M(\omega) \approx 20 \log \frac{1}{\omega RC} \qquad (20)$$

Equation 20 is the high-frequency asymptote. Asymptotes 19 and 20 are plotted in Figure 12.7. Note that they cross at the break frequency

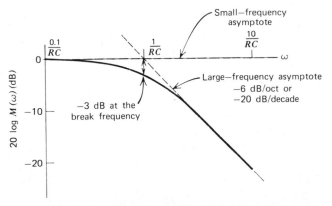

Figure 12.7   Semilog plot of first-order magnitude function.

Table 12.1 *Comparison of Large Frequency Approximation with Actual Magnitude Response*

ω	$20 \log \dfrac{1}{\omega RC}$	$20 \log \left[\dfrac{1}{1+(\omega RC)^2}\right]^{1/2}$
$1/RC$	0.0 dB	-3.01 dB
$2/RC$	-6.0 dB	-6.99 dB
$10/RC$	-20 dB	-20.04 dB

$\omega = 1/RC$. Values of (20) are given in Table 12.1. Note that in (20) and Figure 12.7 if we double the frequency (a one-octave change) the gain decreases by 6 dB; a one-decade (factor of 10) change in ω yields a 20 dB change in the gain.

A semilog plot of the phase is shown in Figure 12.8. Frequently, the phase is approximated by the dashed line on the semilog plot determined by the points $\phi(0.1/RC) = 0°$ and $\phi(10/RC) = 90°$ as shown in Figure 12.8. The error with this approximation does not exceed six degrees.

A two-port with the above magnitude characteristic is called a lowpass filter, because the gain is higher at low frequencies than at high frequencies. The frequency at which the gain is down 3 dB is called the *cutoff frequency* of the filter. The cutoff frequency can be adjusted by changing R or C. The frequency at which the gain of a network function is down 3 dB from its maximum value is also called the *half-power frequency*, because the output current or voltage is $1/\sqrt{2}$ of its maximum value and power is proportional to the square of the magnitude of the output voltage or current phasor which means that *the square of the magnitude of the output is one-half of its maximum value.*

Finally, let us consider the same RC circuit but let the output be the

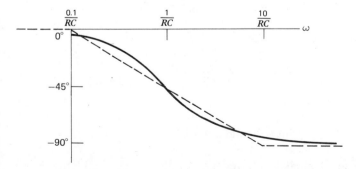

Figure 12.8 Semilog plot of first-order phase function.

voltage across the resistor. From (13)

$$H_r(j\omega) - \frac{j\omega RC}{1 + j\omega RC} \qquad (21)$$

where

$$M(\omega) = \frac{\omega RC}{[1 + (\omega RC)^2]^{1/2}} \qquad (22)$$

and

$$\phi(\omega) = \left(90° - \tan^{-1}\frac{\omega RC}{1}\right) \qquad (23)$$

The magnitude and phase functions (22) and (23) are sketched in Figure 12.9. A two-port with this magnitude function is called a highpass filter. Note that the same circuit can be a highpass or a lowpass filter depending on the response chosen as the output variable.

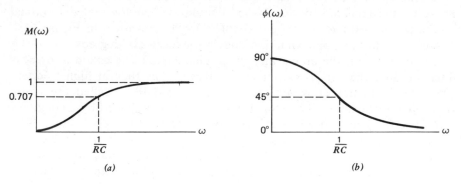

Figure 12.9 Frequency response of high-pass filter. (a) Magnitude. (b) Phase.

In terms of decibels

$$\text{Gain} = 20 \log(\omega RC) - 10 \log[1 + (\omega RC)^2] \qquad (24)$$

and the low-frequency asymptote is

$$\text{Gain} \approx 20 \log(\omega RC) \text{ dB} \qquad (25)$$

and the high-frequency asymptote is

$$\text{Gain} \approx 20 \log(1) = 0 \text{ dB} \qquad (26)$$

The gain function in decibels is plotted on semilog paper in Figure 12.10 along with the low- and high-frequency asymptotes.

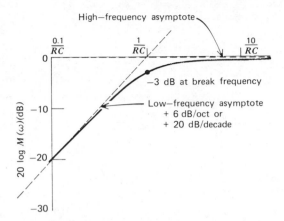

Figure 12.10 Semilog plot of the gain in decibels.

12.1.2 SECOND-ORDER BANDPASS NETWORK FUNCTION

Steeper slopes for the magnitude function $M(\omega)$ can be obtained with higher-order network functions. In this section we consider the second-order bandpass function

$$H(s) = \frac{\left(\dfrac{\omega_r}{Q}\right)s}{s^2 + \dfrac{\omega_r}{Q}s + \omega_r^2} \tag{27}$$

It is fairly common practice among circuit designers to express any second-order complex frequency polynomial $s^2 + b_1 s + b_0$ in the form $s^2 + \omega_r s/Q + \omega_r^2$ where ω_r is called the *resonant frequency* and Q is simply called *the Q or the quality factor of the circuit*.

The poles of this network function are complex for $Q > 1/2$, that is,

$$H(s) = \frac{\left(\dfrac{\omega_r}{Q}\right)s}{\left(s + \dfrac{\omega_r}{2Q} + j\omega_r\sqrt{1 - 1/4Q^2}\right)\left(s + \dfrac{\omega_r}{2Q} - j\omega_r\sqrt{1 - 1/4Q^2}\right)} \tag{28}$$

Thus, this network function has zeros at $s = 0$ and s equal to infinity (since the degree of the denominator is greater than the degree of the numerator). The poles are located at $-(\omega_r/2Q) \pm j\omega_r\sqrt{1 - 1/4Q^2}$. These poles and zeros are plotted in the complex plane in Figure 12.11. The plot has been normalized by the resonant frequency ω_r. A three-dimensional plot of the bandpass function is shown in Figure 12.12. We note that the function is

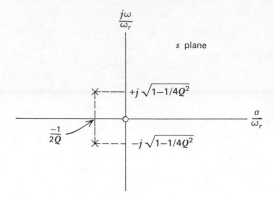

Figure 12.11 Pole-zero locations.

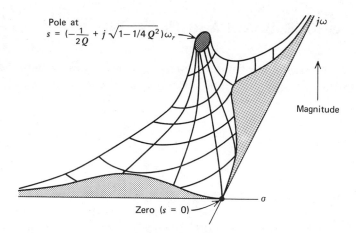

Figure 12.12 Three-dimensional pole-zero plot.

zero at $s = 0$ and peaks along the $j\omega$ axis in the neighborhood of the complex pole. The function goes to zero as s goes to infinity.

In order to study the behavior of the bandpass function along the $j\omega$ axis we substitute $s = j\omega$ into (27). Thus, (27) becomes

$$H(j\omega) = \frac{j\omega\omega_r/Q}{(\omega_r^2 - \omega^2) + j\omega\omega_r/Q} \tag{29}$$

for sinusoidal steady-state analysis. Let us make the substitution $\omega = \omega_r\hat{\omega}$ in (29), which normalizes the resonant frequency to $1\,\text{rad/s}$ for convenience. Thus,

$$H(j\hat{\omega}) = \frac{j\hat{\omega}/Q}{(1 - \hat{\omega}^2) + j\hat{\omega}/Q} \tag{30}$$

where

$$M(\hat{\omega}) = \frac{|\hat{\omega}|/Q}{[(1 - \hat{\omega}^2)^2 + (\hat{\omega}/Q)^2]^{1/2}} \tag{31}$$

and

$$\phi(\hat{\omega}) = 90° - \tan^{-1}\left(\frac{\hat{\omega}}{Q(1 - \hat{\omega}^2)}\right) \tag{32}$$

Both (31) and (32) are plotted in Figure 12.13 as a function of Q and the

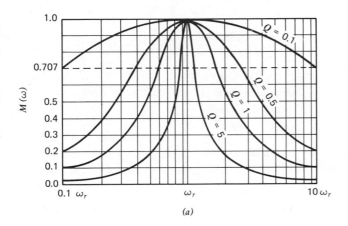

(a)

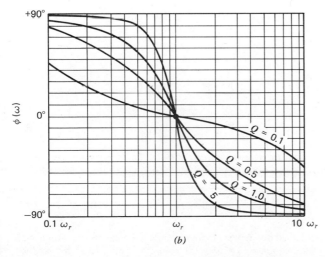

(b)

Figure 12.13 Magnitude and phase characteristic of the bandpass function. (a) Magnitude. (b) Phase.

unnormalized frequency ω. Note that at the resonant frequency $\hat{\omega} = 1$ rad/s ($\omega = \omega_r$ rad/s)

$$M(1) = 1 \tag{33}$$

$$\phi(1) = 0° \tag{34}$$

as $\omega \to 0$

$$H(j\hat{\omega}) \approx j\hat{\omega}/Q \tag{35}$$

so that

$$M(0) = 0 \tag{36}$$

and

$$\phi(0) = 90° \tag{37}$$

finally, as $\omega \to \infty$,

$$H(j\hat{\omega}) \approx \frac{j\hat{\omega}/Q}{-\hat{\omega}^2} = \frac{-j}{Q\hat{\omega}} \tag{38}$$

so that

$$M(\infty) = 0 \tag{39}$$

and

$$\phi(\infty) = -90° \tag{40}$$

The transfer function is independent of Q at these three frequencies. However, the width of the magnitude characteristic in Figure 12.13a is clearly a function of Q. To find the half-power frequencies it is convenient to write (30) in the form

$$H(j\hat{\omega}) = \frac{1}{1 + jQ\left(\hat{\omega} - \dfrac{1}{\hat{\omega}}\right)} \tag{41}$$

Since the maximum value of $M(\hat{\omega})$ is unity then the half-power magnitude $[M(\hat{\omega}) = 1/\sqrt{2}]$ occurs at the frequencies at which

$$Q\left(\hat{\omega} - \frac{1}{\hat{\omega}}\right) = \pm 1 \tag{42}$$

in (41). The solution to equation (42) is

$$\hat{\omega}_{1,2,3,4} = \pm \frac{1}{2Q} \pm \sqrt{\left(\frac{1}{2Q}\right)^2 + 1} \tag{43}$$

Note that there are four frequencies which satisfy (42) but we are only interested in the positive values of (43).[2]

$$\hat{\omega}_2 = \sqrt{\left(\frac{1}{2Q}\right)^2 + 1} + \frac{1}{2Q} \tag{44}$$

[2] Remember that in sinusoidal steady-state analysis there is no need to analyze the circuit with negative frequencies because the response is simply the complex conjugate of the positive frequency response, which implies that $M(\omega) = M(-\omega)$ and $\phi(\omega) = -\phi(\omega)$—that is, $M(\omega)$ is an even function of ω and $\phi(\omega)$ is an odd function of ω.

and

$$\hat{\omega}_1 = \sqrt{\left(\frac{1}{2Q}\right)^2 + 1} - \frac{1}{2Q} \tag{45}$$

The bandwidth of the bandpass filter characteristic in Figure 12.13a is defined as the difference between the 3-dB frequencies.

$$\text{BW} \triangleq \hat{\omega}_2 - \hat{\omega}_1 = \frac{1}{Q} \tag{46}$$

or, in terms of the unnormalized frequency ω,

$$\omega_2 - \omega_1 = \omega_r/Q \tag{47}$$

so that

$$\boxed{Q = \frac{\omega_r}{\omega_2 - \omega_1}} \tag{48}$$

Furthermore, note that

$$\omega_1\omega_2 = \omega_r^2 \tag{49}$$

that is, the resonant frequency is the geometric mean of the half-power frequencies. The variation in the bandwidth as a function of Q is illustrated in Table 12.2.

Frequently the following approximations are used to compute the half-power frequencies. For large Q (44) and (45) become

$$\hat{\omega}_{1,2} \approx 1 \mp \frac{1}{2Q} \text{ (large } Q) \tag{50}$$

or in terms of the unnormalized frequency

$$\boxed{\omega_{1,2} \approx \left(1 \mp \frac{1}{2Q}\right)\omega_r} \quad \text{(large } Q) \tag{51}$$

Table 12.2   Selectivity of the Second-order Bandpass Function

| Q | Half-Power Frequencies | | Bandwidth |
	ω_1	ω_2	$\omega_2 - \omega_1$
0.1	$0.1\,\omega_r$	$10.1\,\omega_r$	$10.0\,\omega_r$
0.5	$0.41\,\omega_r$	$2.41\,\omega_r$	$2.0\,\omega_r$
1	$0.62\,\omega_r$	$1.62\,\omega_r$	$1.0\,\omega_r$
5	$0.905\,\omega_r$	$1.105\,\omega_r$	$0.2\,\omega_r$
10	$0.951\,\omega_r$	$1.051\,\omega_r$	$0.1\,\omega_r$
50	$0.990\,\omega_r$	$1.010\,\omega_r$	$0.02\,\omega_r$
100	$0.995\,\omega_r$	$1.005\,\omega_r$	$0.01\,\omega_r$

When Q is small we write (41) and (42) as

$$\hat{\omega}_{1,2} = \frac{1}{2Q}(\pm 1 + \sqrt{1 + 4Q^2}) \tag{52}$$

and

$$\hat{\omega}_{1,2} \approx \frac{1}{2Q}[\pm 1 + \sqrt{(1 + 2Q^2)^2}] \tag{53}$$

so that

$$\hat{\omega}_2 \approx \frac{1}{Q} \tag{54}$$

or

$$\boxed{\omega_2 \approx \frac{\omega_r}{Q}} \quad \text{(small } Q) \tag{55}$$

and

$$\hat{\omega}_1 \approx Q \tag{56}$$

or

$$\boxed{\omega_1 \approx Q\omega_r} \quad \text{(small } Q) \tag{57}$$

An examination of Table 12.2 reveals that the large Q approximation (51) is very good for $Q \geqslant 5$ and is only in error by approximately 20% for $Q = 1$. Also the small Q approximation is very good for $Q \leqslant 0.2$ and is in error by approximately 20% for $Q = 1/2$.

Example 1

It is desired to design a second-order bandpass filter with a center (resonant) frequency of 1 rad/s and a bandwidth of 0.2 rad/s. From (48) we see that a Q of 5 is required. To realize the bandpass network function we can use the series RLC circuit. From Figure 10.8 with the resistor terminals as the output port, the network function is a bandpass function (see (48) in Chapter Ten).

$$H(s) = \frac{\dfrac{R}{L}s}{s^2 + s\dfrac{R}{L} + \dfrac{1}{LC}} \tag{58}$$

Note that

$$\omega_r^2 = 1/LC = 1 \text{ rad/s} \tag{59}$$

and

$$Q = \frac{\omega_r L}{R} = 5 \tag{60}$$

We have two equations and three unknowns, which means that we have a certain amount of freedom to choose one of the parameters. Let $L = 1$ H, then from (59) $C = 1$ F and from (60) $R = 0.2\ \Omega$. The circuit is illustrated in Figure 12.14.

Recall that this bandpass function can also be realized with the active RC circuit

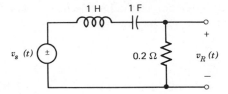

Figure 12.14   RLC bandpass filter.

in Figure 10.13. If we choose $C_1 = C_2 = 1$ F, $G_1 = G_3 = 1/2$ mho, and $G_2 = 1$ mho then Equation 76 in Chapter Ten becomes

$$H(s) = \frac{sA}{s^2 + s(3 - \frac{1}{2}A) + 1} \tag{61}$$

To achieve a Q of 5 we must choose $A = 5.6$, then

$$H(s) = \frac{5.6s}{s^2 + 0.2s + 1} \tag{62}$$

This active RC circuit is shown in Figure 12.15. Transfer function 62 has the same shape as (58) but it has a gain of 28 at the resonant frequency as opposed to a gain of one for the RLC circuit.

Note that if we connect a load to the output port of the RLC circuit in Figure 12.14 such that the output current is not zero, then the transfer function (58) will change. However, in Figure 12.15 a load connected to the output port has a negligible effect on the transfer function (61) because the amplifier is a low impedance voltage source at its output terminals. Many times an RLC filter has been designed to work well by itself, but when it is connected into a circuit its performance changes due to the loading by the circuit to which the output of the filter is connected. This loading must be included in the original design of the filter. In active RC circuits, loading is usually not a problem and second-order filter sections can be cascaded to realize a higher-order filter function. The design

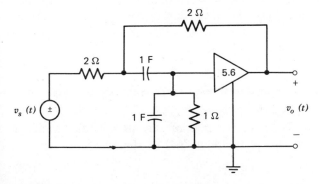

Figure 12.15   Active RC bandpass filter.

procedure for RLC circuits is much more complex due to the loading that occurs when another second-order section is connected to the output of the previous section.

12.2 BODE PLOTS

In the previous section we studied the magnitude and phase characteristics of some simple first- and second-order network functions. In this section we introduce a general method for plotting the magnitude and phase of a network function on a logarithmic scale. The plots make use of the fact that the logarithm of a product of terms is equal to the sum of the logarithms of each term. The method is based on the factored form of the transfer function.

To find the magnitude and phase of the sinusoidal steady state response we write the transfer function in the form

$$H(j\omega) = K \frac{\left|\frac{j\omega}{z_1} - 1\right| \left|\frac{j\omega}{z_2} - 1\right| \cdots \left|\frac{j\omega}{z_m} - 1\right| e^{j(\alpha_1 + \cdots + \alpha_m)}}{\left|\frac{j\omega}{p_1} - 1\right| \left|\frac{j\omega}{p_2} - 1\right| \cdots \left|\frac{j\omega}{p_n} - 1\right| e^{j(\beta_1 + \cdots + \beta_n)}} \tag{63}$$

From (2), the magnitude of the transfer function in decibels is

$$20 \log[M(j\omega)] = 20 \log K + 20 \log\left|\frac{j\omega}{z_1} - 1\right| + \cdots + 20 \log\left|\frac{j\omega}{z_m} - 1\right|$$

$$- 20 \log\left|\frac{j\omega}{p_1} - 1\right| - \cdots - 20 \log\left|\frac{j\omega}{p_n} - 1\right| \tag{64}$$

and the phase

$$\phi(\omega) = (\alpha_1 + \cdots + \alpha_m - \beta_1 - \cdots - \beta_n) \tag{65}$$

Note that the analysis of the frequency response of the network has been broken into a simple summation of pole or zero dependent terms. To make a graph of the magnitude and phase we only need to consider the following four types of terms.

(a) Constants.   This is a term such as

$$20 \log K \tag{66}$$

if K is positive the phase is 0°, and if K is negative the phase is 180°. The magnitude of this term is plotted in Figure 12.16 for a value of $K = 100$.

(b) Roots at the Origin.   This means a pole or zero at $s = 0$ that yields a factor $j\omega$ in (63) so that we have a factor in (64) of the form

$$20 \log |j\omega|^{\pm 1} \tag{67}$$

where the + sign implies that the root is a zero and the minus sign implies a

Figure 12.16   Constant term.

pole. The phase is 90° for the zero and $-90°$ for the pole. The magnitudes are plotted in Figure 12.17. Note that the 0-dB point occurs at $\omega = 1$ rad/s if the coefficient of ω is unity, and the slope of the magnitude is $+6$ dB/oct for a zero and -6 dB/oct for a pole.

In the case of a multiple zero or pole of multiplicity n we have

$$20 \log[|j\omega|^{\pm n}] \tag{68}$$

Thus the 0-dB point is still at 1 rad/s, but the slope of the magnitude is $\pm 6n$ dB/oct and the phase is $\pm 90°n$ where the positive sign is taken in the case of a multiple zero and the negative sign for a multiple pole.

(c) Real Roots. This means a factor of the form $[(s/\sigma_i) + 1]$ in the transfer function where the root $-\sigma_i$ is located on the real axis. This factor contributes the following term to (64).

$$20 \log \left| \frac{j\omega}{\sigma_i} + 1 \right|^{\pm 1} \tag{69}$$

For small values of ω the above term is approximated by the asymptote

$$20 \log(1) = 0 \text{ dB} \tag{70}$$

and the phase is zero degrees. For large values of ω the magnitude

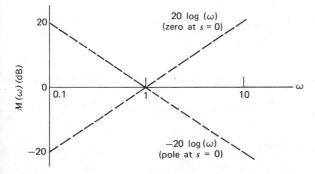

Figure 12.17   Roots at the origin.

asymptote is

$$\pm 20 \log \left| \frac{\omega}{\sigma_i} \right| \tag{71}$$

The sign is positive if the factor is a zero, and negative if the factor is a pole. Note that this asymptote crosses the 0-dB axis at $\omega = |\sigma_i|$, which is called the break frequency. At the break frequency the magnitude of the real root term is ± 3 dB and the phase is $\pm 45°$. The magnitude in decibels and the phase are plotted in Figure 12.18. In the phase plot it is assumed that $\sigma_i > 0$. If $\sigma_i < 0$ simply interchange the two phase plots.

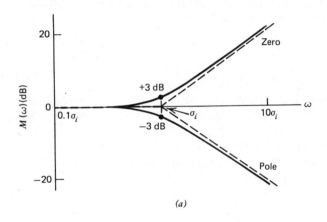

(a)

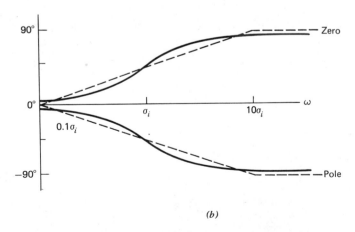

(b)

Figure 12.18 Real roots. (a) Magnitude. (b) Phase ($\sigma_i > 0$).

(d) Complex Roots. Complex roots are always grouped in pairs with their conjugates, and each conjugate pair is written in the form

$$\frac{s^2}{\omega_r^2} + \frac{1}{Q\omega_r}s + 1 \tag{72}$$

If the complex roots are poles, then ω_r is the resonant frequency of the pole pair and Q is called the Q or quality factor of the pole pair. In (64) this term is expressed as

$$20 \log \left| 1 - \frac{\omega^2}{\omega_r^2} + j\frac{\omega}{Q\omega_r} \right|^{\pm 1} \tag{73}$$

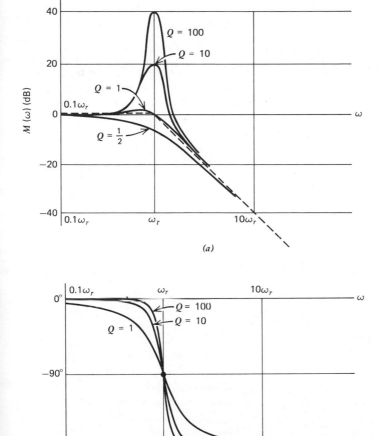

(a)

(b)

Figure 12.19 Complex roots. (a) Magnitude (pole). (b) Phase (pole).

The small frequency asymptote is

$$20 \log(1) = 0 \, \text{dB} \tag{74}$$

and the phase is zero for sufficiently small ω. The large frequency asymptote is

$$20 \log \left| \frac{\omega^2}{\omega_r^2} e^{j180°} \right|^{\pm 1} = \pm 20 \log \left(\frac{\omega}{\omega_r} \right)^2 \tag{75}$$

This asymptote has a slope of $\pm 12 \, \text{dB/oct}$ and crosses the 0-dB axis at the resonant frequency as shown in Figure 12.19a. In this figure the slope is negative since we have assumed that the roots are poles of the transfer function. For large values of ω the phase is approximately $\pm 180°$. At the break frequency ω_r the magnitude is

$$\pm 20 \log \left| j \frac{1}{Q} \right| = \pm 20 \log \left(\frac{1}{Q} \right) \tag{76}$$

The magnitude and phase functions are plotted in Figure 12.19 as a function of Q. The roots are assumed to be poles. In the case of complex zeros simply invert the plots in Figure 12.19. Since the roots are complex only for $Q > 1/2$, only values of $Q \geqslant 1/2$ were considered in these plots because case (c) applies for real roots ($Q \leqslant 1/2$).

Frequency response plots computed from (64) are called *Bode diagrams* or *Bode plots* in honor of H. W. Bode who originally proposed their use.[3]

Example 2

Consider the bandpass function

$$H(s) = \frac{10.1s}{s^2 + 10.1s + 1} \tag{77}$$

Sketch the magnitude and phase of this transfer function using the Bode diagram.

Upon factoring (77) we obtain

$$H(s) = \frac{10.1s}{(s + 0.1)(s + 10)} = \frac{10.1s}{\left(\dfrac{s}{0.1} + 1 \right)\left(\dfrac{s}{10} + 1 \right)} \tag{78}$$

Note that we have two real poles at -0.1 and -10. This is a low Q circuit ($Q = 1/10.1$). We write

$$20 \log|H(j\omega)| = 20 \log 10.1 + 20 \log|\omega| - 20 \log \left| 1 + \frac{j\omega}{0.1} \right| \tag{79}$$

$$- 20 \log \left| 1 + \frac{j\omega}{10} \right|$$

[3] H. W. Bode, *Network Analysis and Feedback Amplifier Design.* Van Nostrand Co., Princeton, N.J., 1945.

and

$$\phi(\omega) = 90° - \tan^{-1} 10\omega - \tan^{-1} 0.1\,\omega \qquad (80)$$

The logarithm of the magnitude of the frequency response is simply the summation of the four terms in (79) as shown in Figure 12.20a. The solid line indicates the sum of these four terms and the 3-dB difference at the breakpoints is considered. Recall that in the previous section we said that in low Q circuits the cutoff frequencies were $\omega_1 \approx Q$ and $\omega_2 \approx 1/Q$, which yields $\omega_1 \approx 0.1$ rad/s and $\omega_2 \approx 10$ rad/s for this circuit. Note that in Figure 12.20a the response is indeed down 3 dB at these frequencies. The phase is sketched in Figure 12.20b.

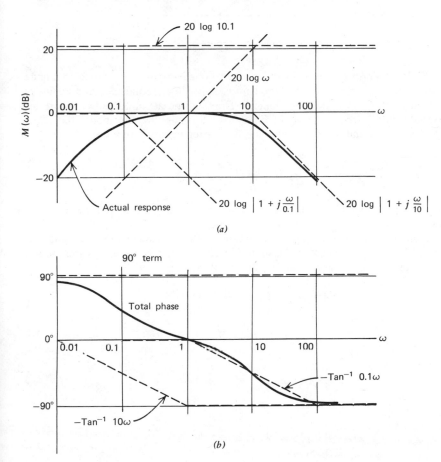

Figure 12.20 Bandpass function ($Q = 1/10.1$). (a) Magnitude. (b) Phase.

Example 3

The gain of an operational amplifier is not constant with respect to frequency due to the junction and other parasitic capacitances of the transistors in the amplifier. A

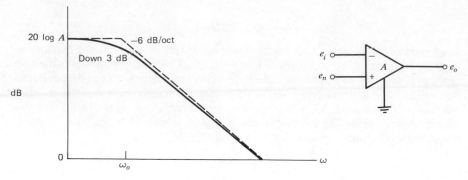

Figure 12.21 Operational amplifier frequency response.

typical Bode diagram of the gain of an operational amplifier is shown in Figure 12.21. From this graph we see that the gain is A for low frequencies and begins to decrease at 6 dB/oct at higher frequencies. Drawing the asymptotes, we locate the approximate cutoff frequency ω_o in Figure 12.21. Thus, we can model the input-output behavior of the amplifier with the first-order network function

$$\frac{\mathbf{E}_o(s)}{\mathbf{E}_n(s) - \mathbf{E}_i(s)} = \frac{A}{\dfrac{s}{\omega_o} + 1} \tag{81}$$

In the 741 op amp $A \approx 10^5$ (100 dB) and $\omega_o \approx 100$ rad/s. Thus, the op amp is a lowpass filter with a very small cutoff frequency.

Let us now compute the transfer function of the typical inverting amplifier in Figure 12.22. If we assume that the input impedance of the op amp is infinite and the output impedance is zero, the circuit equations are

$$\left(\frac{1}{R_s} + \frac{1}{R_f}\right)\mathbf{E}_i(s) - \frac{1}{R_f}\mathbf{E}_o(s) = \frac{\mathbf{V}_s}{R_s} \tag{82}$$

and

$$\mathbf{E}_o(s) = -A(s)\mathbf{E}_i(s) \tag{83}$$

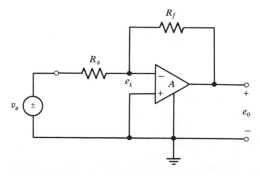

Figure 12.22 Inverting amplifier.

where

$$A(s) = \frac{A}{\dfrac{s}{\omega_o} + 1}$$

Upon the elimination of $\mathbf{E}_i(s)$ we obtain

$$\frac{\mathbf{E}_o(s)}{\mathbf{V}_s} = \frac{-\dfrac{R_f}{R_s}}{1 + \dfrac{1}{A} + \dfrac{R_f}{R_s A} + \dfrac{s}{\omega_o A}\left(1 + \dfrac{R_f}{R_s}\right)} \tag{84}$$

and usually the low frequency gain A is large enough so that

$$\frac{\mathbf{E}_o(s)}{\mathbf{V}_s} \approx \frac{-\dfrac{R_f}{R_s}}{1 + \dfrac{s}{\omega_o A R_s/(R_s + R_f)}} \tag{85}$$

Note that at low frequencies the gain is $-R_f/R_s$ as we computed in Chapter Seven, but at high frequencies the gain decreases 6 dB/oct. The break frequency is $\omega_b = \omega_o A R_s/(R_s + R_f)$. For example, if $R_f = 9\,\text{k}\Omega$ and $R_s = 1\,\text{k}\Omega$ so that the low frequency gain is 9, and if the op amp is the 741 with $A = 10^5$, and $\omega_o = 10^2\,\text{rad/s}$, then

$$\omega_b = (10^2)(10^5)\left(\frac{1}{10}\right) = 10^6\,\text{rad/s} \tag{86}$$

The frequency response of the amplifier in Figure 12.22 is plotted in Figure 12.23. Note that the break frequency of the amplifier with feedback is $\omega_o[AR_s/(R_s + R_f)]$ or $10^4\,\omega_o$ in this example. Thus, reducing the gain increases the bandwidth.

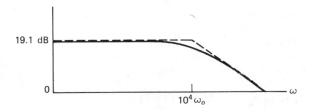

Figure 12.23   Inverting amplifier frequency response.

12.3 SCALING

Often the circuit examples in this book have probably seemed unrealistic in the sense that we have had element values of one farad or breakpoint frequencies of 1 rad/s, and so on. However, these circuits can readily be scaled in order to realize more practical element values and frequency responses.

To begin let us express KCL for the complex frequency network in the form

$$\mathbf{A}_n\mathbf{I}(s) = 0 \tag{87}$$

where the rows of the matrix $\mathbf{A}_n$ correspond to the nodes of the circuit and the columns correspond to the branches of the circuit and $\mathbf{I}(s)$ is a vector representing the complex amplitudes of the branch currents. The KVL equation has the form

$$\mathbf{M}\mathbf{V}(s) = 0 \tag{88}$$

where the rows of $\mathbf{M}$ correspond to the loops of the circuit and the columns of $\mathbf{M}$ correspond to the branches of the circuit and $\mathbf{V}(s)$ is a vector representing the complex amplitudes of the branch voltages. These two equations are similar to Equations 47 and 48 in Chapter Three, only they involve the complex amplitudes of the currents and voltages. The elements in the matrices $\mathbf{A}_n$ and $\mathbf{M}$ are either $+1$, -1, or 0. For example, if branch k is connected between nodes m and n and the current is referenced to flow from node m to node n, then $a_{mk} = +1$ and $a_{nk} = -1$. If branch k is not connected to node m then $a_{mk} = 0$. Similarly, if branch k is in loop j, then $m_{jk} = \pm 1$ depending on the polarity of the branch. If branch k is not in loop j, then $m_{jk} = 0$. In addition, to the Kirchhoff equations we have the branch constraints of the circuit. Some typical branch constraints are listed in column 1 of Table 12.3. The last four constraints are the four types of controlled source constraints.

Now suppose that we wish to conserve power by reducing the current amplitudes by a. This means that all of the branch currents must be divided

Table 12.3 Magnitude and Frequency Scaling of Circuit Elements

Unscaled	Magnitude Scaled	Frequency Scaled
$\mathbf{V}(s) = R\mathbf{I}(s)$	$\mathbf{V}(s) = (aR)(\mathbf{I}(s)/a)$	$\mathbf{V}(s) = R\mathbf{I}(s)$
$\mathbf{V}(s) = sL\mathbf{I}(s)$	$\mathbf{V}(s) = s(aL)(\mathbf{I}(s)/a)$	$\mathbf{V}(s) = s\left(\dfrac{L}{b}\right)\mathbf{I}(s)$
$\mathbf{V}(s) = \dfrac{1}{sC}\mathbf{I}(s)$	$\mathbf{V}(s) = \dfrac{1}{s(C/a)}(\mathbf{I}(s)/a)$	$\mathbf{V}(s) = \dfrac{1}{s(C/b)}\mathbf{I}(s)$
$\mathbf{V}_x(s) = \mu\mathbf{V}_y(s)$	$\mathbf{V}_x(s) = \mu\mathbf{V}_y(s)$	$\mathbf{V}_x(s) = \mu\mathbf{V}_y(s)$
$\mathbf{V}_x(s) = r\mathbf{I}_y(s)$	$\mathbf{V}_x(s) = (ar)(\mathbf{I}_y(s)/a)$	$\mathbf{V}_x(s) = r\mathbf{I}_y(s)$
$\mathbf{I}_x(s) = g_m\mathbf{V}_y(s)$	$\dfrac{\mathbf{I}_x(s)}{a} = \left(\dfrac{g_m}{a}\right)\mathbf{V}_y(s)$	$\mathbf{I}_x(s) = g_m\mathbf{V}_y(s)$
$\mathbf{I}_x(s) = \beta\mathbf{I}_y(s)$	$\dfrac{\mathbf{I}_x(s)}{a} = \beta\dfrac{\mathbf{I}_y(s)}{a}$	$\mathbf{I}_x(s) = \beta\mathbf{I}_y(s)$

by a. Equation 87 becomes

$$A \left[\frac{1}{a} \mathbf{I}(s) \right] = 0 \qquad (89)$$

In addition, the branch constraints must be modified as shown in column 2 of Table 12.3, that is, to reduce the currents in our circuit by a, we must increase all resistances and inductances by a and reduce all conductances, capacitances, and current sources by a. This scaling leaves all voltages unchanged.

The next type of scaling is called frequency scaling. The frequency response of a circuit is scaled by replacing the complex frequency variable s by s/b. In an actual circuit this type of scaling is accomplished by dividing every inductor and every capacitor in the circuit by b as shown in column 3 of Table 12.3. The frequency response obtained at $s = j\omega_o$ in the unscaled circuit now is obtained at $s = j(b\omega_o)$ in the frequency scaled circuit. However, not only is the frequency response of the circuit scaled but the transient response is also scaled by b. For example, the transfer function (7) becomes

$$H(s) = Kb^{n-m} \frac{(s - bz_1)(s - bz_2) \cdots (s - bz_m)}{(s - bp_1)(s - bp_2) \cdots (s - bp_n)} \qquad (90)$$

so that the transient response is now

$$y_t(t) = \sum_{i=1}^{n} k_i e^{-p_i b t} \qquad (91)$$

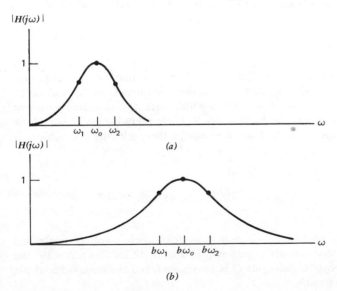

(a)

(b)

Figure 12.24 Effect of frequency scaling on the frequency response.

For example, if b is greater than 1, than all of the capacitors and inductors are *reduced* by the factor b, the breakpoint frequencies of the frequency response are *increased* by b, and the transient response is *faster* by a factor b (time constants are reduced by b). Figures 12.24 and 12.25 graphically illustrate the effect of scaling on the frequency and time domain responses. Below we give an example of scaling.

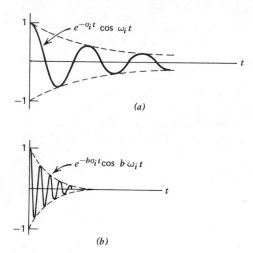

Figure 12.25 Effect of frequency scaling on the transient response. (*a*) Unscaled circuit. (*b*) Scaled circuit.

Example 4

In the bandpass filters discussed in Section 12.1.2 suppose that we want to reduce the currents by $a = 10^3$ and we want a resonant frequency of 10^4 rad/s. To accomplish this frequency scaling we set $b = 10^4$ since the unscaled resonant frequency is 1 rad/s. Let us scale both the series RLC circuit in Figure 12.14 and the active RC circuit in Figure 12.15. The elements in these circuits are scaled as follows.

$$R \rightarrow aR$$
$$L \rightarrow aL/b$$
$$C \rightarrow C/ab$$

The voltage amplifier is unchanged since it is modeled with a voltage-controlled voltage source. The scaled circuits are shown in Figure 12.26. Since $b = 10^4$ the resonant frequency is now 10^4 rad/s but Q is unchanged, and the bandwidth of the filter is now $0.2\omega_r = 2000$ rad/s.

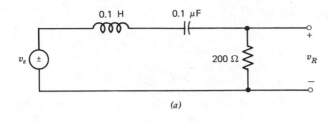

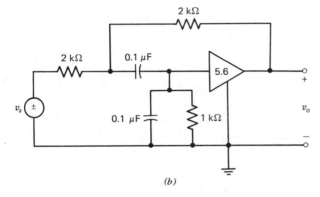

Figure 12.26   Magnitude and frequency scaling. (*a*) RLC circuit. (*b*) Active RC circuit.

12.4 COMPLEX LOCI

Sometimes it is useful to have a pictorial diagram of the variation of the real and imaginary part of a complex quantity such as a phasor or network function with respect to some parameter in the circuit such as the frequency of the input signal. For example, the admittance of a series RLC circuit is

$$Y(j\omega) = \frac{1}{R + j\left(\omega L - \frac{1}{\omega C}\right)} \tag{92}$$

At the resonant frequency $\omega_r = 1/\sqrt{LC}$ we have

$$Y(j\omega_r) = \frac{1}{R} = G \tag{93}$$

and for $\omega > \omega_r$ the circuit is inductive so that $\text{Im}[Y(j\omega)] < 0$ and for $\omega < \omega_r$ the circuit is capacitive so that $\text{Im}[Y(j\omega)] > 0$. Furthermore,

$$\text{Re}[Y(j\omega)] = \frac{R}{R^2 + \left(\omega L - \frac{1}{\omega C}\right)^2} \tag{94}$$

and

$$\text{Im}[Y(j\omega)] = \frac{-\left(\omega L - \dfrac{1}{\omega C}\right)}{R^2 + \left(\omega L - \dfrac{1}{\omega C}\right)^2} \tag{95}$$

From (94) we see that $\text{Re}[Y(j\omega)] > 0$ for $R > 0$. The plot of $Y(j\omega)$ in the complex plane as a function of ω is shown in Figure 12.27. The locus is a circle with diameter G.

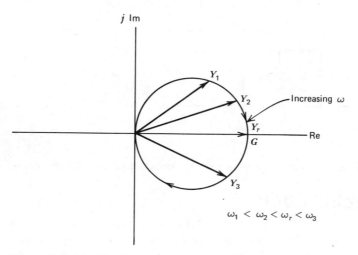

Figure 12.27 Locus of admittance function series RLC circuit.

To prove that the locus is a circle we write

$$\mathbf{I}(j\omega) = Y(j\omega)\mathbf{V}(j\omega) \tag{96}$$

and let the input phasor $\mathbf{V}(j\omega) = 1$ so that

$$\mathbf{I}(j\omega) = Y(j\omega) \tag{97}$$

Also, note that $\mathbf{V}_R = R\mathbf{I}$ and $\mathbf{V}_{LC} = j[\omega L - (1/\omega C)]\mathbf{I}$ where

$$\mathbf{V}_R + \mathbf{V}_{LC} = 1 \tag{98}$$

that is, the voltage across the resistor plus the voltage across the LC combination must equal the input voltage, a constant equal to 1. Furthermore, note that $\mathbf{V}_R$ is always 90° out of phase with $\mathbf{V}_{LC}$. If two vectors are always 90° out of phase and their sum is a constant, then the locus of the

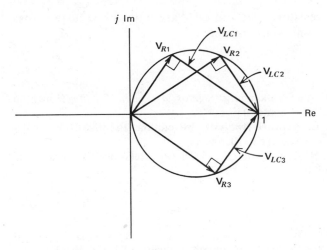

Figure 12.28 Locus of the resistor voltage $\mathbf{V}_R$.

phasor $\mathbf{V}_R$ must lie on the circle shown in Figure 12.28. Since

$$\mathbf{I}(j\omega) = \frac{1}{R}\mathbf{V}_R = Y(j\omega) \tag{99}$$

it follows that the locus of $Y(j\omega)$ must be a circle.

As a final example let us plot the complex locus of the admittance of a series RC circuit.

$$Y(j\omega) = \frac{1}{R + \dfrac{1}{j\omega C}} \tag{100}$$

The locus is the semicircle shown in Figure 12.29. It is a semicircle because

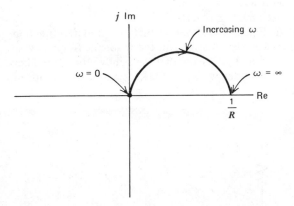

Figure 12.29 Admittance locus of series RC circuit.

the voltage across the resistor is 90° out of phase with the voltage across the capacitor and their sum is a constant.

REFERENCES

1. Cruz, J. B. and M. E. Van Valkenburg. *Signals in Linear Circuits*, Houghton Mifflin Co., Boston, 1974, Chapter 11.
2. Van Valkenburg, M. E. *Network Analysis*, 3rd edition, Prentice-Hall, Englewood Cliffs, N.J., 1974, Chapters 9, 10, and 13.
3. Desoer, C. A. and E. S. Kuh. *Basic Circuit Theory*, McGraw-Hill, New York, 1969, Chapters 15 and 16.

PROBLEMS

1. Sketch the magnitude and phase of the network function $\mathbf{V}_C(s)/\mathbf{V}_s(s)$ and $\mathbf{V}_R(s)/\mathbf{V}_s(s)$ versus ω for the series RC circuit in Figure 12.3. Let $R = 1\,\text{k}\Omega$ and $C = 1\,\mu\text{F}$. What is the effect of (a) increasing R or C by a factor of 10, or (b) decreasing R or C by a factor of 10?

2. If $R = 1\,\text{k}\Omega$ and $L = 1\,\text{mH}$ in Figure P12.2, sketch the magnitude and phase of the network functions $\mathbf{V}_L(s)/\mathbf{V}_s(s)$ and $\mathbf{V}_R(s)/\mathbf{V}_s(s)$ versus ω.

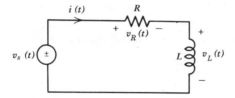

Figure P12.2

3. Repeat Problem 2 for $R = 100\,\Omega$ and $L = 1\,\text{mH}$.
4. Repeat Problem 2 for $R = 1\,\text{k}\Omega$ and $L = 10\,\mu\text{H}$.
5. For the series RLC circuit in Figure P12.5 verify that the transfer function $\mathbf{V}_R(s)/\mathbf{V}_s(s)$ is a bandpass function. Express Q and the resonant frequency ω_r in terms of the circuit parameters R, L, C. Show that the bandwidth of the filter is R/L. Finally, design a bandpass filter with $f_r = 100\,\text{kHz}$ and a bandwidth of $10\,\text{kHz}$. Assume that $R = 1\,\text{k}\Omega$.

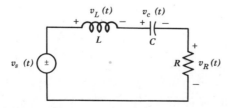

Figure P12.5

6. In the series RLC circuit show that for an input voltage $V_s/\underline{0°}$ the voltage drop across the inductor is $jQ \cdot V_s$ and the voltage drop across the capacitor is $-jQ \cdot V_s$ when the frequency of the input voltage is equal to the resonant frequency of the circuit.

7. In the series resonant circuit, Figure P12.5, show that in the steady state the *stored* energy

$$w(t) = \frac{1}{2}Li^2(t) + \frac{1}{2}Cv_c^2(t)$$

is constant and equal to $(1/2)L(V/R)^2$ at the resonant frequency $\omega_r = 1/\sqrt{LC}$. It is assumed that the input phasor $V_s = V/\underline{0°}$. At resonance the total energy *dissipated* per cycle is $(1/2)(V^2/R)T$ where $T = 2\pi/\omega_r$. Show that

$$\frac{2\pi \text{ (total stored energy at resonance)}}{\text{(energy dissipated per cycle at resonance)}} = Q$$

8. For the parallel resonant circuit in Figure P12.8 verify that the transfer function $\mathbf{I}_R(s)/\mathbf{I}_s(S)$ is a bandpass function. Express Q, ω_r, and the bandwidth in terms of the circuit parameters R, L, and C. Design a bandpass filter with $f_r = 1$ MHz and a bandwidth of 10 kHz. Let $R = 10$ kΩ.

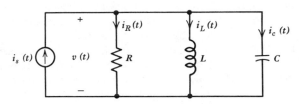

Figure P12.8

9. In the parallel resonant circuit in Figure P12.8 show that for an input current $I_s/\underline{0°}$ the current through the capacitor is $jQ \cdot I_s$ and the current through the inductor is $-jQ \cdot I_s$ when the frequency of the input current is equal to the resonant frequency of the circuit.

10. Sketch the Bode plots (magnitude and phase) for the following network functions:

(a) $\dfrac{10}{(10^{-3}s + 1)(10^{-4}s + 1)}$ (b) $\dfrac{10s}{(s + 1)(0.1s + 1)}$

(c) $\dfrac{50(10^{-2}s + 1)}{(0.1s + 1)(10^{-3}s + 1)}$ (d) $\dfrac{1000}{s(0.1s + 1)}$

11. Repeat Problem 10 for the following two transfer functions:

(a) $\dfrac{20}{10^{-4}s^2 + 10^{-3}s + 1}$ (b) $\dfrac{0.01s^2 + 0.02s + 1}{(s + 1)(0.01s + 1)}$

12. Repeat Problem 10 for the transfer functions below.

(a) $\dfrac{10(s + 10)}{(s + 1)(s + 100)}$ (b) $\dfrac{s^2}{s^2 + 10^2s + 10^6}$.

13. In Example 3 suppose that the gain-bandwidth of the op amp $\omega_o A = 2\pi \times 1$ MHz. What is the bandwidth of the inverting amplifier in Figure 12.22 if $R_s = 1$ kΩ and (a) $R_f = 100$ kΩ, (b) $R_f = 10$ kΩ, and (c) $R_f = 1$ kΩ?

14. Repeat Problem 13 only for the noninverting amplifier in Figure P12.14.

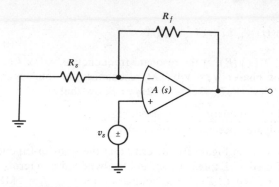

Figure P12.14

15. In Problem 12 replace the complex variable s by $s/10^3$. What is the effect on the pole-zero locations, the frequency response, and the time domain response?

16. In the example in Figure 12.14, scale the elements as follows: $L \to 10^3 L$, $R \to 10^3 R$, and $C \to C/10^3$. What effect has this scaling had on the pole-zero locations, time domain response and the frequency domain response? Explain your results. Suppose that $L \to L/10^6$ and $C \to C/10^6$. What is the effect now on the pole-zero locations, the frequency response and the time domain response? Show that this latter scaling is equivalent to replacing s by $s/10^6$.

17. Design a lowpass filter with a transfer function of the form

$$\frac{K\omega_r^2}{s^2 + \frac{\omega_r}{Q}s + \omega_r^2}$$

Use the series RLC circuit with $R = 100$ Ω and $v_c(t)$ the response. Find L and C such that $Q = 1$ and $f_r = 1$ kHz. Sketch the magnitude and phase of $V_C(s)/V_s(s)$ and $V_L(s)/V_s(s)$ versus ω.

18. Find the transfer function $V_o(s)/V_s(s)$ for the circuit in Figure P12.18 where r denotes the resistance of the inductor coil. Neglect r and design this circuit to

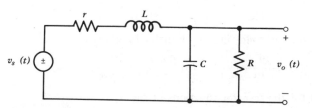

Figure P12.18

have a $Q = 1$ and a resonant frequency $\omega_r = 1 \, \text{rad/s}$. Assume $R = 100 \, \Omega$. If $r = 1 \, \Omega$, compute the change in the resonant frequency, Q, and the gain at resonance. Repeat the above analysis with $Q = 100$. Frequency scale both of the above designs so that the resonant frequency is $1 \, \text{kHz}$.

19. Show that the circuit in Figure P12.19 yields the transfer function

$$\frac{V_o(s)}{V_s(s)} = \frac{Y_1 Y_3}{Y_1 Y_3 + Y_1 Y_4 + Y_2 Y_3 + Y_2 Y_4 + Y_3 Y_4}$$

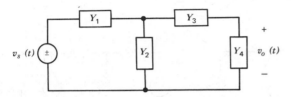

Figure P12.19

20. In Figure P12.19 each branch represents a resistor or capacitor. Select the elements to realize the lowpass transfer function

$$H(s) = \frac{K\omega_r^2}{s^2 + \dfrac{\omega_r}{Q} s + \omega_r^2}$$

Let each resistor be R ohms and each capacitor C farads. What is the Q of the circuit? Let $R = 1 \, \Omega$ and $C = 1 \, \text{F}$, then scale the impedance up by 10^3 and the frequency response up by 10^4. Sketch the magnitude and phase of this transfer function versus ω.

21. Repeat Problem 20, only find the RC circuit that realizes the second-order bandpass transfer function

$$H(s) = \frac{K\omega_r s}{s^2 + \dfrac{\omega_r}{Q} s + \omega_r^2}$$

22. Repeat Problem 20, only find the RC circuit that realizes the second-order highpass transfer function

$$H(s) = \frac{s^2}{s^2 + \dfrac{\omega_r}{Q} s + \omega_r^2}$$

23. Impedance scale the circuits in Figure P12.23 such that all branch currents are reduced by 10^2 and the voltages are unchanged.

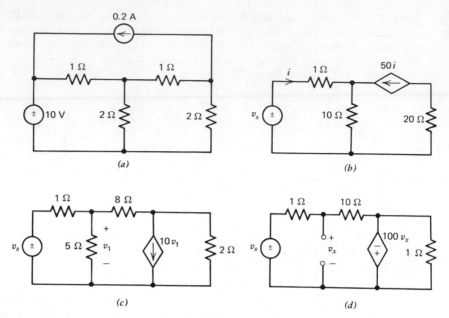

Figure P12.23

24. Scale the frequency response up by a factor of 10^5 and increase the impedance by 50 in Figure P12.24.

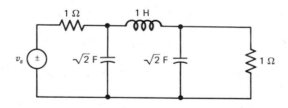

Figure P12.24

25. Show that the transfer function for the circuit in Figure P12.25 is

$$\frac{V_o(s)}{V_s(s)} = \frac{AY_1Y_2}{Y_1Y_2 + Y_1Y_3 + Y_2Y_3 + Y_3Y_f + Y_2Y_f(1-A)}$$

26. Assume that the amplifier in Figure P12.25 is an inverting amplifier, $A < 0$. A second-order bandpass function can be realized by choosing $Y_1 = sC_1$, $Y_2 = G_2$, $Y_3 = sC_3$, and $Y_f = G_f$. Let the element values be unity and find the gain $A < 0$ such that $Q = 5$. Scale the impedance up by 10^3 and the frequency such that $f_r = 2$ kHz.

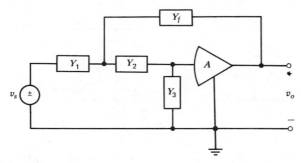

Figure P12.25

27. Repeat Problem 26 only let $Y_1 = G_1$, $Y_2 = sC_2$, $Y_3 = G_3$, and $Y_f = sC_f$.
28. The circuit in Figure P12.28 is called a twin T or notch filter. Show that if $C_1 = C_2 = C$, $C_3 = 2C$, $G_1 = G_2 = G$, and $G_3 = 2G$, then

$$H(s) = \frac{V_o(s)}{V_s(s)} = \frac{s^2 + \left(\dfrac{1}{RC}\right)^2}{s^2 + \dfrac{4}{RC}s + \left(\dfrac{1}{RC}\right)^2}$$

If $R = 1\,k\Omega$, find the value of C such that the transfer function is zero (the steady state output is zero) at the frequency 120 Hz, that is, $H(j2\pi \times 120) = 0$. Sketch the magnitude and phase of $H(j\omega)$ versus ω.

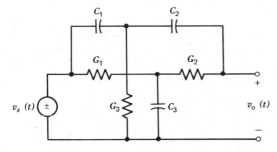

Figure P12.28

29. A small-signal model of a common-emitter transistor amplifier is shown in Figure P12.29. The capacitor C_π represents the effects of the pn junction capacitance and other parasitic capacitive effects. Find the transfer function $H(s) = V_o(s)/V_s(s)$. If $r_\pi = 2\,k\Omega$, $g_m = 0.04$ mho, $R_s = 100\,\Omega$, $R_L = 2\,k\Omega$, and $C_\pi = 100$ pF, sketch $|H(j\omega)|$ versus ω. At what frequency does the gain decrease to 3 dB below the low frequency gain. Typically this frequency determines the effective bandwidth of the amplifier.

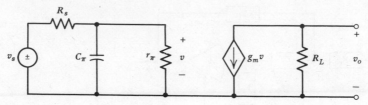

Figure P12.29

30. In Figure P12.30 under what conditions will $\text{Im}[Z(j\omega)] = 0$ for all ω? Under what conditions will $Z(j\omega) = Y(j\omega)$ for all ω?

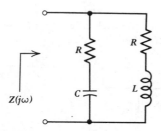

Figure P12.30

31. In Figure P12.31 find the transfer function $H(s) = V_o(s)/V_{in}(s)$ and select R such that $v_o(t) = 0.1 v_{in}(t)$ for low frequencies. What is the bandwidth of this circuit? Place a capacitor in parallel with the resistor R and find the value of capacitance such that $V_o(s)/V_{in}(s) = 0.1$ for all ω. The addition of the capacitor compensates for the reduction in bandwidth due to the large value of R. This compensation technique is used in X10 attenuation probes for oscilloscopes. The 1-MΩ resistor and 10-pF capacitor model the input impedance characteristics of the oscilloscope.

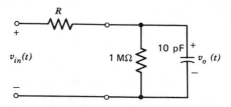

Figure P12.31

32. Plot the complex locus of the impedance of the parallel resonant circuit in Figure P12.8.

33. The small-signal model of the bipolar transistor for the base-emitter terminals is illustrated in Figure P12.33. In general, $r_b \ll r_\pi$. Explain how one can use the plot of the complex locus of $Z(j\omega)$ to determine r_π and r_b.

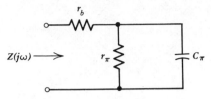

Figure P12.33

Chapter Thirteen
The Transformer

We have already shown that the operation of the electric generator is based on the principle of magnetic induction, which is mathematically expressed by Faraday's law. Another useful device based on the principle of magnetic induction is the transformer, which consists of N magnetically coupled coils. Since each coil has a pair of terminals, the transformer is an N-port. Our discussion will be restricted primarily to transformers with only two magnetically coupled coils (a two-port). Basically the operation of the transformer is as follows. A time-varying current in one coil produces a time-varying flux that links the second coil. The voltage induced in the second coil is proportional to the time-rate-of-change of the flux and the number of turns in the second coil that couple the flux. Therefore, the voltage across the second coil is essentially $(n_2/n_1)v_1$ where v_1 is the voltage across the first coil, n_1 is the number of turns in the first coil, and n_2 is the number of turns in the second coil. Since the transformer consists only of a pair of coupled coils, it cannot deliver more energy at one port than it takes in at the other port. It follows that the current in the second coil should be proportional to $(n_1/n_2)i_1$. Thus, if the voltage across the second coil is double the voltage across the first coil, then the current in the second coil is approximately one-half the current in the first coil. As we will soon see, this ability to change the current and voltage levels has valuable applications in communication systems and in the transmission of power from a generator to a load. First let us develop some mathematical models for the transformer.

13.1 TRANSFORMER CONSTRAINTS

Consider the coupled coils in Figure 13.1. If we neglect the series resistance and capacitive effects of the coils, then Faraday's law yields

$$v_1 = n_1 \frac{d\phi_{11}}{dt} + n_1 \frac{d\phi_{12}}{dt}$$

and

$$v_2 = n_2 \frac{d\phi_{21}}{dt} + n_2 \frac{d\phi_{22}}{dt}$$

(1)

where the flux ϕ_{11} is the flux produced by i_1, which links coil 1; ϕ_{12} is the flux produced by i_2, which links coil 1; ϕ_{21} is the flux produced by i_1, which links coil 2; and ϕ_{22} is the flux produced by i_2, which links coil 2. Let n_1 and n_2 represent the number of turns of coil 1 and coil 2, respectively. By definition, the fluxes ϕ_{11} and ϕ_{22} will always be positive. However, ϕ_{12} and ϕ_{21} can be positive (flux from the coils in the same direction) or negative (flux from the coils in opposite directions). In Figure 13.1 note that for the given current assignment the flux produced by i_1 aids the flux produced by i_2 so that ϕ_{12} and ϕ_{21} are positive. However, if one of the coils was wound in the reverse direction on the core, or if the polarity of the current and voltage at one of the ports was reversed, then ϕ_{12} and ϕ_{21} would be negative. Please note that we are only setting up reference directions for the voltage and current. If our assigned direction is opposite to the actual flow or polarity, then we will obtain a negative answer.

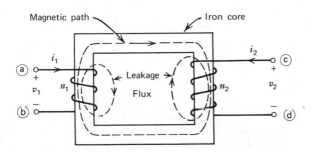

Figure 13.1 Two coupled coils.

Let us assume that the above fluxes are linear functions of the current producing them, that is,

$$n_1\phi_{11} = L_1 i_1$$
$$n_1\phi_{12} = \pm M_{12} i_2$$
$$n_2\phi_{21} = \pm M_{21} i_2$$
$$n_2\phi_{22} = L_2 i_2$$

(2)

so that (1) becomes

$$v_1 = L_1 \frac{di_1}{dt} \pm M_{12} \frac{di_2}{dt}$$

and

$$v_2 = \pm M_{21} \frac{di_1}{dt} + L_2 \frac{di_2}{dt}$$

(3)

The inductances L_1 and L_2 are called the self-inductances of the coil, and M_{12} and M_{21} are called the mutual inductances of the coil. Because of certain symmetry properties of coupled coils it can be shown that $M_{12} = M_{21} = M$ in linear coils.[1] *It is convenient to assume that M is always a positive number and its sign in (2) and (3) is positive if for the assigned current directions the fluxes ϕ_{11} and ϕ_{12}, and ϕ_{22} and ϕ_{21} aid each other, otherwise the negative sign is chosen.*

The typical circuit symbol for the magnetically coupled coils is shown in Figure 13.2a. The dots are used to indicate the directions of the windings on the core and are *independent* of the assigned current directions. *The convention is that if the two port currents are both assigned flowing into the dots or out of the dots then the fluxes aid each other. However, if one port current is assigned into the dot and the other port current is assigned flowing away from the dot then the fluxes oppose each other.* From this convention verify from Figure 13.1 that the dots are correctly placed in Figure 13.2a. Of course, in practice we may not know how the coils are wound on the core, but the location of the dots for our circuit diagram can be determined by measurement techniques. For example, open circuit one port and apply an increasing current at the other port. From Lenz's law we conclude that the dots belong at the input terminal at which the input current is increasing and the output terminal at which the voltage is positive. One could also use sinusoidal techniques and observe the phase difference between the input current and the output voltage.

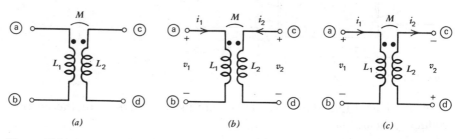

Figure 13.2   (a) Transformer symbol for two coupled coils. (b) and (c) Two of the four possible choices for the assignment of current and voltage in the standard reference system.

[1] S. Ramo, J. R. Whinnery, and T. Van Duzer, *Fields and Waves in Communication Electronics*, Wiley, 1965, pp. 284–285.

In Figure 13.2*b* both currents are assigned flowing into the dots; thus for this current assignment the branch constraints are

$$v_1 = L_1 \frac{di_1}{dt} + M \frac{di_2}{dt}$$

$$v_2 = M \frac{di_1}{dt} + L_2 \frac{di_2}{dt}$$

(4)

However, for the current assignment in Figure 13.2*c* the branch constraints are

$$v_1 = L_1 \frac{di_1}{dt} - M \frac{di_2}{dt}$$

$$v_2 = - M \frac{di_1}{dt} + L_2 \frac{di_2}{dt}$$

(5)

Finally the coil at the input port of the transformer is called the *primary winding* and the coil at the output port of the transformer is called the *secondary winding*.

13.2 ENERGY CONSIDERATIONS

The purpose of the following discussion is to use energy arguments to show that $L_1 L_2 \geq M^2$ in any two-port transformer. The total power in the two port at any time is

$$v_1 i_1 + v_2 i_2 = \left(L_1 \frac{di_1}{dt} + M \frac{di_2}{dt} \right) i_1 + \left(M \frac{di_1}{dt} + L_2 \frac{di_2}{dt} \right) i_2$$

(6)

and the total energy taken by the transformer is

$$w(t) = \int_{-\infty}^{t} \left[\left(L_1 \frac{di_1}{dt} + M \frac{di_2}{dt} \right) i_1 + \left(M \frac{di_1}{dt} + L_2 \frac{di_2}{dt} \right) i_2 \right] dt$$

(7)

If we assume that $i_1(-\infty)$ and $i_2(-\infty)$ are zero, the above expression simplifies to

$$w(t) = \frac{1}{2} L_1 i_1^2(t) + M i_1(t) i_2(t) + \frac{1}{2} L_2 i_2^2(t)$$

(8)

Since a pair of coupled coils is passive, that is, the total energy $w(t) \geq 0$ for all time, then

$$\frac{1}{2} L_1 i_1^2 + M i_1 i_2 + \frac{1}{2} L_2 i_2^2 \geq 0.$$

(9)

If we divide inequality (9) by i_2^2 and let $x = i_1/i_2$, then

$$f(x) \triangleq \frac{1}{2} L_1 x^2 + Mx + \frac{1}{2} L_2 \geq 0$$

(10)

This result implies that the roots of the equation

$$\frac{1}{2} L_1 x^2 + Mx + \frac{1}{2} L_2 = 0 \tag{11}$$

are imaginary, or at most equal (see Figure 13.3). Note that if $f(x)$ lies below the axis, a violation of (10), then there are two real roots. From (11)

$$x = \frac{-M \pm \sqrt{M^2 - L_1 L_2}}{L_1} \tag{12}$$

The restriction $w(t) \geqslant 0$ then implies

$$L_1 L_2 \geqslant M^2 \tag{13}$$

The ratio

$$\boxed{k = \frac{M}{\sqrt{L_1 L_2}} \leqslant 1} \tag{14}$$

is defined as the *coefficient of coupling* for the transformer. The result is identical if we had used $-M$ in (6) to (12).

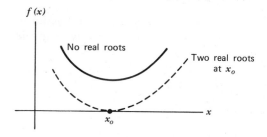

Figure 13.3 Energy constraint forces $f(x) \geq 0$.

13.3 PERFECTLY COUPLED TRANSFORMER MODEL

It is interesting to observe the case when $|\phi_{12}| = \phi_{22}$ and $|\phi_{21}| = \phi_{11}$, that is, all of the flux produced by the current i_2 links both coil 2 and coil 1, and all of the flux produced by the current i_1 links both coil 1 and coil 2 so that there is no leakage, then we say the coils are *perfectly coupled*. Substituting this condition for perfect coupling into (1) we obtain

$$v_1 = n_1 \frac{d\phi_{11}}{dt} \pm n_1 \frac{d\phi_{22}}{dt} \tag{15}$$

$$v_2 = \pm n_2 \frac{d\phi_{11}}{dt} + n_2 \frac{d\phi_{22}}{dt} \tag{16}$$

so that the multiplication of (16) by $\mp (n_1/n_2)$ and the addition of (15) and (16) yields

$$v_1 = \pm \frac{n_1}{n_2} v_2 \tag{17}$$

where the plus sign again indicates that with the assigned polarity the fluxes aid each other. *Thus, in the perfectly coupled transformer the input and output voltages are simply related by the turns ratio.* Next we show that the coefficient of coupling is unity.

From (2) recall that

$$n_1 \phi_{11} = L_1 i_1 \tag{18}$$

$$n_1 |\phi_{12}| = M i_2 \tag{19}$$

$$n_2 |\phi_{21}| = M i_1 \tag{20}$$

$$n_2 \phi_{22} = L_2 i_2 \tag{21}$$

but if the coils are perfectly coupled so that $\phi_{11} = |\phi_{21}|$ and $\phi_{22} = |\phi_{12}|$ then from (18) to (21) it follows that

$$\frac{n_1}{n_2} = \frac{L_1}{M} = \frac{M}{L_2} \tag{22}$$

and (22) yields

$$L_1 L_2 = M^2 \tag{23}$$

Hence, the coefficient of coupling is "1" for the perfectly coupled transformer.

13.4 THE IDEAL TRANSFORMER MODEL

Finally, we would like to further simplify our equation to arrive at the most elementary model of the transformer, the *ideal transformer*. Suppose we have a perfectly coupled transformer. Let us divide (3) by L_1 and L_2 respectively and utilize (22) to obtain

$$\frac{v_1}{L_1} = \frac{di_1}{dt} + \frac{M}{L_1} \frac{di_2}{dt} = \frac{di_1}{dt} \pm \frac{n_2}{n_1} \frac{di_2}{dt}$$
$$\frac{v_2}{L_2} = \frac{M}{L_2} \frac{di_1}{dt} + \frac{di_2}{dt} = \pm \frac{n_1}{n_2} \frac{di_1}{dt} + \frac{di_2}{dt} \tag{24}$$

If we assume that the permeability μ of the iron core of the transformer is infinite then L_1 and L_2 are infinite so that (24) yields

$$i_1 = \mp \frac{n_2}{n_1} i_2 \tag{25}$$

The ideal transformer is a perfectly coupled transformer (17), which, in addition, satisfies (25). If, for the assigned current directions, the mutual coupling is positive then the upper signs in (17) and (25) are used, and the lower signs are used for negative coupling. Finally note that in the ideal transformer the instantaneous power

$$p(t) = v_1 i_1 + v_2 i_2 \equiv 0$$

for all t. This can easily be proved by the substitution of (17) and (25) into this expression.

13.5 APPLICATION OF THE TRANSFORMER IN CIRCUITS

The transformer finds very wide application in two significant areas of electrical systems: communication systems and power systems. In communication systems, the transformer is used to change impedance levels in order to match impedances so that maximum power can be delivered to a load. In power systems, transformers are used to change voltage and current levels in order to reduce transmission line losses and to make available a specified voltage. For example, in North America the specified residential rms voltage is 120 V and home appliances are designed to operate with this voltage. If a higher or lower voltage is needed then a transformer must be used to change the voltage level.

Let us begin with its application in changing impedance levels. Consider the circuit in Figure 13.4. The generator v_g has a fixed internal resistance

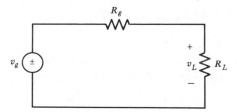

Figure 13.4 Generator driving a resistive load.

R_g. The power delivered to the load R_L is

$$P_L = \left(\frac{v_g}{R_L + R_g}\right)^2 R_L \tag{26}$$

Recall from Chapter Eleven that $R_L = R_g$ yields maximum power transfer. When the generator and load resistors are equal we say that the load is matched to the generator.

Now let us examine the circuit in Figure 13.5. The 90-Ω load resistor is

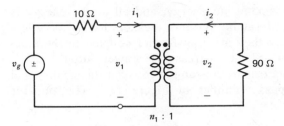

Figure 13.5   Transformation of voltage, current, and resistance levels.

transformer coupled to the generator, which has a 10-Ω internal resistance. We will now show that by a proper choice of the turns ratio n_1, we can make the 90-Ω resistor look like a 10-Ω resistor to the generator so that maximum power transfer is achieved. Let us first compute the DP resistance at the input port of the transformer assuming that the transformer is ideal. The equations are

$$\frac{v_1}{v_2} = \frac{n_1}{1} \tag{27}$$

$$\frac{i_1}{i_2} = -\frac{1}{n_1} \tag{28}$$

and

$$90 i_2 = - v_2 \tag{29}$$

Upon the elimination of i_2 and v_2 from the above equations we obtain

$$v_1 = (n_1{}^2 90) i_1 \tag{30}$$

The resistance seen by the generator is $n_1{}^2 \times 90\ \Omega$. If $n_1 = 1/3$, that is, the secondary coil (output) of the transformer has three turns to every turn of the primary coil (input), then

$$\frac{v_1}{i_1} = 10\ \Omega \tag{31}$$

The equivalent circuit is shown in Figure 13.6, and note that with the above turns ratio, maximum power is delivered to the equivalent load, but since

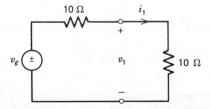

Figure 13.6   Load matched to generator when $n_1 = 1/3$.

the transformer is ideal it dissipates no energy, and all of the energy is delivered to the 90-Ω resistor. In audio systems speakers have a very low resistance, therefore frequently they are transformer coupled to the electronic circuits driving them in order to increase the power output.[2]

As a second impedance transformation example consider the small-signal model of a transistor bandpass amplifier in Figure 13.7. The transfer

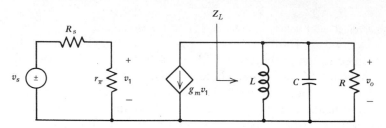

Figure 13.7   Bandpass amplifier model.

function for this amplifier is

$$\frac{\mathbf{V}_o(s)}{\mathbf{V}_s} = H(s) = - g_m \left(\frac{r_\pi}{r_\pi + R_s} \right) Z_L(s) \tag{32}$$

where

$$Z_L(s) = \frac{s/C}{s^2 + \dfrac{1}{RC} s + \dfrac{1}{LC}} = \frac{Ks}{s^2 + \dfrac{\omega_r}{Q} s + \omega_r^2} \tag{33}$$

Equating coefficients in (33) we find that

$$C = \frac{Q}{\omega_r R} \tag{34}$$

and

$$L = \frac{R}{\omega_r Q} \tag{35}$$

For example, suppose that $R = 50\ \Omega$ and that the bandpass filter must have a resonant frequency of 1-MHz and a bandwidth of 10 kHz, then, from Chapter Twelve

$$Q = \frac{1\ \text{MHz}}{10\ \text{kHz}} = 100$$

From (34) and (35) we find that $C = 0.318\ \mu\text{F}$ and $L = 79.6\ \text{nH}$.

---

[2] Due to the nonlinear flux-current characteristic of the iron core inductor, the transformer is a good source of distortion. Therefore, in modern high fidelity systems the transformer is eliminated by using low impedance power amplifiers to drive the speakers.

In practice the above design may prove to be very poor because we have neglected the series resistance of the inductor. For example, the impedance function including this resistance (see Figure 13.8) is

$$Z_L(s) = \frac{\frac{1}{C}\left(s + \frac{r_s}{L}\right)}{s^2 + s\left(\frac{1}{RC} + \frac{r_s}{L}\right) + \frac{1}{LC}\left(1 + \frac{r_s}{R}\right)} \tag{36}$$

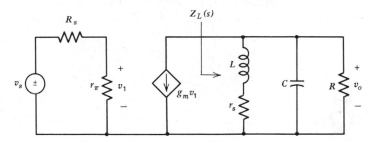

Figure 13.8 Amplifier model with coil resistance.

In the denominator typically $r_s/R \ll 1$ so that in order to neglect r_s we need the condition

$$\frac{1}{RC} \gg \frac{r_s}{L} \tag{37}$$

Upon substitution of (34) and (35) into (37) we obtain a clearer understanding of this requirement.

$$R \gg Q^2 r_s \tag{38}$$

The above condition is not always satisfied, particularly when Q is high and R is low.

If inequality (38) is not satisfied the problem can be rectified by placing a transformer between R and the LC circuit as shown in Figure 13.9, so that

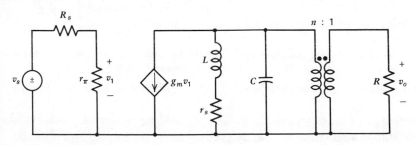

Figure 13.9 Amplifier with transformer coupling.

new resistance seen by the LC circuit is n^2R. From (34) we see that the capacitance is reduced by n^2, and from (35) the inductance is increased by n^2. Since the inductance is proportional to the square of the turns ratio this means that the increase in the coil size is proportional to n, thus r_s increases only by n and (38) becomes

$$n^2R \gg Q^2nr_s \tag{39}$$

The transformer yields an overall improvement proportional to n.

As our final example, consider the power transmission system in Figure 13.10, which consists of a generator with impedance Z_g, two transformers,

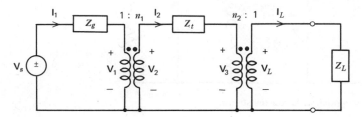

Figure 13.10   Power transmission system.

a transmission line with impedance Z_t, and a load Z_L. Let us find the Thévenin equivalent circuit seen by the load. The first two mesh equations are

$$-\mathbf{V}_s + Z_g\mathbf{I}_1 + \mathbf{V}_1 = 0$$
$$-\mathbf{V}_2 + Z_t\mathbf{I}_2 + \mathbf{V}_3 = 0 \tag{40}$$

and the transformer constraints are

$$\mathbf{V}_1 = \frac{\mathbf{V}_2}{n_1} \qquad \mathbf{I}_1 = n_1\mathbf{I}_2$$
$$\mathbf{V}_3 = n_2\mathbf{V}_2 \qquad \mathbf{I}_2 = \frac{\mathbf{I}_L}{n_2} \tag{41}$$

Upon elimination of all the variables except $\mathbf{V}_L$ and $\mathbf{I}_L$ we obtain

$$\mathbf{V}_L = \frac{n_1}{n_2}\mathbf{V}_s - \left[\frac{Z_t}{n_2^2} + Z_g\left(\frac{n_1}{n_2}\right)^2\right]\mathbf{I}_L \tag{42}$$

From this equation the Thévenin equivalent circuit is obtained. See Figure 13.11.

In the design of this system Z_g, Z_t, and Z_L will be assumed given. It is desired to minimize the power lost in the transmission line Z_t and also the customer expects a specified voltage $\mathbf{V}_L$. The average power lost in the

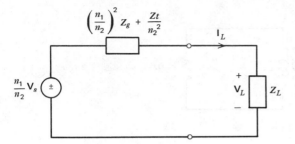

Figure 13.11   Thévenin equivalent of the power system.

transmission line is

$$P_{Z_t} = |\mathbf{I}_2|^2 \operatorname{Re}(Z_t) = \frac{|\mathbf{I}_L|^2}{n_2^2} \operatorname{Re}(Z_t) \tag{43}$$

and from Figure 3.11 the output voltage is

$$\mathbf{V}_L = \frac{Z_L}{Z_L + Z_g\left(\dfrac{n_1}{n_2}\right)^2 + \dfrac{Z_t}{n_2^2}} \cdot \left(\frac{n_1}{n_2}\mathbf{V}_s\right) \tag{44}$$

From these two equations we see that n_2 should be large in order to significantly reduce the transmission line loss (43), and the ratio n_1/n_2 determines the voltage $\mathbf{V}_L$.

In conclusion we should note one other desirable feature about the power system. It is desirable to have good voltage regulation, that is, under various load conditions the voltage $\mathbf{V}_L$ should remain relatively constant, which means that

$$Z_L \gg Z_g\left(\frac{n_1}{n_2}\right)^2 + \frac{Z_t}{n_2^2}$$

in (44). Thus, a good impedance match for maximum power transfer is not desirable since the voltage $\mathbf{V}_L$ would vary by a factor of 2 between open circuit condition and matched load condition. One can only achieve good voltage regulation and maximum power transfer if the load is relatively constant. Also, under the matched condition only one-half of the generated power is delivered to the load. This poor efficiency is intolerable in power distribution systems.

13.6 FREQUENCY RESPONSE OF THE NONIDEAL TRANSFORMER

In the previous section we assumed that the transformer was ideal. Let us now analyze the response of a nonideal transformer. The mesh equations

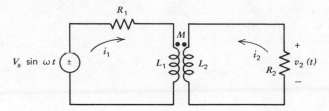

Figure 13.12   Transformer-coupled load.

for the circuit in Figure 13.12 in terms of the complex frequency s are

$$R_1\mathbf{I}_1(s) + sL_1\mathbf{I}_1(s) + sM\mathbf{I}_2(s) = \mathbf{V}_s$$
$$sM\mathbf{I}_1(s) + sL_2\mathbf{I}_2(s) + R_2\mathbf{I}_2(s) = 0 \tag{45}$$

The transfer function

$$\frac{\mathbf{I}_2(s)}{\mathbf{V}_s} = \frac{-sM}{s^2(L_1L_2 - M^2) + s(L_1R_2 + R_1L_2) + R_1R_2} \tag{46}$$

Since $\mathbf{V}_2(s) = -R_2\mathbf{I}_2(s)$, the input voltage-output voltage transfer function is

$$\frac{\mathbf{V}_2(s)}{\mathbf{V}_s} = H_v(s) = \frac{sMR_2}{s^2(L_1L_2 - M^2) + s(L_1R_2 + R_1L_2) + R_1R_2} \tag{47}$$

This circuit is a second-order system and we can express (47) in the resonant form

$$H_v(s) = \frac{Ks}{s^2 + \dfrac{s\omega_r}{Q} + \omega_r^2} \tag{48}$$

where

$$\omega_r^2 = \frac{R_1R_2}{L_1L_2 - M^2}$$

$$Q = \frac{(\sqrt{L_1L_2 - M^2})(\sqrt{R_1R_2})}{L_1R_2 + L_2R_1}$$

and

$$K = MR_2/(L_1L_2 - M^2)$$

The maximum value of $|H_v(j\omega)|$ occurs at the resonant frequency ω_r and

$$|H_v(j\omega_r)| = \frac{MR_2}{L_1R_2 + R_1L_2} \tag{49}$$

In the above circuit let us choose the ratio of our coil windings such that

$$R_2 = \left(\frac{n_2}{n_1}\right)^2 R_1 \tag{50}$$

Since the self-inductance of a coil is proportional to the square of its turns ratio, then L_1 and L_2 are related by the equation

$$L_2 = \left(\frac{n_2}{n_1}\right)^2 L_1 \tag{51}$$

Upon substitution of (50) and (51) into (47) and (48) we obtain

$$H_v(s) = \frac{sMR_1/L_1^2}{s^2(1-k^2) + s(2R_1/L_1) + R_1^2/L_1^2} \tag{52}$$

and

$$H_v(j\omega_r) = \frac{1}{2} M/L_1 \tag{53}$$

where

$$\omega_r^2 = \left(\frac{R_1}{L_1}\right)^2 \frac{1}{(1-k^2)} \tag{54}$$

and

$$Q = \frac{1}{2}\sqrt{1-k^2} \tag{55}$$

Note that $Q < 1/2$ for $k < 1$, which means the circuit is overdamped. Also, note that if $k = 1$, then from (22), $M/L_1 = n_2/n_1$, so that (53) becomes

$$\frac{V_2(j\omega_r)}{V_s} = \frac{1}{2}\left(\frac{n_2}{n_1}\right) \tag{56}$$

The factor of 1/2 is present since half the voltage drop is across R_1 and the other half across R_2.

Assume that the transformer is to operate at 60 Hz and suppose that $R_1 = 10\,\Omega$ and $k = 0.9$, then (54) becomes

$$2\pi \times 60 = \frac{10}{L_1} \frac{1}{(1-(0.9)^2)^{1/2}}$$

or

$$L_1 = \frac{10}{(2\pi)(60)\sqrt{0.19}} = 61 \text{ mH} \tag{57}$$

Therefore, to obtain maximum power transfer at the frequency 60 Hz, the primary winding should have a self-inductance of 61 mH. The secondary self-inductance depends on R_2. See (50) and (51). From (55) note that with the given k

$$Q = 0.22 \tag{58}$$

Therefore from our small Q bandwidth approximation in Chapter Twelve the bandwidth of the above transformer is approximately the difference between

$$f_2 = 60 \text{ Hz}/0.22 = 273 \text{ Hz} \tag{59}$$

and

$$f_1 = (0.22)(60 \text{ Hz}) = 13.2 \text{ Hz} \tag{60}$$

where

$$\sqrt{f_1 f_2} = 60 \text{ Hz}$$

If we designed the transformer so that $k = 0.99$, then $Q = 0.07$ and

$$f_2 = 60 \text{ Hz}/0.07 = 857 \text{ Hz} \tag{61}$$

and

$$f_1 = (0.07)(60 \text{ Hz}) = 4.2 \text{ Hz} \tag{62}$$

Thus, increasing the coefficient of coupling broadens the bandwidth, but from (54) the self-inductance L_1 must increase to keep ω_r constant.

The above analysis clearly illustrates that transformer responses are frequency dependent and, to achieve maximum efficiency, they should be operated within the designed bandwidth and with the specified terminating resistors. If Q is very small, then this operating range can be fairly large. Also, we should remember that in the case of iron core inductors the flux-current relationship is nonlinear so that the incremental inductance depends on the current in the coil or load on the transformer. This nonlinear characteristic plus the hysteresis phenomena further complicate the design of iron core transformers.

13.7 TRANSFORMER MODELS

The perfectly coupled transformer and the ideal transformer are the result of drastic model simplifications. However, in practice, transformers can be made highly efficient with coefficients of coupling close to 1. Thus, these assumptions can be used to give a first approximation to the circuit behavior. If a more exact analysis is needed, then better models must be used as was illustrated in the last section. In this section we present a transformer model that more clearly illustrates the effect of both the perfectly coupled and ideal transformer assumptions.

Consider the transformer model in Figure 13.13. We would like to express L_a, L_b, and L_c in terms of the coupling coefficient and self-

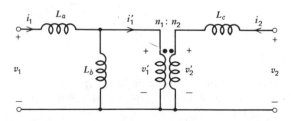

Figure 13.13 Equivalent transformer model.

inductances of the transformer. Thus,

$$v_1 = L_a \frac{di_1}{dt} + v_1' \tag{63}$$

$$v_2 = L_c \frac{di_2}{dt} + v_2' \tag{64}$$

However,

$$v_1' = \frac{n_1}{n_2} v_2' \tag{65}$$

$$i_1' = -\frac{n_2}{n_1} i_2 \tag{66}$$

and

$$v_1' = L_b \frac{d}{dt} (i_1 - i_1') \tag{67}$$

Substituting (66) into (67) and this result into (63), we obtain

$$v_1 = (L_a + L_b) \frac{di_1}{dt} + \frac{n_2}{n_1} L_b \frac{di_2}{dt} \tag{68}$$

Similarly, using (65) to (67) to eliminate v_2' in (64), we obtain

$$v_2 = \frac{n_2}{n_1} L_b \frac{di_1}{dt} + \left[L_c + \left(\frac{n_2}{n_1} \right)^2 L_b \right] \frac{di_2}{dt} \tag{69}$$

From (68) and (69) we conclude that

$$L_1 = L_a + L_b \tag{70}$$

$$L_2 = L_c + \left(\frac{n_2}{n_1} \right)^2 L_b \tag{71}$$

$$M = \frac{n_2}{n_1} L_b \tag{72}$$

Thus

$$L_a = L_1 - \frac{n_1}{n_2} M \tag{73}$$

$$L_c = L_2 - \frac{n_2}{n_1} M \tag{74}$$

But,

$$\frac{M}{k\sqrt{L_1 L_2}} = 1 \tag{75}$$

and from (51) it follows that (75) can be written as

$$\frac{M}{k L_2} = \frac{n_1}{n_2} \tag{76}$$

so that

$$L_a = L_1(1-k) \tag{77}$$

$$L_c = L_2(1-k) \tag{78}$$

$$L_b = M^2/kL_2 = kL_1 \tag{79}$$

Thus, inductances L_a, L_b, and L_c can be expressed in terms of L_1, L_2, and k. Note that $L_a = L_c = 0$ when $k = 1$ (the perfectly coupled case). Furthermore, the model reduces to the ideal transformer model if L_b is infinite. The model in Figure 13.13 more clearly illustrates the process of model reduction through various assumptions. The inductances L_a and L_c are called leakage inductance since they are zero when the coils are perfectly coupled. Since the inductor is a short circuit for dc signals, the transformer is a short circuit to dc signals due to the shunt inductor L_b unless L_1 and L_2 are infinite, which means that L_b is an open circuit.

The above model is still not very accurate since we neglected loss in the transformer due to the resistance of the coils, hysteresis loss, and eddy currents in the core (currents induced in the core due to the time-varying magnetic field in the core). Also, stray capacitance was neglected. A more complete model is shown in Figure 13.14. The capacitances C_1 and C_2 represent the stray capacitance of the coils, and C_3 represents the stray capacitance between the coils. These capacitances seriously affect the frequency response of the transformer, as do the inductors. The resistors take into account the fact that heat is dissipated in the transformer due to the above mentioned losses. Of course, if the power rating of the

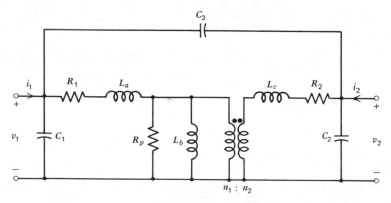

Figure 13.14   Transformer model including hysteresis and eddy current losses and capacitive effects.

transformer is exceeded, the resulting smoke will add to the world's pollution problems.

13.8 THREE COUPLED COILS

Let us conclude this chapter on transformers with a discussion of the circuit equations for a transformer with three or more coils.

In case three coils are coupled, Figure 13.15, it becomes necessary to generate more symbols to determine whether or not the fluxes aid or oppose. In Figure 13.15a, for the assigned current directions the flux

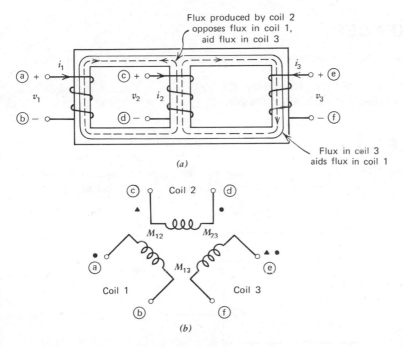

Figure 13.15   Three winding transformer. (a) Illustration of magnetic coupling. (b) Circuit symbol.

produced by coil 1 aids that produced by coil 3 and opposes that produced by coil 2. Hence, the dots are assigned as in Figure 13.15b. However, the flux produced by coil 2 aids the flux produced by coil 3. Thus, we need a new symbol, the triangle in Figure 13.15b, to illustrate this coupling. Hence, the picture is complete. Note that if any of the port current-voltage polarities are reversed, the magnetic symbol convention for the flux flow is not disturbed. Also from energy arguments one can again conclude that $M_{21} = M_{12}$, $M_{31} = M_{13}$ and $M_{23} = M_{32}$ in linear circuits. Thus, the equations for the coupled coils in Figure 13.15 are

$$v_1 = L_1 \frac{di_1}{dt} - M_{12} \frac{di_2}{dt} + M_{13} \frac{di_3}{dt}$$

$$v_2 = -M_{12} \frac{di_1}{dt} + L_2 \frac{di_2}{dt} + M_{23} \frac{di_3}{dt} \tag{80}$$

$$v_3 = +M_{13} \frac{di_1}{dt} + M_{23} \frac{di_2}{dt} + L_3 \frac{di_3}{dt}$$

This result can easily be extended to any number of multiple windings.

REFERENCES

1. Balabanian, N. *Fundamentals of Circuit Theory*, Allyn and Bacon, Boston, 1961, Chapter 9.
2. Hayt, W. H. Jr. and J. E. Kemmerly. *Engineering Circuit Analysis*, 2nd edition, McGraw-Hill, New York, 1971, Chapter 15.

PROBLEMS

1. According to our convention have the dots been properly assigned to the coils in Figure P13.1*b*? Is the dot assignment in Figure P13.1*c* correct also? Refer to Figure P13.1*a*.

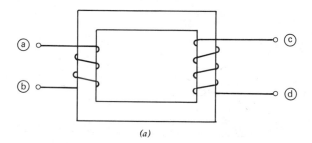

(a)

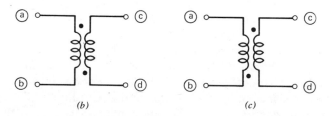

(b)   *(c)*

Figure P13.1

2. Show that the pair of coupled coils is equivalent to the corresponding three uncoupled inductor models in Figure P13.2.

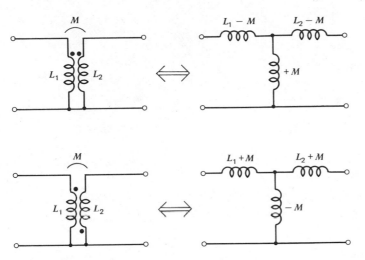

Figure P13.2

- 3. Terminals Ⓐ and Ⓑ are the primary terminals and Ⓒ and Ⓓ are the secondary terminals of a transformer. With the secondary open and a voltage $v_{ab} = 10 \cos 2\pi 10^3 t$ applied to the primary it is found that in the steady state $i_{ab} = 0.1 \sin 2\pi 10^3 t$ and $v_{cd}(t) = -0.9 \cos 2\pi 10^3 t$. When the secondary is shorted it is found that $i_{cd} = 0.9 \sin 2\pi 10^3 t$. Calculate L_1, L_2, M and the coefficient of coupling k. At which terminals should dots be placed?

4. If $L_1 = 100$ mH, $L_2 = 1$ mH and $M = 8$ mH for a pair of coupled coils, what is the coefficient of coupling k? If the self-inductance of a coil is proportional to the square of the number of turns, what is the turns ratio n_1/n_2 for these coils?

5. Assume that the transformer in Figure P13.5 is ideal and calculate the turns ratio so that the resistance seen on the primary side is $1800 \, \Omega$.

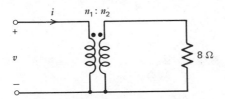

Figure P13.5

6. Assume that the coupled coils in Figure P13.6 are ideal and show that the input resistance is $[(n_1 + n_2)/n_2]^2 R$.

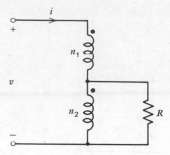

Figure P13.6

7. Assume that the coupled coils in Figure P13.7 are ideal and show that the input resistance is

$$\left(\frac{n_1 - n_2}{n_2}\right)^2 R$$

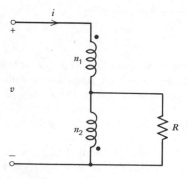

Figure P13.7

8. Calculate the power delivered to the 8-Ω load in Figure P13.8 with $n = 1$. What is the maximum power available from the source? Choose n so that maximum power is delivered to the 8-Ω resistor.

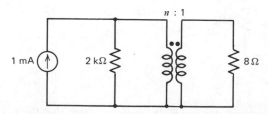

Figure P13.8

9. In Figure 13.7 calculate the gain at resonance from (32). Assume $g_m = 0.4$, $r_\pi \gg R_s$ and use the L and C parameter values for $f_r = 1\,\text{MHz}$, $BW = 10\,\text{kHz}$, $R = 50$. If $r_s = 0.01\,\Omega$ calculate from (36) the new resonant frequency and Q. Use the same values for R, L, and C. What is the gain at resonance now? Finally, in Figure 13.9 use the ideal transformer model and let $n = 20$. Recalculate L and C from (34) and (35). Assume that $r_s = 0.2\,\Omega$ now (increased by n) and recalculate the effect of r_s on f_r, Q, and gain for this transformer circuit.

10. Assuming perfect coupling, show that the two circuits in Figure P13.10 have the same driving-point impedances where

$$R_{eq} = \left(\frac{\sqrt{L_2} + \sqrt{L_1}}{\sqrt{L_2}}\right)^2 R \quad \text{and}$$

$$L_{eq} = (L_1 + L_2)\left(1 + \frac{2\sqrt{L_1 L_2}}{L_1 + L_2}\right) = (\sqrt{L_1} + \sqrt{L_2})^2$$

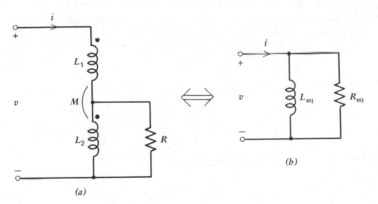

(a)   (b)

Figure P13.10

11. Design the resonant circuit in Figure P13.11 (find L_1, L_2, and C) such that $f_r = 500\,\text{kHz}$, $BW = 10\,\text{kHz}$. Use the equivalent model in Figure P13.10b and assume that $R_{eq} = 5\,\text{k}\Omega$. This circuit combines the inductor and transformer into one tapped coil.

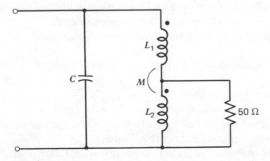

Figure P13.11

- 12. In the power transmission system, Figure 13.10, let $n_1 = 1$, $Z_g = 0.01\,\Omega$, $Z_t = 1\,\Omega$, $Z_L = 0.1\,\Omega$, and $V_s = 200$ V rms. Find n_2 such that V_L (open circuit) = 120 V rms. What is the voltage V_L, the power dissipated in the load, and the power dissipated in the line with Z_L connected. Repeat the above calculations with $n_1 = 100$.

13. In the power transmission sytem, Figure 13.10, suppose that $V_s = 200$ V rms and $\text{Re}(Z_g) = 0.01\,\Omega$. Can the generator supply 1 MW of power to the load?

- 14. In Figure 13.10 what is the equivalent impedance seen by the generator?

15. An audio transformer is to be built to transform $4\,\Omega$ to $3600\,\Omega$ on the primary side. Assume a coupling coefficient of 0.9 and calculate the self-inductance L_1 of the primary coil so that $\omega_r = 2\,\pi \times 1500$ rad/s. Calculate Q and the bandwidth of the audio transformer.

16. Repeat Problem 15 for a coupling coefficient of 0.98.

17. Find the transfer function $V_o(s)/V_s$ of the circuit in Figure P13.17 and sketch the magnitude and phase of its frequency response for $R_s = 1\,\text{k}\Omega$, $R_L = 10\,\Omega$, $L_1 = 100$ mH, and $L_2 = 1$ mH. Assume that the coils are perfectly coupled and use the transformer model in Figure 13.13.

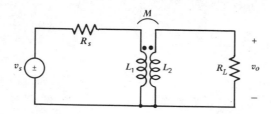

Figure P13.17

18. Find the step response of the circuit in Problem 17.

19. Given three coupled coils show that if they are perfectly coupled, then

$$v_1 = \pm \frac{n_1}{n_2} v_2 \quad \text{and} \quad v_1 = \pm \frac{n_1}{n_3} v_3$$

where the plus sign implies that the fluxes aid and a minus sign implies that the coils are wound such that the respective fluxes are in opposition.

Show that the ideal transformer assumption leads to the constraint

$$n_1 i_1 = \mp n_2 i_2 \mp n_3 i_3$$

Show that the instantaneous power is always zero.

20. In Figure P13.20 let $v_s = 170 \sin 2\pi 60 t$, $n_1 = 10$, and $n_2 = 1$. Assume that the transformer and diodes are ideal and sketch $v_o(t)$. In the reverse bias direction, what is the largest voltage that appears across the diodes?

21. In Problem 20 let $R_L = 1\,\text{k}\Omega$ and place a 10-μF capacitor in parallel with R_L. Sketch $v_o(t)$. Repeat with a 100-μF capacitor.

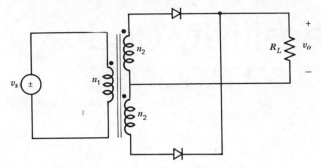

Figure P13.20

22. The transformer equations in impedance form are

$$\begin{bmatrix} V_1(s) \\ V_2(s) \end{bmatrix} = s \begin{bmatrix} L_1 & \pm M \\ \pm M & L_2 \end{bmatrix} \begin{bmatrix} I_1(s) \\ I_2(s) \end{bmatrix}$$

Show that the transformer equations in admittance form are

$$\begin{bmatrix} I_1(s) \\ I_s(s) \end{bmatrix} = \frac{1}{s(L_1 L_2 - M^2)} \begin{bmatrix} L_2 & \mp M \\ \mp M & L_1 \end{bmatrix} \begin{bmatrix} V_1(s) \\ V_2(s) \end{bmatrix}$$

and develop an equivalent circuit model.

Chapter Fourteen
Sensitivity and
Noise Analysis

The changes in the responses of a circuit due to changes in its parameter values are frequently computed and used in the evaluation of the circuit. For example, in the manufacturing process the parameter values of a circuit can only be controlled to within certain tolerances. The more narrow the tolerances must be in order to meet performance specifications, the more costly the manufacturing process. Also, one often must consider the variation of parameter values with respect to temperature and other environmental conditions. Typically, the sensitivity function that characterizes a response variation with respect to the circuit parameters is nonlinear. However, if the parameter variations are not too large this nonlinear sensitivity function can be approximated with a linear function. Typically the linear function is simply the first derivative of the response with respect to a particular parameter evaluated at the nominal set of parameter values. We call this function the first-order sensitivity function. In this chapter we define the first-order sensitivity function, show how to compute it, and demonstrate its applications in circuit design.

14.1 FIRST-ORDER SENSITIVITY

The change in the response of a circuit due to a parameter change can be determined simply by reanalyzing the circuit with the new parameter value

436

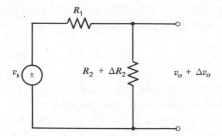

Figure 14.1 Parameter variation analysis.

or set of parameter values. For example, for the circuit in Figure 14.1:

$$v_o + \Delta v_o = \frac{R_2 + \Delta R_2}{R_1 + R_2 + \Delta R_2} \cdot v_s \tag{1}$$

where the nominal value of the network function is

$$\frac{v_o}{v_s} = H = \frac{R_2}{R_1 + R_2} \tag{2}$$

We conclude from (1) and (2) that

$$v_o + \Delta v_o = v_o \left(\frac{1 + \Delta R_2 / R_2}{1 + \Delta R_2 / (R_1 + R_2)} \right) \tag{3}$$

Equation 3 represents the change in Δv_o due to large changes in R_2. If, $\Delta R_2 / R_2 \ll 1$, then

$$\frac{\Delta v_o}{v_o} = \frac{\Delta R_2}{R_2} + \left(1 + \frac{\Delta R_2}{R_2}\right) \left[-\frac{\Delta R_2}{R_1 + R_2} + \left(\frac{\Delta R_2}{R_1 + R_2}\right)^2 - \left(\frac{\Delta R_2}{R_1 + R_2}\right)^3 + \cdots \right]$$

or

$$\frac{\Delta v_o}{v_o} \approx \frac{\Delta R_2}{R_2} - \frac{\Delta R_2}{R_1 + R_2} = \frac{R_1}{R_2(R_1 + R_2)} \cdot \Delta R_2 \tag{4}$$

and the fractional change of the response with respect to a fractional change in a parameter is approximately

$$\frac{\Delta v_o}{v_o} \approx \frac{R_1}{R_1 + R_2} \cdot \frac{\Delta R_2}{R_2} \tag{5}$$

Note that $R_1 / (R_1 + R_2) = R_2 / H \cdot \partial H / \partial R_2$, therefore:

$$\frac{\Delta v_o}{v_o} \approx S_{R_2}^H \cdot \frac{\Delta R_2}{R_2} \tag{6}$$

where

$$S_{R_2}^H \triangleq \frac{R_2}{H} \cdot \frac{\partial H}{\partial R_2} \tag{7}$$

is called the first-order sensitivity of the network function H with respect to the parameter R_2. In case $H = 0$ we use the definition

$$\hat{S}_{R_s}^{H} \triangleq R_2 \frac{\partial H}{\partial R_2} \tag{8}$$

In some cases the first-order variations are not accurate enough to predict the response change due to large variations in the parameters. In such cases Monte Carlo analysis is used. This is a brute force technique that analyzes the circuit a large number of times, each time letting the elements assume values within some prescribed distribution that is determined by the manufacturing process. This technique is costly. However, in many cases the first-order variations (7) and (8) are useful in the evaluation of a circuit and are cheaper to compute. Also, in the design of circuits with optimization techniques one defines an error function that is the measure of the distance between some desired response and the actual response. The optimization algorithm varies the circuit parameters in a direction that minimizes the error. Frequently, the partial derivative of the error function with respect to the adjustable parameters is required by the algorithm in order to determine the most likely direction of the parameter vector that will minimize the error function in that neighborhood. The parameters are adjusted in that direction.

The next example shows how the first-order sensitivity function (8) can be applied to a tolerance analysis problem in which the changes are sufficiently small so that a large parameter change analysis is not necessary.

Example 1

Consider the Wheatstone bridge in Figure 14.2. The bridge is said to be balanced when $I_m = 0$, which is obtained when

$$R_x = \frac{R_2 R_3}{R_1} \tag{9}$$

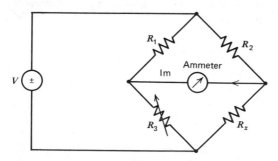

Figure 14.2 Wheatstone bridge.

An unknown resistor can be determined by calibrating the dial on the variable resistor R_3 and adjusting R_3 until the current $I_m = 0$. The resistor R_1, R_2, and R_3 are assumed known. The dial of R_3 can be calibrated so that the value of R_x is read directly from this dial. The Thévenin equivalent circuit for this bridge is shown in Figure 14.3 where

$$R_p = \frac{R_1 R_3}{R_1 + R_3} + \frac{R_2 R_x}{R_2 + R_x} \tag{10}$$

and

$$V_t = \frac{R_2 R_3 - R_1 R_x}{(R_1 + R_3)(R_2 + R_x)} V \tag{11}$$

From (11) and Figure 14.3 we see that $I_m = 0$ when (9) is satisfied so that $V_t = 0$.

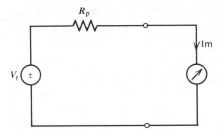

Figure 14.3   Thévenin equivalent.

Now suppose that the ammeter can only detect I_m to within $\pm 1 \, \mu$A. To within what tolerance can we measure R_x? To answer this question let us use the first-order sensitivity function. If we neglect the resistance of the ammeter, then

$$I_m = \frac{V_t}{R_p} \tag{12}$$

or, from (10) and (11),

$$I_m = \frac{(R_2 R_3 - R_1 R_x) V}{R_1 R_3 (R_2 + R_x) + R_2 R_x (R_1 + R_3)} \tag{13}$$

From (13) we obtain

$$\frac{\partial I_m}{\partial R_x} = \frac{-R_1 V}{[R_1 R_3 (R_2 + R_x) + R_2 R_x (R_1 + R_3)]}$$
$$- \frac{(R_1 R_3 + R_1 R_2 + R_2 R_3)(R_2 R_3 - R_1 R_x) V}{[R_1 R_3 (R_2 + R_x) + R_2 R_x (R_1 + R_3)]^2} \tag{14}$$

If the bridge is balanced, then

$$R_x \frac{\partial I_m}{\partial R_x} = \frac{-R_1 R_x V}{R_1 R_3 (R_2 + R_x) + R_2 R_x (R_1 + R_3)} \Bigg|_{I_m = 0} \tag{15}$$

and, on substitution of (9) into (15), we obtain the first-order sensitivity function

$$\hat{S}^{I_m}_{R_x} = \frac{-R_1 V}{R_1^2 + R_1 R_3 + R_1 R_2 + R_2 R_3} \Bigg|_{I_m = 0} \tag{16}$$

However,

$$\frac{\Delta R_x}{R_x} \approx (\hat{S}_{R_x}^{I_m})^{-1} \Delta I_m \qquad (17)$$

If $R_1 = R_2 = 1 \text{ k}\Omega$ and $V = 10 \text{ V}$, then

$$\frac{\Delta R_x}{R_x} \approx \left(\frac{2 \times 10^3 + 2R_3}{-10}\right)(\pm 10^{-6} \text{ A}) \qquad (18)$$

so that

$$\left|\frac{\Delta R_x}{R_x}\right| \approx (2 \times 10^{-4} + 2R_3 \times 10^{-7}) \qquad (19)$$

If R_x is in the neighborhood of $1 \text{ k}\Omega$, then $R_3 \approx 1 \text{ k}\Omega$, and, from (19), we can measure R_x to better than one part in 2500 or 0.04%. Of course, we must also add in the tolerances on the reference resistors, R_1, R_2, and R_3. If they are only accurate to within 1%, then our measurement of R_x will also be accurate to within approximately 3% worst case. With precise resistors the bridge null method can be an extremely accurate measurement tool.

14.2 THE SENSITIVITY CIRCUIT

In the previous section we saw that, given the transfer function of a circuit, we could compute the first-order sensitivity with respect to a parameter by taking the partial derivative of that network function with respect to the parameter. However, in many cases it is tedious to find the network function in symbolic form, and computer algorithms for this task are too costly. In such cases one could find an approximation to the first-order sensitivity with respect to changes in the parameter values by separately changing each parameter and recomputing the response. The first-order sensitivity is approximately the difference between the original response and the response obtained with the new parameter value divided by the difference between the original and new parameter values. This method requires a separate analysis for each sensitivity function. Also, the method is subject to large errors since the calculation involves computing small differences between large numbers. A much faster and more accurate approach uses the concept of a sensitivity circuit[1], which we present in this section.

Let us consider the simple voltage divider in Figure 14.4. The Kirchhoff equations are

$$-v_s + v_1 + v_2 = 0 \qquad (20)$$
$$i_s + i_1 = 0 \qquad (21)$$
$$-i_1 + i_2 = 0 \qquad (22)$$

[1] M. L. Bihovski, "The Accuracy of Electric Network Analyzers," Izv. ANSSR Otd. Tekhn, Nauk, No. 8, 1948 (in Russian). In English see also J. V. Leeds, Jr., "Transient and Steady-State Sensitivity Analysis," *IEEE Transactions on Circuit Theory*, Vol. CT-13, No. 3, Sept., 1966, pp. 288–289.

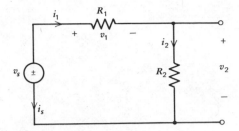

Figure 14.4   Voltage divider circuit.

and the branch constraints are

$$R_1 i_1 = v_1 \tag{23}$$

$$R_2 i_2 = v_2 \tag{24}$$

In order to calculate the first-order variation of the above voltages and currents with respect to changes in R_2, let us take the partial derivative of (20) to (24) with respect to R_2. The new equations are

$$-0 + \frac{\partial v_1}{\partial R_2} + \frac{\partial v_2}{\partial R_2} = 0 \tag{20b}$$

$$\frac{\partial i_s}{\partial R_2} + \frac{\partial i_1}{\partial R_2} = 0 \tag{21b}$$

$$-\frac{\partial i_1}{\partial R_2} + \frac{\partial i_2}{\partial R_2} = 0 \tag{22b}$$

and

$$R_1 \frac{\partial i_1}{\partial R_2} = \frac{\partial v_1}{\partial R_2} \tag{23b}$$

$$i_2 + R_2 \frac{\partial i_2}{\partial R_2} = \frac{\partial v_2}{\partial R_2} \tag{24b}$$

In (20b) $\partial v_s/\partial R_2 - 0$ since the value of the independent source does not depend on R_2. To compute the sensitivity, first determine i_2 from (20) to (24) and substitute i_2 into (24b). Then solve (20b) to (24b) for

$$\frac{\partial v_1}{\partial R_2}, \quad \frac{\partial v_2}{\partial R_2}, \quad \frac{\partial i_s}{\partial R_2}, \quad \frac{\partial i_1}{\partial R_2}, \quad \text{and} \quad \frac{\partial i_2}{\partial R_2}$$

The circuit model for (20b) to (24b) is shown in Figure 14.5a. Note the similarity between Figures 14.4 and 14.5a, except the voltage of the independent voltage source is zero in the sensitivity circuit and the resistor R_2, with respect to which the first-order changes are to be computed, is in series with a voltage source whose value is i_2, the current through the resistor R_2 in the original circuit. Note that the currents and voltages in the

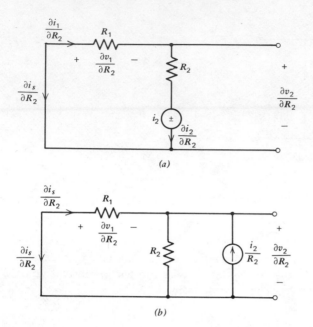

(a)

(b)

Figure 14.5   Sensitivity circuit for R_2 variation.

sensitivity circuit now have units of A/Ω and V/Ω, respectively. Figure 14.5b illustrates the sensitivity circuit with a Norton equivalent of the R_2 branch. Thus, in any circuit, when we differentiate the Kirchhoff equations with respect to any parameter in the circuit, the *form of the Kirchhoff equations does not change. This means that the original circuit and the sensitivity circuit have the same topology, only in the sensitivity circuit the currents and voltages are replaced by partial derivatives with respect to that parameter.* Several examples are given below that illustrate the generation and use of the sensitivity circuit in analysis.

Example 2

Consider the four-bit D/A converter in Figure 14.6. The LSB is $2^{-4}V_r = V_r/16$. It is very important that the tolerances on the resistor be such that the error in the output voltage e_o be less than $\pm\frac{1}{2}$LSB, that is, $\pm\frac{1}{32}V_r$. Let us use the sensitivity circuit to compute the sensitivity of e_o with respect to the resistor R_1 of the most significant bit. The nominal value of the resistor R_1 is $2R$, identical to the other shunt resistors.

The sensitivity circuit is shown in Figure 14.7. By simple series parallel combinations in Figure 14.7 it is easy to show that

$$\frac{\partial e_o}{\partial R_1} = \frac{1}{2}i_1 \qquad (25)$$

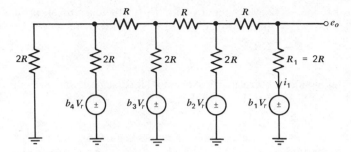

Figure 14.6   Four-bit D/A converter.

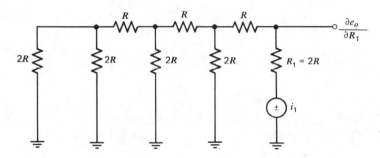

Figure 14.7   A sensitivity circuit for D/A converter.

The maximum value of i_1 in Figure 14.6 occurs when $b_1 = 1$, and $b_2 = b_3 = b_4 = 0$. We obtain

$$\max|i_1| = \left|\frac{-V_r}{4R}\right| \tag{26}$$

therefore

$$\left|\frac{\partial e_o}{\partial R_1}\right| \leq \frac{V_r}{8R} \tag{27}$$

We use (27) to write

$$\max|\Delta e_o| \approx \left(\frac{V_r}{8R}\right)\left(\frac{\Delta R_1}{2R}\right)(2R)$$

or

$$\max|\Delta e_o| \approx \left(\frac{V_r}{4}\right)\left(\frac{\Delta R_1}{2R}\right) \tag{28}$$

If we neglect the other parameter tolerances then since we require $|\Delta e_o| \leq \frac{1}{32} V_r$, from (28) we require

$$\left(\frac{V_r}{4}\right)\left(\frac{\Delta R_1}{2R}\right) \leq \frac{1}{32} V_r$$

or

$$\frac{\Delta R_1}{2R} \leq \frac{1}{8} = 2^{-3} \tag{29}$$

which means that a tolerance of 12.5% is required for R_1. In the case of an n-bit converter in which $\frac{1}{2}\text{LSB} = 2^{-(n+1)}$ the tolerance on the MSB $2R$ resistor would be

$$\frac{\Delta R_1}{2R} - 2^{-(n-1)} \tag{30}$$

Thus, a 10-bit converter would require a MSB resistor accuracy of one part in $2^9 = 512$, or the resistor would have to have a tolerance of less than 0.2%.

Example 3

Next let us compute the sensitivity of the voltage across a diode in Figure 14.8 with respect to the temperature of the diode. Recall that the voltage across the

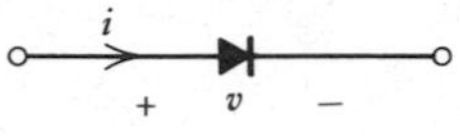

Figure 14.8   Diode.

diode is given by the equation

$$v = \frac{kT}{q}\ln(i/I_s + 1) \tag{31}$$

Not only are v and i dependent on temperature, but the saturation current I_s is also temperature dependent. This dependence is approximated by the equation[2]

$$I_s(T) = I_s(T_o)\left(\frac{T}{T_o}\right)^3 \exp\left[-E_g\left(\frac{1}{T}-\frac{1}{T_o}\right)/k\right]$$

where E_g is the band gap, which is 1.1 eV or $1.1 \times 1.6 \times 10^{-19} = 1.76 \times 10^{-19}$ J for silicon and $k = 1.23 \times 10^{-23}$ J/K. Thus

$$I_s(T) = I_s(T_o)\left(\frac{T}{T_o}\right)^3 \exp\left[-1.43 \times 10^4 \left(\frac{1}{T}-\frac{1}{T_o}\right)\right] \tag{32}$$

The derivative of (31) with respect to T is

$$\frac{dv}{dT} = \frac{k}{q}\ln\left(\frac{i}{I_s}+1\right) + \frac{kT}{q}\frac{1}{I_s}\frac{\partial i}{\partial T}\frac{1}{\frac{i}{I_s}+1}$$

$$-\frac{kT}{q}\frac{i}{I_s^2}\frac{\partial I_s}{\partial T}\frac{1}{\frac{i}{I_s}+1} \tag{33}$$

Let us assume that the diode is forward biased at $v = V_o > 0$ and $i = I_o > 0$ and that $I_o/I_s \gg 1$ so that

$$\frac{dv}{dT} \approx \frac{V_o}{T} + \frac{kT}{q}\frac{1}{I_o}\frac{\partial i}{\partial T} - \frac{kT}{q}\frac{1}{I_s}\frac{\partial I_s}{\partial T} \tag{34}$$

---

[2] T. E. Idleman, F. S. Jenkins, W. J. McCalla, and D. O. Pederson, "SLIC—A Similator for Linear Integrated Circuits," *Journal of Solid-State Circuits*, Vol. SC-6, August 1971, pp. 188–203.

Note that in the forward-bias region of the diode the incremental diode conductance

$$\frac{di}{dv} = \frac{q}{KT} I_s e^{qv/kT} \tag{35}$$

so that at the operating point (I_o, V_o) we write

$$g_o = \frac{di}{dv}\bigg|_{I_o, V_o} = \frac{q}{kT} I_s e^{qV_o/kT} \tag{36}$$

and if $I_o/I_s \gg 1$, then

$$g_o \approx \frac{qI_o}{kT} \tag{37}$$

and we can write (34) as

$$\frac{dv}{dT} = \frac{1}{g_o}\frac{\partial i}{\partial T} + \left(\frac{V_o}{T} - \frac{kT}{qI_s}\frac{\partial I_s}{\partial T}\right) \tag{38}$$

The equivalent circuit for (38) is shown in Figure 14.9.

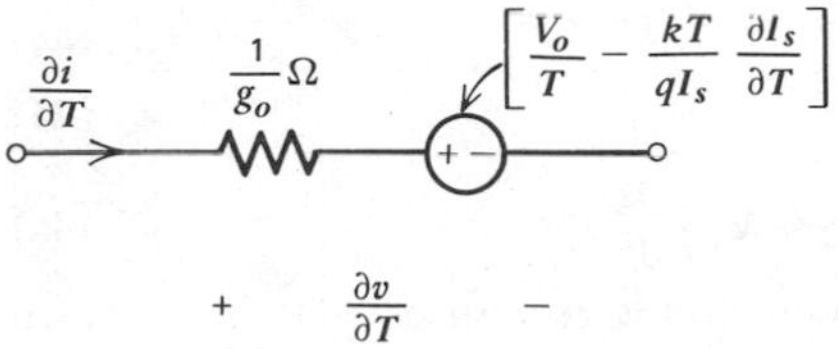

Figure 14.9 Sensitivity branch constraint of diode.

Let us assume that $V_o = 700\text{ mV}$ and $T_o = 300\text{ K}$, then from (32)

$$\frac{\partial I_s}{\partial T}\bigg|_{T_o} = I_s(T_o)\left(\frac{3}{T_o} + \frac{1.43 \times 10^4}{T_o^2}\right) \tag{39}$$

or

$$\frac{\partial I_s}{\partial T}\bigg|_{300\text{ K}} = I_s(T_o)(0.01 + 0.159) \tag{40}$$

so that (38) becomes

$$\frac{dv}{dT}\bigg|_{I_o, V_o, T_o} = \frac{1}{g_o}\frac{\partial i}{\partial T}\bigg|_{I_o, V_o, T_o} + \left[\frac{700\text{ mV}}{300\text{ K}} - 26\text{ mV}(0.169/\text{K})\right]$$

or

$$\frac{dv}{dT}\bigg|_{I_o, V_o, T_o} = \frac{1}{g_o}\frac{\partial i}{\partial T}\bigg|_{I_o, V_o, T_o} - 2.2\text{ mV/K} \tag{41}$$

Thus, if we assume that $\partial i/\partial T \approx 0$, then, in the forward-biased state,

$$\frac{\Delta v}{\Delta T} \approx -2.2\text{ mV/K} \quad \text{or} \quad -2.2\text{ mV/°C} \tag{42}$$

This is an important sensitivity parameter in the design of integrated circuits.

Example 4

Let us now consider a circuit with energy storage elements, for example, the series RLC circuit in Figure 14.10. Suppose that we are interested in the sensitivity

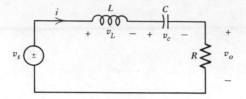

Figure 14.10   Series RLC circuit.

of the steady-state response phasor $\mathbf{V}_o$ (see Figure 14.11) with respect to parameter variations. The equations for this circuit are

$$\mathbf{V}_L = j\omega L \mathbf{I} \tag{43}$$

$$\mathbf{I} = j\omega C \mathbf{V}_C \tag{44}$$

$$\mathbf{V}_o = R\mathbf{I} \tag{45}$$

and the KVL equation is

$$\mathbf{V}_L + \mathbf{V}_C + \mathbf{V}_o = \mathbf{V}_s \tag{46}$$

In order to find the sensitivity of the above circuit responses with respect to changes in the inductor we differentiate (43) to (46) with respect to L and obtain

$$\frac{\partial \mathbf{V}_L}{\partial L} = j\omega \mathbf{I} + j\omega L \frac{\partial \mathbf{I}}{\partial L} \tag{47}$$

$$\frac{\partial \mathbf{I}}{\partial L} = j\omega C \frac{\partial \mathbf{V}_C}{\partial L} \tag{48}$$

$$\frac{\partial \mathbf{V}_o}{\partial L} = R \frac{\partial \mathbf{I}}{\partial L} \tag{49}$$

and

$$\frac{\partial \mathbf{V}_L}{\partial L} + \frac{\partial \mathbf{V}_C}{\partial L} + \frac{\partial \mathbf{V}_o}{\partial L} = 0 \tag{50}$$

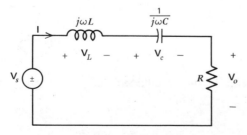

Figure 14.11   Phasor model of series RLC circuit.

Note that the parameters R and C and the input voltage phasor $\mathbf{V}_s$ are independent of changes in the inductance. The circuit model for (47) to (50) is illustrated in Figure 14.12.

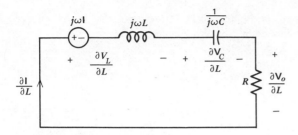

Figure 14.12   Sensitivity circuit.

An analysis of this sensitivity circuit yields

$$\frac{\partial \mathbf{V}_o}{\partial L} - \frac{R}{R + j\left(\omega L - \dfrac{1}{\omega C}\right)}(-j\omega \mathbf{I}) \tag{51}$$

but, from Figure 14.11, we obtain

$$\mathbf{I} = \frac{\mathbf{V}_s}{R + j\left(\omega L - \dfrac{1}{\omega C}\right)} \tag{52}$$

and

$$\mathbf{V}_o = R\mathbf{I} \tag{53}$$

Therefore, from (51) to (53), the sensitivity of $\mathbf{V}_o$ with respect to L is

$$S_L^{V_o}(j\omega_o) = \frac{L}{\mathbf{V}_o}\frac{\partial \mathbf{V}_o}{\partial L} = \frac{-j\omega L}{R + j\left(\omega L - \dfrac{1}{\omega C}\right)} \tag{54}$$

Note that the sensitivity function is dependent on frequency. At the resonant frequency $\omega_o = 1/\sqrt{LC}$ we obtain

$$S_L^{V_o}(j\omega_o) = \frac{-j\omega_o L}{R} = -Q \tag{55}$$

Thus, the higher the circuit Q the more sensitive the response is to changes in the inductance.

Finally, let us examine the sensitivity of the time responses of this circuit with respect to parameter variations. Assume that we are interested in the response for $t \geqslant 0$ and that $v_c(0) = v_{co}$ and $i(0) = i_{Lo}$. The circuit equations are

$$v_L(t) + v_c(t) + v_o(t) = v_s(t) \tag{56}$$

$$i(t) = i_{Lo}u(t) + \frac{1}{L}\int_0^t v_L \, dt \tag{57}$$

$$v_c(t) = v_{co}u(t) + \frac{1}{C}\int_0^t i\,dt \tag{58}$$

and

$$v_o(t) = Ri(t) \tag{59}$$

The equivalent circuit is shown in Figure 14.13. Note that the initial energy conditions are introduced as independent sources for convenience.

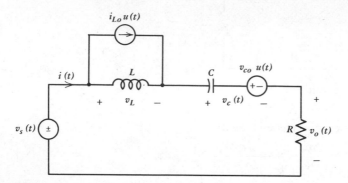

Figure 14.13   Series RLC circuit with initial conditions.

Let us find the sensitivity of the time response of this circuit with respect to changes in the initial condition v_{co}. From (56) to (59) we obtain

$$\frac{\partial v_L}{\partial v_{co}} + \frac{\partial v_c}{\partial v_{co}} + \frac{\partial v_o}{\partial v_{co}} = 0 \tag{60}$$

$$\frac{\partial i}{\partial v_{co}} = \frac{1}{L}\int_0^t \frac{\partial v_L}{\partial v_{co}}\,dt \tag{61}$$

$$\frac{\partial v_c}{\partial v_{co}} = u(t) + \frac{1}{C}\int_0^t \frac{\partial i}{\partial v_{co}}\,dt \tag{62}$$

and

$$\frac{\partial v_o}{\partial v_{co}} = R\frac{\partial i}{\partial v_{co}} \tag{63}$$

The circuit model for these sensitivity equations is shown in Figure 14.14. The steady-state value of $\partial v_o/\partial v_{co}$ is zero, hence the solution is

$$\frac{\partial v_o}{\partial v_{co}} = k_1 e^{p_1 t} + k_2 e^{p_2 t} + 0 \tag{64}$$

where the natural frequencies p_1 and p_2 are roots of the circuit characteristic equation

$$s^2 + \frac{R}{L}s + \frac{1}{LC} = 0 \tag{65}$$

and in writing (64) it is assumed that p_1 and p_2 are distinct. In order to evaluate k_1

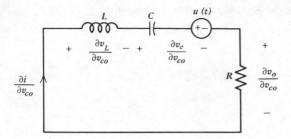

Figure 14.14   Sensitivity circuit for variation in v_{co}.

and k_2 we need to consider the initial condition circuit in Figure 14.15. From this circuit we obtain

$$\frac{\partial v_o(0)}{\partial v_{co}} = 0 \tag{66}$$

and

$$\frac{\partial v_L(0)}{\partial v_{co}} = -1 \tag{67}$$

but

$$\frac{\partial v_L}{\partial v_{co}} = L\frac{d}{dt}\left(\frac{\partial i}{\partial v_{co}}\right) \tag{68}$$

and

$$R\frac{\partial i}{\partial v_{co}} = \frac{\partial v_o}{\partial v_{co}} \tag{69}$$

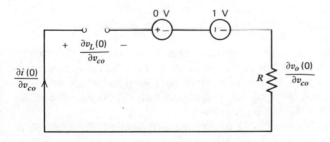

Figure 14.15   Initial conditions in the sensitivity circuit.

From (68) and (69) we obtain

$$\frac{\partial v_L}{\partial v_{co}} = \frac{L}{R}\frac{d}{dt}\left(\frac{\partial v_o}{\partial v_{co}}\right) \tag{70}$$

so that, from (67),

$$\frac{L}{R}\frac{d}{dt}\left(\frac{\partial v_o}{\partial v_{co}}\right)\bigg|_{t=0} = -1 \tag{71}$$

with the two initial condition constraints (66) and (71), we can solve for k_1 and k_2 in (64).

$$k_1 + k_2 = 0 \tag{72}$$

and

$$\frac{L}{R}(p_1 k_1 + p_2 k_2) = -1 \tag{73}$$

The solution to (72) and (73) is

$$k_1 = \frac{R}{L(p_2 - p_1)} \tag{74}$$

$$k_2 = \frac{R}{L(p_1 - p_2)} \tag{75}$$

If we assume that $R = 1\,\Omega$, $C = 1\,F$, and $L = \frac{1}{2}H$, then $s_{1,2} = -1 \pm j$, $k_1 = j$, $k_2 = -j$, and

$$\frac{\partial v_o}{\partial v_{co}} = -2\,e^{-t}\sin t \tag{76}$$

or the approximate change in the output due to a small change in the initial condition v_{co} is

$$\Delta v_o \approx -2(\Delta v_{co})\,e^{-t}\sin t \tag{77}$$

Obviously initial condition perturbations have less and less effect on the response as time increases.

PROBLEMS

1. Determine the first-order sensitivity $S_{R_1}^H$ in Figure 14.1.
2. If there is a 1% change in the parameter p, use the sensitivity function (7) to find the approximate percentage change in the transfer function H if (a) $S_p^H = 0.1$, (b) $S_p^H = 1$, and (c) $S_p^H = 10$.
3. In the Wheatstone bridge example suppose that $R_x \approx 1\,k\Omega$ and $R_1 = 100\,\Omega$, $R_2 = 200\,\Omega$, and $R_3 = 500\,\Omega$. What is the tolerance (17) on the measurement of R_x? Suppose that the input voltage is reduced to $1\,V$. What effect does this reduction have on the tolerance of the measurement of R_x? Neglect the effect of the tolerances on R_1, R_2, and R_3.
4. In example 1 find the tolerance (17) on the measurement of R_x, if (a) $R_x = 100\,\Omega$, (b) $R_x = 10\,k\Omega$, and (c) $R_x = 1\,M\Omega$. Assume that $V = 10\,V$, $R_1 = R_2 = 1\,k\Omega$, and $R_3 = R_x$. Neglect the effect of the tolerances on R_1, R_2, and R_3.
5. From (9) find $S_{R_k}^{R_x}$ for $k = 1, 2,$ and 3. If the resistors R_1, R_2, and R_3 have a 1% tolerance, what is the tolerance on the measurement of R_x? Assume worst-case and assume that the error caused by the tolerance on the ammeter is negligible in comparison to the errors caused by the tolerances on R_1, R_2, and R_3.
6. In the bridge circuit in Figure P14.6 determine $S_{L^m}^{I_m}$. Let $R_2 = R_3 = 1\,k\Omega$, $V = 10\,V$, $f = 10\,kHz$, and suppose that $\Delta_{I_m} = \pm 1\,\mu A$. What is the tolerance on the measurement of L due to the tolerance on ΔI_m for a (a) $L = 1\,mH$, (b) $L = 15\,mH$, and (c) $L = 100\,mH$?

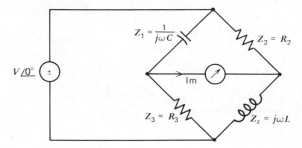

Figure P14.6

7. Given S_R^H show that $S_G^H = -S_R^H$ where $R = 1/G$.
8. Show that $S_L^{H(j\omega)} = S_{Z_L}^{H(j\omega)}$ and $S_C^{H(j\omega)} = -S_{Z_C}^{H(j\omega)}$ where $Z_L = j\omega L$ and $Z_C = 1/j\omega C$.
9. In the inverting amplifier circuit in Figure P14.9 find S_A^H where $H = e_o/e_i$. Assume that the op amp has infinite input impedance, zero output impedance, and a gain $A = 10^3$. Use this sensitivity function to find the percent change in the output voltage for a 20% change in A for (a) $R_f/R_s = 10$, (b) $R_f/R_s = 100$, and (c) $R_f/R_s = 10^3$.

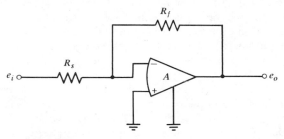

Figure P14.9

10. Repeat Problem 9 for $A = 10^5$.
11. In chapter 12, Example 1, the transfer function of the active bandpass filter is

$$H(s) = \frac{sA}{s^2 + s\left(3 - \frac{1}{2}A\right) + 1}$$

Find $S_A^{H(s)}$. For $s = j\omega$, at what frequency is the first-order variation in $H(j\omega)$ greatest? Express this maximum in terms of the Q of the filter.

12. In Figure P14.12 we wish to find $S_p^{v_o}$ and S_p^i where p is the 1-Ω resistor.
    (a) Draw the sensitivity circuit model.
    (b) Analyze the circuit in Figure P14.12 and the sensitivity circuit and compute $S_p^{v_o}$ and S_p^i.

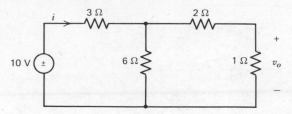

Figure P14.12

13. Repeat Problem 12 for the 6-Ω resistor in Figure P14.12.

14. In Figure 14.4 suppose that we wish to compute $R_2(\partial y/\partial R_2)$ where y denotes a voltage or current in the circuit. Show that the sensitivity circuit model is similar to the model in Figure 14.5a except the voltage source is now $R_2 i_2$.

15. In the D/A converter example the MSB resistor in a 10-bit converter had to have a tolerance better than 0.2%. Estimate the tolerance required for the LSB resistor in this converter.

16. We wish to compute the sensitivity $I_s(\partial y/\partial I_s)$ where y denotes a response and I_s denotes the saturation current of a semiconductor diode $(i = I_s(e^{v/V_T} - 1)$. Develop a sensitivity circuit model for the diode branch similar to the model in Figure 14.9.

17. For the series RLC circuit in Figure 14.10, Example 4, derive the sensitivity circuit model for computing the first-order sensitivity of the phasors with respect to the parameter C. Compute $S_C^{V_c}$ and $S_C^{V_o}$ from this model. What is the value of these sensitivities at the resonant frequency of the circuit? Check your answers by computing the above sensitivities from the transfer functions $H_c(j\omega) = V_c/V_s$ and $H_o(j\omega) = V_o/V_s$.

18. For the series RLC circuit in Figure 14.13 derive the sensitivity circuit model for computing first-order variations in the responses due to small changes in the initial condition i_{L0}. If $R = 1\,\Omega$, $C = 1\,F$, and $L = \frac{1}{2}H$, compute $\partial v_o/\partial i_{L0}$.

19. In Figure 14.13 derive the sensitivity circuit for computing $\partial v_o(t)/\partial L$.

20. Show that $S_p^{|H(j\omega)|} = \text{Re}[S_p^{H(j\omega)}]$ where p is a real circuit parameter. Hint: write $|H(j\omega)| = [H(j\omega)H^*(j\omega)]^{1/2}$.

Chapter Fifteen
Reciprocity and
Interreciprocity

In this concluding chapter some general properties of networks are discussed. At first it may seem that these properties, although interesting, are not very useful. However, it is shown in later sections that the reciprocity and interreciprocity properties can be used to significantly reduce the amount of computation required in many sensitivity and noise analysis problems.

15.1 TELLEGEN'S THEOREM

Tellegen's theorem is valid for any lumped network that obeys Kirchhoff's laws. To begin let us assign a branch voltage v_k and current i_k to each branch with the standard reference system so that the power absorbed by the branch at any time t is $v_k(t)i_k(t)$. Tellegen's theorem states that if all the branch voltages v_k satisfy KVL and all the branch currents satisfy KCL, then

$$\sum_{k=1}^{b} v_k(t) \cdot i_k(t) = 0 \tag{1}$$

where the summation includes every branch of the lumped circuit and b is the number of branches.

To prove Tellegen's theorem we write

$$S = \sum_{k=1}^{b} v_k i_k = \frac{1}{2} \sum_{p=0}^{n-1} \sum_{q=0}^{n-1} (e_p - e_q) i_{pq} \tag{2}$$

where n is the number of nodes in the circuit, e_p and e_q are node voltages and "0" represents the ground node so that $e_o = 0$ and recall that any branch voltage v_k can be represented as the difference between two node voltages. The current i_{pq} represents the total current flowing from node ⓟ to node ⓠ through any branch or collection of parallel branches whose terminals terminate at nodes ⓟ and ⓠ. Note that the factor $\frac{1}{2}$ appears in (2) since terms of the form $(e_p - e_q)i_{pq}$ and $(e_q - e_p)i_{qp}$ are equal since $i_{pq} = -i_{qp}$ so that the double summation in (2) counts everything twice. Next, write (2) as

$$S = \sum_{p=0}^{n-1} \sum_{q=0}^{n-1} e_p i_{pq} - \sum_{p=0}^{n-1} \sum_{q=0}^{n-1} e_q i_{pq}$$

or

$$S = \sum_{p=0}^{n-1} e_p \left(\sum_{q=0}^{n-1} i_{pq} \right) - \sum_{q=0}^{n-1} e_q \left(\sum_{p=0}^{n-1} i_{pq} \right) \tag{3}$$

Since the branch currents satisfy KCL,

$$\sum_{q=0}^{n-1} i_{pq} = \sum_{p=0}^{n-1} i_{pq} = 0 \tag{4}$$

Hence, Equation 1 is verified.

Equation 1 probably appears obvious, since it states that *the sum of the power in the branches of the network at any time t is zero.* This result implies a conservation of energy principle. Of course, KCL and KVL are based on the principle of conservation of energy. However, a much more surprising result is the following. Let us take two circuits with the same topology, that is, if the first circuit has a set of branch voltages $v_1(t)$, $v_2(t), \ldots v_b(t)$ and branch currents $i_1(t)$, $i_2(t) \ldots i_b(t)$ with polarities assigned under the standard reference system, then the second circuit has a set of branch voltages $\hat{v}_1(\tau)$, $\hat{v}_2(\tau), \ldots, \hat{v}_b(\tau)$ and branch currents $\hat{i}_1(\tau)$, $\hat{i}_2(\tau), \ldots, \hat{i}_b(\tau)$ with polarities assigned as above and there is a one-to-one correspondence between the nodes and the branches of the two circuits, that is, if branch k is connected between nodes ⓟ and ⓠ, then the second circuit has a corresponding branch k (not necessarily the same element type or value) connected between nodes ⓟ and ⓠ. Circuits 1 and 2 can be the same circuits, but perhaps with different excitations. If these branch voltages and currents satisfy Kirchhoff's laws in their respective circuits, then

$$\sum_{k=1}^{b} v_k(t) \cdot \hat{i}_k(\tau) = 0 \tag{5}$$

and

$$\sum_{k=1}^{b} \hat{v}_k(\tau) i_k(t) = 0 \tag{6}$$

It is interesting to note that one can use the branch responses at time t in one circuit, and the branch responses at time τ in the second circuit. The proof of (5) and (6) is analogous to that used to verify (1) and is left as an exercise.

Finally, it should be noted that in the sinusoidal steady-state analysis of linear time-invariant circuits, Tellegen's theorem for the phasor circuit model has the form

$$\mathbf{P} = \sum_{k=1}^{b} \mathbf{V}_k(j\omega)\mathbf{I}_k^*(j\omega) = 0 \qquad (7)$$

From the proof of Tellegen's theorem one can easily show that (7) is valid without the conjugate of the current, that is,

$$\sum_{k=1}^{b} \mathbf{V}_k(j\omega)\mathbf{I}_k(j\omega) = 0 \qquad (8)$$

However, the real part of (8) no longer represents the real average power. Expressions 5 and 6 can also be expressed in terms of the phasors of both circuits.

15.2 RECIPROCITY

Basically, reciprocity implies that for a given two-port the input and the output ports of a system can be interchanged and the transfer immittances will be unchanged. Specifically, a reciprocal network has the following properties. For simplicity the discussion will be limited to the linear time-invariant sinusoidal steady-state phasor circuit model.

Case I

If the two-port N in Figure 15.1 is *reciprocal*, then given that the independent sources are equal,

$$\mathbf{I}_s(j\omega) = \hat{\mathbf{I}}_o(j\omega) \qquad (9)$$

it follows that the responses are equal,

$$\hat{\mathbf{V}}_s(j\omega) = \mathbf{V}_o(j\omega) \qquad (10)$$

Note that in Figure 15.1a we have denoted the branch variables as $\{\mathbf{V}_k, \mathbf{I}_k\}$ and in Figure 15.1b as $\{\hat{\mathbf{V}}_k, \hat{\mathbf{I}}_k\}$, because, even though the networks N are

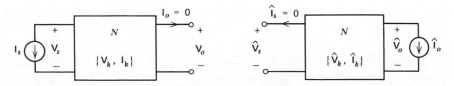

Figure 15.1   Input current—open circuit output voltage.

identical, they are excited differently at their ports so that the branch variables can have different values in the two circuits. It is important to note that reciprocity *does not* imply that the transfer functions V_o/V_s and $\hat{V}_s/\hat{V}_o$ are equal.

Case II

If the two-port N is *reciprocal* in Figure 15.2, then, given that the independent sources are equal,

$$V_s(j\omega) = \hat{V}_o(j\omega) \tag{11}$$

it follows that the responses are equal,

$$\hat{I}_s(j\omega) = I_o(j\omega) \tag{12}$$

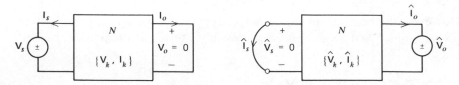

Figure 15.2   Input voltage—short circuit output current.

Case III

If the two-port N is *reciprocal* in Figure 15.3, then given that the independent sources satisfy the equation

$$V_s(j\omega) = -a\hat{I}_o(j\omega) \tag{13}$$

it follows that the responses are related by the equation

$$V_o(j\omega) = a\hat{I}_s(j\omega) \tag{14}$$

where a is an arbitrary constant.

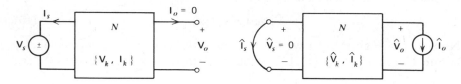

Figure 15.3   Input voltage—open circuit output voltage.

Case IV

If the two-port N is reciprocal in Figure 15.4, then given that the independent sources satisfy the equation

$$I_s(j\omega) = -a\hat{V}_o(j\omega) \tag{15}$$

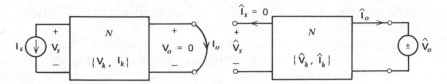

Figure 15.4   Input current—short circuit output current.

it follows that the responses are related by the equation

$$\mathbf{I}_o(j\omega) = a\hat{\mathbf{V}}_s(j\omega) \tag{16}$$

15.2.1 DEFINITION OF RECIPROCITY

Now that we have enunciated some of the prominent properties of reciprocal circuits, let us define reciprocity for lumped circuits. First, recall that from Tellegen's theorem we can write for the two-port N that

$$\sum_{k=1}^{b} [\mathbf{V}_k(j\omega)\hat{\mathbf{I}}_k(j\omega)] = -[\mathbf{V}_s(j\omega)\hat{\mathbf{I}}_s(j\omega) + \mathbf{V}_o(j\omega)\hat{\mathbf{I}}_o(j\omega)] \tag{17}$$

and

$$\sum_{k=1}^{b} [\hat{\mathbf{V}}_k(j\omega)\mathbf{I}_k(j\omega)] = -[\hat{\mathbf{V}}_s(j\omega)\mathbf{I}_s(j\omega) + \hat{\mathbf{V}}_o(j\omega)\mathbf{I}_o(j\omega)] \tag{18}$$

where b denotes the total number of branches in the lumped circuit N and the right side of (17) and (18) includes the branches that terminate the two-port. Let us now define reciprocity for linear time-invariant sinusoidal steady-state circuits. The phasors are usually frequency dependent but, for convenience, we drop the $j\omega$ argument below.

A linear time-invariant sinusoidal steady-state circuit is reciprocal if

$$\sum_{k=1}^{b} (\mathbf{V}_k\hat{\mathbf{I}}_k - \hat{\mathbf{V}}_k\mathbf{I}_k) \equiv 0 \tag{19}$$

for all permissible phasors $\{\mathbf{V}_k, \mathbf{I}_k\}$ and $\{\hat{\mathbf{V}}_k, \hat{\mathbf{I}}_k\}$.

Permissible phasors are those that satisfy the circuit equations. Now if (19) is true then, from (17) and (18), it follows that

$$\mathbf{V}_s\hat{\mathbf{I}}_s + \mathbf{V}_o\hat{\mathbf{I}}_o - \hat{\mathbf{V}}_s\mathbf{I}_s - \hat{\mathbf{V}}_o\mathbf{I}_o = 0 \tag{20}$$

Note that (19) and (20) were formed by taking the difference between (17) and (18). Given (20) the properties (9) to (16) can easily be verified. For example, in Case I $\mathbf{I}_o = 0$ and $\hat{\mathbf{I}}_s = 0$ so that (20) becomes

$$\mathbf{V}_o\hat{\mathbf{I}}_o - \hat{\mathbf{V}}_s\mathbf{I}_s = 0 \tag{21}$$

Clearly if $\mathbf{I}_s = \hat{\mathbf{I}}_o$, then $\mathbf{V}_o = \hat{\mathbf{V}}_s$. The other three properties can be verified just as easily.

The prime question that should be in one's mind is the following. *What types of linear time-invariant lumped circuits satisfy* (19)? This question leads us to the following theorem.

In the sinusoidal steady state any linear time-invariant circuit consisting of only resistors, capacitors, inductors, and transformers satisfies (19); hence, is a reciprocal circuit.

To prove the above theorem for two-terminal elements such as resistors, capacitors, and inductors, we write the branch constraints for the first network as

$$\mathbf{V}_k = Z_k \mathbf{I}_k \tag{22}$$

and for the second network as

$$\hat{\mathbf{V}}_k = Z_k \hat{\mathbf{I}}_k \tag{23}$$

The corresponding term in (19) is

$$\mathbf{V}_k \hat{\mathbf{I}}_k - \hat{\mathbf{V}}_k \mathbf{I}_k$$

Upon substitution of (22) and (23) into the above term, we obtain

$$Z_k \mathbf{I}_k \hat{\mathbf{I}}_k - Z_k \hat{\mathbf{I}}_k \mathbf{I}_k \equiv 0 \tag{24}$$

In the case of the transformer the constraints for the first circuit are

$$\begin{aligned}
\mathbf{V}_k &= j\omega L_1 \mathbf{I}_k + j\omega M \mathbf{I}_{k+1} \\
\mathbf{V}_{k+1} &= j\omega M \mathbf{I}_k + j\omega L_2 \mathbf{I}_{k+1}
\end{aligned} \tag{25}$$

and for the second circuit simply place a carat above the phasors. We must associate two branches with the transformer element. The appropriate terms in (19) are

$$\mathbf{V}_k \hat{\mathbf{I}}_k - \hat{\mathbf{V}}_k \mathbf{I}_k + \mathbf{V}_{k+1} \hat{\mathbf{I}}_{k+1} - \hat{\mathbf{V}}_{k+1} \mathbf{I}_{k+1} \tag{26}$$

Upon substitution of (25) into (26) and a similar expression for the constraints of the second circuit into the above expression, we obtain

$$\begin{aligned}
&(j\omega L_1 \mathbf{I}_k + j\omega M \mathbf{I}_{k+1})\hat{\mathbf{I}}_k - (j\omega L_1 \hat{\mathbf{I}}_k + j\omega M \hat{\mathbf{I}}_{k+1})\mathbf{I}_k \\
&+ (j\omega M \mathbf{I}_k + j\omega L_2 \mathbf{I}_{k+1})\hat{\mathbf{I}}_{k+1} - (j\omega M \hat{\mathbf{I}}_k + j\omega L_2 \hat{\mathbf{I}}_{k+1})\mathbf{I}_{k+1} \equiv 0
\end{aligned} \tag{27}$$

We conclude that the *linear time-invariant resistor, capacitor, inductor, and transformer are reciprocal elements*. One can show that the dependent source and the independent source are *not* reciprocal elements.

Another way to view reciprocity is from the matrix representation of the network equations. For example, in Figure 15.1 suppose that the circuit is an $n+1$ node circuit that is described by the nodal equations

$$
\begin{bmatrix}
y_{11} & y_{12} \cdots y_{1n} \\
y_{21} & y_{22} \cdots y_{2n} \\
\cdot & \cdot \\
\cdot & \cdot \\
\cdot & \cdot \\
y_{n1} & y_{n2} \cdots y_{nn}
\end{bmatrix}
\begin{bmatrix}
E_1 \\
E_2 \\
\cdot \\
\cdot \\
\cdot \\
E_n
\end{bmatrix}
=
\begin{bmatrix}
I_1 \\
I_2 \\
\cdot \\
\cdot \\
\cdot \\
I_n
\end{bmatrix}
\tag{28}
$$

which we write as

$$\mathbf{YE} = \mathbf{I} \tag{29}$$

where $\mathbf{Y}$ is the admittance array, $\mathbf{E}$ is the vector of node voltages, and $\mathbf{I}$ is the vector of independent source currents. The circuit is said to be reciprocal if

$$\mathbf{Y} = \mathbf{Y}^T \tag{30}$$

that is, the admittance array is symmetric.

To verify that $\mathbf{V}_o/\mathbf{I}_s = \hat{\mathbf{V}}_s/\hat{\mathbf{I}}_o$ in case I, which is the only case involving only current source inputs, suppose that $\mathbf{I}_s$ is connected between node ① and the reference node and that $\mathbf{V}_o$ is measured from node ⓝ to ground. Under these conditions

$$
\mathbf{V}_o = \frac{
\begin{vmatrix}
y_{11} & y_{12} \cdots \mathbf{I}_s \\
y_{21} & y_{22} \cdots 0 \\
\cdot & \cdot \\
\cdot & \cdot \\
\cdot & \cdot \\
y_{n1} & y_{n2} \cdots 0
\end{vmatrix}
}{|\mathbf{Y}|}
\tag{31}
$$

or

$$\frac{\mathbf{V}_o}{\mathbf{I}_s} = \frac{(-1)^{n-1}|\mathbf{Y}_{1n}|}{|\mathbf{Y}|} \tag{32}$$

where $|\mathbf{Y}_{1n}|$ denotes the $(1, n)$ minor of the array $\mathbf{Y}$. With the current source connected to the output port, the voltage at the input port is

$$
\hat{\mathbf{V}}_s = \frac{
\begin{vmatrix}
0 & y_{12} \cdots y_{1n} \\
0 & y_{22} \cdots y_{2n} \\
\cdot \\
\cdot \\
\cdot \\
\hat{\mathbf{I}}_o & y_{n2} \cdots y_{nn}
\end{vmatrix}
}{|\mathbf{Y}|}
\tag{33}
$$

or

$$\frac{\hat{\mathbf{V}}_s}{\hat{\mathbf{I}}_o} = \frac{(-1)^{n-1}|\mathbf{Y}_{n1}|}{|\mathbf{Y}|} \tag{34}$$

In order for (34) and (32) to be identical we require that

$$|\mathbf{Y}_{1n}| = |\mathbf{Y}_{n1}| \tag{35}$$

Equation 35 is satisfied if the admittance matrix is symmetric. The admittance matrix of an RLC transformer circuit can be shown to be symmetric by means of Equation 17 in Chapter Four and the transformer model in Problem 22, Chapter Thirteen.

If one expresses the network equations by means of the mesh formulation

$$\mathbf{Z}\mathbf{I}_m = \mathbf{V} \tag{36}$$

where $\mathbf{Z}$ is an impedance array, $\mathbf{I}_m$ is the vector of mesh currents and $\mathbf{V}$ denotes the independent voltage sources, then it can be shown that

$$\frac{\mathbf{I}_o}{\mathbf{V}_s} = \frac{\hat{\mathbf{I}}_s}{\hat{\mathbf{V}}_o}$$

in Figure 15.2 provided that the impedance array $\mathbf{Z}$ is symmetric, and the impedance matrix of an RLC transformer circuit is symmetric.

15.3 APPLICATIONS OF RECIPROCITY

Reciprocity has two basic applications. The first deals with the fact that transmission in the reciprocal two-port is the same in either direction under the conditions of properties (9) to (16). The second important application deals with the computation of transfer functions in a multiple-input but single-output circuit. In this situation the reciprocity property can be used to significantly reduce the computational effort.

15.3.1 RECIPROCAL TWO-PORTS

Figure 15.5a illustrates a filter that could be found in a communication system. Typically R_1 represents the generator resistance and R_2 represents the load resistance. This circuit, excluding port terminations, is a reciprocal circuit. Upon application of property (13) and (14) we find that the circuit in Figure 15.5b yields an equivalent response, only the excitation and response currents are scaled by the factor a. By means of a source transformation, and by taking the output response to be the voltage drop across the resistor R_1, we obtain the circuit in Figure 15.5c. Note that the voltage transfer function for this circuit is $R_1\mathbf{V}_o/R_2\mathbf{V}_s$. Thus the transfer functions for the circuits in Figures 15.5a and 15.5c are identical except for the scale factor R_1/R_2.

The above exercise has a definite purpose because in some applications Figure 15.5c may be preferable to Figure 15.5a. For example, suppose that $R_1 = R_2$, but in practice the load resistor has a parasitic capacitance

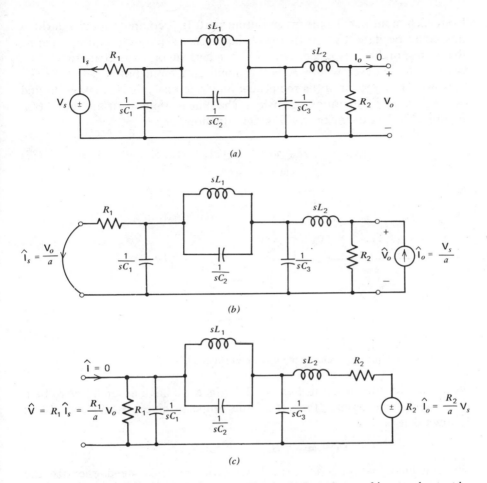

Figure 15.5 Communication circuit filter. (*a*) Circuit. (*b*) Interchange of input and output by reciprocity. (*c*) Conversion of (*b*) to voltage input-voltage output.

associated with it that affects the designed response of the circuit in Figure 15.5*a*. By using Figure 15.5*c* the parasitic load capacitance can be included in the value of C_1. Thus, the designed response is achieved. Also, in cases in which $R_1 \neq R_2$, the exchange would be necessary in cases where the generator resistance is R_2 and the load resistance is R_1.

15.3.2 MULTIPLE INPUT-SINGLE OUTPUT TRANSFER FUNCTION CALCULATIONS

Suppose that we have a linear circuit with several inputs and only one output. Suppose further that we need to compute the transfer function from each input port to the given output port. In a computational algorithm this could be done by exciting each input port separately and calculating the corresponding response. This is a standard superposition approach that requires K analyses where K is the number of input ports.

However, if the circuit is reciprocal only one analysis is required to find the K transfer functions. Consider the linear time-invariant circuit in Figure 15.6. Let us write the transfer functions as

$$\frac{\mathbf{V}_o}{\mathbf{I}_k} = H_{ok}(j\omega) \qquad k = 1, 2, \ldots, K \tag{37}$$

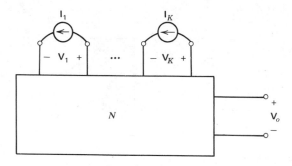

Figure 15.6   Multiple-input single-output reciprocal circuit.

Now consider the circuit in Figure 15.7. Since we have assumed N to be a linear reciprocal circuit, then from the property given by (9) and (10) it follows that

$$\hat{\mathbf{V}}_k = H_{ko} = H_{ok} \qquad k = 1, 2, \ldots, K \tag{38}$$

We conclude that in a reciprocal network one only need energize the circuit at the output port and measure or calculate the responses at the other ports in order to determine the above transfer functions. To more clearly see the advantage of this result, let us discuss a typical application in sensitivity analysis.

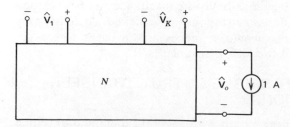

Figure 15.7   Transfer function determination.

15.3.3 SENSITIVITY CALCULATIONS

For example, consider the low-pass filter circuit in Figure 15.8. Suppose that we wish to compute $\partial V_o/\partial R_1$, $\partial V_o/\partial C_2$, and $\partial V_o/\partial L_3$. One approach would be to find the transfer function V_o/V_s and differentiate this function with respect to the parameters R_1, C_2, and L_3, respectively. This approach would not be too difficult in this example because the circuit is relatively simple. However, in a large circuit this approach would not be feasible and we would have to resort to computer analysis and the sensitivity circuit method in the previous chapter.

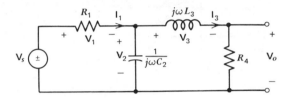

Figure 15.8 Low-pass filter.

The three sensitivity circuits that must be analyzed are shown in Figure 15.9. In the first circuit we have differentiated the Kirchhoff equations and branch constraints with respect to R_1. For example, the branch constraint

$$I_1 = \frac{V_1}{R_1} \tag{39}$$

becomes

$$\frac{\partial I_1}{\partial R_1} = \frac{1}{R_1}\frac{\partial V_1}{\partial R_1} - \frac{1}{R_1^2}V_1 \tag{40}$$

which is a resistor in shunt with a current source V_1/R_1^2 in Figure 15.9a. Similarly the sensitivity circuit in Figure 15.9b was obtained by differentiating the circuit equations with respect to C_2. Finally, Figure 15.9c is the sensitivity circuit for the inductance L_3. Note that all three circuits in Figure 15.9 are identical excluding the current sources, and they are reciprocal. Therefore, we can use the reciprocity property to analyze the *one* circuit in Figure 15.10 rather than the *three* circuits in Figure 15.9. The circuit in Figure 15.10 is sometimes called the adjoint circuit. By means of reciprocity in this circuit the voltage $\hat{V}_1$ across the resistor R_1 is the sensitivity $\partial V_o/\partial R_1$ provided that we weight $\hat{V}_1$ by the value of the current source in Figure 15.9a, that is,

$$\frac{\partial V_o}{\partial R_1} = -\frac{1}{R_1^2}V_1\hat{V}_1 \tag{41}$$

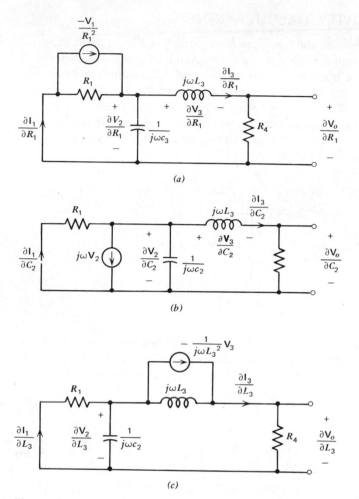

Figure 15.9 Sensitivity circuits. (a) Sensitivity circuit for R_1. (b) Sensitivity circuit for C_2. (c) Sensitivity circuit for L_3.

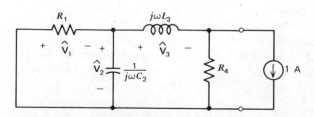

Figure 15.10 Adjoint circuit.

Similarly, we conclude that

$$\frac{\partial V_o}{\partial C_2} = j\omega V_2 \hat{V}_2 \tag{42}$$

and

$$\frac{\partial V_o}{\partial L_3} = \frac{-1}{j\omega L^2} V_3 \hat{V}_3 \tag{43}$$

It is very straightforward to compute these expressions. For example, to compute (43) we note that

$$I_3 = \frac{1}{j\omega L_3} V_3 \tag{44}$$

so that

$$\frac{\partial I_3}{\partial L_3} = \frac{-1}{j\omega L_3^2} V_3 + \frac{1}{j\omega L_3} \frac{\partial V_3}{\partial L_3} \tag{45}$$

From (45) we see that the inductor is shunted by a current source whose value is $(-1/j\omega L_3^2)V_3$. Now the voltage $\hat{V}_3$ in Figure 15.10 is the transfer admittance from the inductor L_3 to the output port, hence the product of this transfer admittance times the value of the current source shunting the inductance in the sensitivity circuit is the sensitivity of the output voltage with respect to L_3 as given by (43). To compute these sensitivities we only need to analyze the original circuit in Figure 15.8 and the adjoint circuit in Figure 15.10 and substitute the results into (41) to (43).

15.4 INTERRECIPROCITY

In the previous section it was shown that reciprocal networks have some useful properties that can simplify multiple-input single-output transfer function calculations. This property was particularly useful in sensitivity circuit analysis. It would be extremely advantageous if the reciprocity properties could be applied to nonreciprocal circuits, in particular circuits with controlled sources. Then we could simplify multiple-input single-output calculations for electronic circuits. In order to see how to handle the controlled source branches so that the reciprocity properties can be used, consider the two networks in Figure 15.11. Suppose that both of these

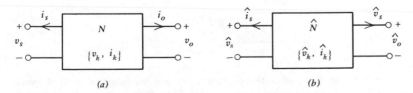

(a)   (b)

Figure 15.11   Interreciprocal networks.

circuits have b branches, which are in one-to-one correspondence. We say that these two-ports are interreciprocal to each other if the reciprocity properties in (9) to (16) are satisfied. In order for these properties to apply, recall that we require

$$\sum_{k=1}^{b} (\mathbf{V}_k \hat{\mathbf{I}}_k - \hat{\mathbf{V}}_k \mathbf{I}_k) \equiv 0 \tag{19}$$

We showed that R, L, and C branches satisfied this identity, and each RLC branch of the circuit N, Figure 15.11a, will continue to satisfy (19) on a branch-by-branch basis if, for each resistor, capacitor, inductor branch, and transformer multiport in N, there is a corresponding resistor, capacitor, inductor branch, or transformer multiport, respectively, in $\hat{N}$ connected to identical nodes and with identical parameter values.

Let us now see how to handle controlled sources so that (19) is satisfied. Examine the voltage controlled current source branch in Figure 15.12a. The branch constraints in the frequency domain are

$$\mathbf{I}_k = 0 \tag{46}$$

and

$$\mathbf{I}_{k+1} = g_m \mathbf{V}_k \tag{47}$$

The controlled source is a two-port so that we must associate two branches in (19) with it, a controlling branch and the controlled branch. Thus, the

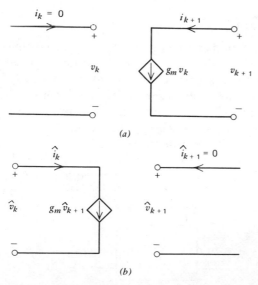

(a)

(b)

Figure 15.12 Relationship between a branch constraint and its interreciprocal network branch constraint. (a) Voltage controlled current source. (b) Interreciprocal network branch constraint.

terms in (19) associated with this controlled source are

$$\mathbf{V}_k\hat{\mathbf{I}}_k - \hat{\mathbf{V}}_k\mathbf{I}_k + \mathbf{V}_{k+1}\hat{\mathbf{I}}_{k+1} - \hat{\mathbf{V}}_{k+1}\mathbf{I}_{k+1} \equiv 0 \qquad (48)$$

Upon substitution of (46) and (47) into (48) we obtain

$$\mathbf{V}_k\hat{\mathbf{I}}_k + \mathbf{V}_{k+1}\hat{\mathbf{I}}_{k+1} - \hat{\mathbf{V}}_{k+1}g_m\mathbf{V}_k \equiv 0 \qquad (49)$$

Equation 49 is satisfied provided that

$$\hat{\mathbf{I}}_k = g_m\hat{\mathbf{V}}_{k+1} \qquad (50)$$

and

$$\hat{\mathbf{I}}_{k+1} = 0 \qquad (51)$$

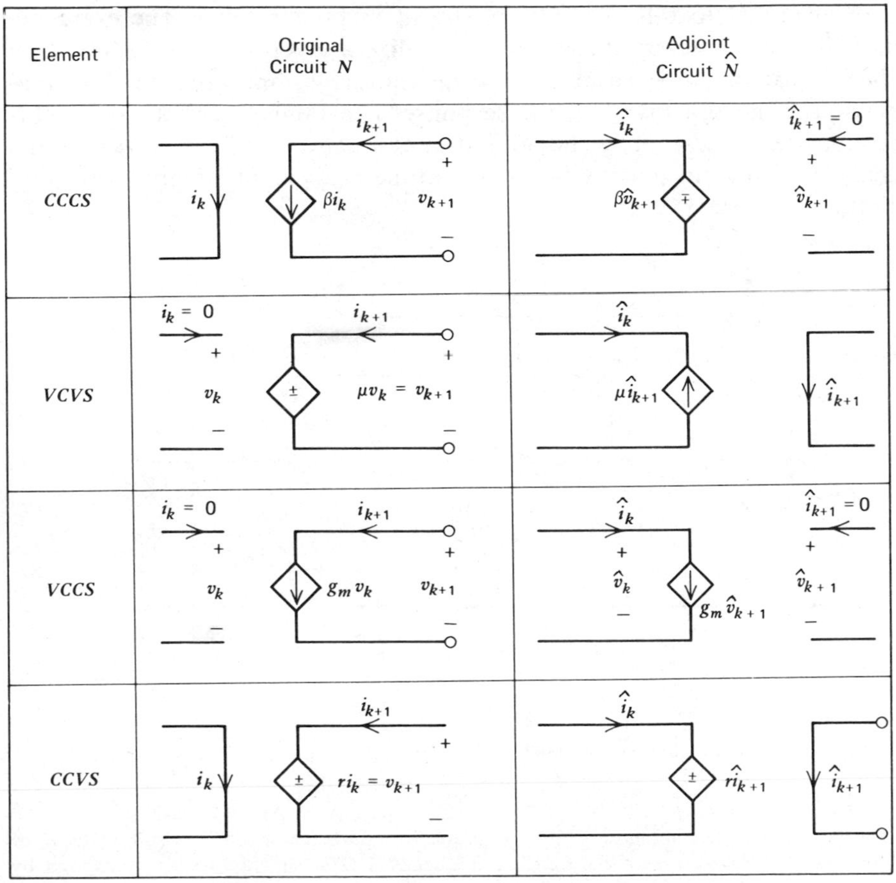

Element	Original Circuit N	Adjoint Circuit $\hat{N}$
CCCS		
VCVS		
VCCS		
CCVS		

Figure 15.13 Controlled source branch constraints and their interreciprocal network constraints.

Thus, these equations tell us to replace the VCCS's by the circuit model in Figure 15.12*b* in order to satisfy (19).

The network $\hat{N}$ is said to be *interreciprocal* to N if (19) is satisfied. A linear time-invariant network N consisting of R, L, C, transformer, and dependent source elements is interreciprocal to a network $\hat{N}$ and vice versa if the network $\hat{N}$ is formed by replacing every dependent source in N by the corresponding model in Figure 5.13. This circuit $\hat{N}$ is again referred to as the *adjoint* of N or the *adjoint circuit*.

15.4.1 APPLICATION OF INTERRECIPROCITY IN SENSITIVITY ANALYSIS

Consider the active RC filter in Figure 15.14. The network that is interreciprocal to this two-port is shown in Figure 15.15. There are numerous active filter configurations to realize a given transfer function. In choosing among the various active filter designs, one criterion for comparison is the sensitivity of the response to parameter variations. In this section we illustrate how the adjoint circuit can be utilized to reduce the analysis of several sensitivity circuits to the analysis of simply the adjoint circuit in Figure 15.15.[1]

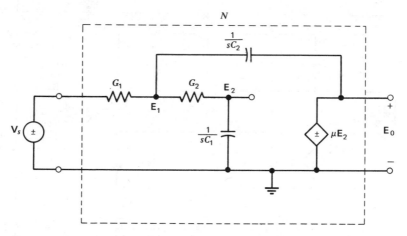

Figure 15.14   Active RC low-pass filter.

[1] For the simple circuit in Figure 15.14 usually one finds the transfer function E_o/V_s in terms of the circuit parameters and the complex frequency s. The sensitivities can be found by differentiating this transfer function with respect to the parameter of interest. The adjoint circuit approach is introduced because in the computer analysis of circuits it is often a good method for computing sensitivities.

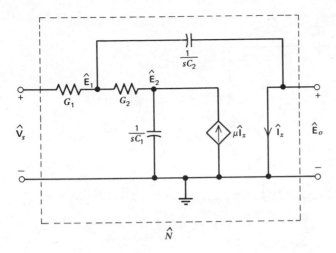

Figure 15.15 Adjoint network.

The node equations for the active filter are

$$(G_1 + G_2 + sC_2)\mathbf{E}_1 - (G_2 + sC_2\mu)\mathbf{E}_2 = G_1\mathbf{E}_1 \tag{52}$$

and

$$- G_2\mathbf{E}_1 + (G_2 + sC_1)\mathbf{E}_2 \tag{53}$$

plus the controlled source constraint

$$\mathbf{E}_o = \mu\mathbf{E}_2 \tag{54}$$

Suppose that we are interested in the sensitivity of the output voltage $\mathbf{E}_o$ with respect to the gain μ of the amplifier. By means of the sensitivity circuit method we write

$$(G_1 + G_2 + sC_2)\frac{\partial\mathbf{E}_1}{\partial\mu} - (G_2 + sC_2\mu)\frac{\partial\mathbf{E}_2}{\partial\mu} - sC_2\mathbf{E}_2 = 0 \tag{55}$$

$$- G_2\frac{\partial\mathbf{E}_1}{\partial\mu} + (G_2 + sC_1)\frac{\partial\mathbf{E}_2}{\partial\mu} \tag{55}$$

and

$$\frac{\partial\mathbf{E}_o}{\partial\mu} = e_2 + \mu\frac{\partial\mathbf{E}_2}{\partial\mu} \tag{57}$$

The sensitivity circuit described by (55) to (57) is shown in Figure 15.16. In the frequency domain in order to find $\partial\mathbf{E}_o/\partial\mu$ we need to find the transfer function from the source E_2 to the output port in Figure 15.16, and weight this transfer function by $\mathbf{E}_2$, the input phasor in the sensitivity circuit. From (13) to (14) and Figure 15.3 we see that we can use the adjoint circuit in Figure

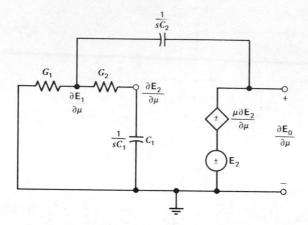

Figure 15.16   Sensitivity circuit for μ variation.

15.17 to find this transfer function which is

$$\frac{\partial \mathbf{E}_o / \partial \mu}{\mathbf{E}_2} = -\hat{\mathbf{I}}_x \tag{58}$$

Therefore,

$$\frac{\partial \mathbf{E}_o}{\partial \mu} = -\hat{\mathbf{I}}_x \mathbf{E}_2 \tag{59}$$

Thus, in order to find the sensitivity of the output voltage $\mathbf{E}_o$ with respect to the amplifier gain μ in the frequency domain, we need to analyze the phasor circuit model of Figure 15.14 to find $\mathbf{E}_2$ and the phasor circuit model of Figure 15.17 to find $\hat{\mathbf{I}}_x$.

Why not simply analyze the phasor model of the sensitivity circuit in

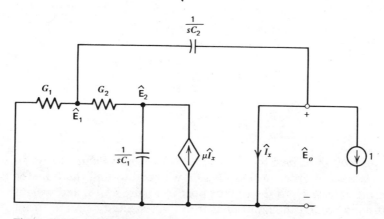

Figure 15.17   Adjoint circuit analysis.

Figure 15.16? If we are interested only in the sensitivity with respect to μ then this would be the best approach. But suppose we also wanted to find the first-order sensitivities $\partial E_o/\partial G_1$, $\partial E_o/\partial G_2$, $\partial E_o/\partial C_1$, and $\partial E_o/\partial C_2$. Note that only the output port voltage variation is of interest. Since conductance branch constraints are of the form

$$\mathbf{I}_a = G_a \mathbf{V}_a \tag{60}$$

then the sensitivity constraint is

$$\frac{\partial \mathbf{I}_a}{\partial G_a} = \mathbf{V}_a + G_a \frac{\partial \mathbf{V}_a}{\partial G_a} \tag{61}$$

We see that the conductance in the sensitivity circuit model with respect to which the derivative is taken is shunted by a current source whose value is $\mathbf{V}_a$. Thus, we need the transfer admittance from the port shunting that conductance to the output. The sensitivity is this transfer admittance weighted by $\mathbf{V}_a$, the value of the current source in the sensitivity circuit. Thus, by means of the adjoint circuit in Figure 15.17,

$$\frac{\partial \mathbf{E}_o}{\partial G_1} = \hat{\mathbf{E}}_1(\mathbf{E}_1 - \mathbf{V}_s) \tag{62}$$

and

$$\frac{\partial \mathbf{E}_o}{\partial G_2} = (\hat{\mathbf{E}}_2 - \hat{\mathbf{E}}_1)(\mathbf{E}_2 - \mathbf{E}_1) \tag{63}$$

where $\hat{\mathbf{E}}_1$ and $(\hat{\mathbf{E}}_2 - \hat{\mathbf{E}}_1)$ are the transfer admittances. Note that one only needs to analyze the circuit and its adjoint to compute (59), (62), and (63).
  Similarly, for the capacitor

$$\mathbf{I}_a = j\omega C_a \mathbf{V}_a \tag{64}$$

and

$$\frac{\partial \mathbf{I}_a}{\partial C_a} = j\omega \mathbf{V}_a + j\omega C \frac{\partial \mathbf{V}_a}{\partial C} \tag{65}$$

The capacitor with respect to which the derivative is taken is shunted by a current source $j\omega \mathbf{V}_a$ in the sensitivity circuit. Thus, the sensitivity $\partial E_o/\partial C_a$ is the transfer function from this current source to the output port weighted by $j\omega \mathbf{V}_a$. The interreciprocity property tells us that this transfer function is simply the voltage across the capacitance in the adjoint circuit that is excited at its output port with a 1-A source. Thus, in our example,

$$\frac{\partial \mathbf{E}_o}{\partial C_1} = j\omega \mathbf{E}_2 \hat{\mathbf{E}}_2 \tag{66}$$

and

$$\frac{\partial \mathbf{E}_o}{\partial C_2} = j\omega (\mathbf{E}_1 - \mathbf{E}_o)(\hat{\mathbf{E}}_1 - \hat{\mathbf{E}}_o) \tag{67}$$

where $\hat{\mathbf{E}}_2$ and $(\hat{\mathbf{E}}_1 - \hat{\mathbf{E}}_o)$ are the transfer admittances. Note that to compute (59), (62), (63), (66), and (67) did not require the analysis of five sensitivity circuits, but only the analysis of the adjoint circuit. Of course, we cannot use the adjoint circuit in Figure 15.17 to compute $\partial\hat{\mathbf{E}}_1/\partial\mu$, or $\partial\hat{\mathbf{E}}_1/\partial G_1$, and so on. If we must compute the sensitivity of more than one response with respect to a number of parameters, we must decide if it is better to analyze the sensitivity circuits directly or use the adjoint circuit concept in the calculation of the sensitivities.

Finally it should be noted that the reciprocity and interreciprocity properties are also very useful in the noise analysis of circuits.[2]

REFERENCES

1. Desoer, C. A. and E. S. Kuh. *Basic Circuit Theory*, Chapters 9, 16, McGraw-Hill Book co., 1969.
2. Director, S. W. and R. A. Rohrer. *IEEE Transactions on Circuit Theory*, Vol. CT-16, Aug. 1969, pp. 318–323 and 330–337; Correction: Vol. CT-19, July 1972, pp. 367–370.

PROBLEMS

1. Prove Equation 5.
2. Verify properties (9) to (16) for the reciprocal two-port in Figure P15.2, that is, for case I find the transfer impedances $Z_{os} = \mathbf{V}_o/\mathbf{I}_s|_{\mathbf{I}_o=0}$ and $Z_{so} = \mathbf{V}_s/\mathbf{I}_o|_{\mathbf{I}_s=0}$ and verify that $Z_{os} = Z_{so}$. Make similar calculations for the other three cases.

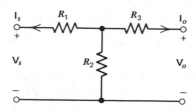

Figure P15.2

3. Repeat Problem 2 for the reciprocal two-port in Figure P15.3.

[2] R. Rohrer, L. Nagel, R. Meyer, and L. Weber, "Computationally Efficient Electronic-Circuit Noise Calculations," *IEEE Journal on solid-State Circuits*, Vol. SC-6, Aug. 1971, pp. 204–213.

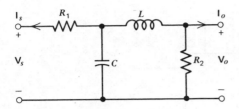

Figure P15.3

4. Show that the admittance and impedance matrices of the two-port in Figure P15.2 are symmetric.
5. Repeat Problem 4 for the two-port in Figure P15.3.
6. Use (13) and (14) and find the transfer functions v_o/v_1, v_o/v_2, v_o/v_3, and v_o/v_4 in Figure P15.6a by means of an analysis of the circuit in Figure P15.6b.

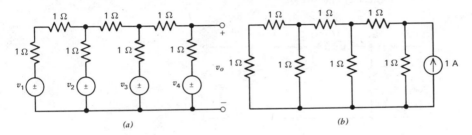

(a)   (b)

Figure P15.6

7. Use source transformations and properties (9) and (10) to find the relationship between V_o/V_s in Figure P15.7a and $\hat{V}_o/V_s$ in Figure P15.7b.

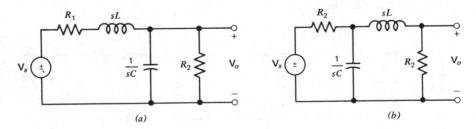

(a)   (b)

Figure P15.7

8. In Figure P15.8 it is found that for a given independent current source $1\underline{/0°}$ A at the frequency ω_o rad/s, the amplitude and phase of the steady-state responses are $V_1 = 10\underline{/30°}$, $V_2 = 3\underline{/-20°}$, and $V_3 = 0.8\underline{/-70°}$. At the frequency ω_o what is the value of the transfer functions V_o/I_1, V_o/I_2, and V_o/I_3 when $I_o = 0$. Explain your answer.

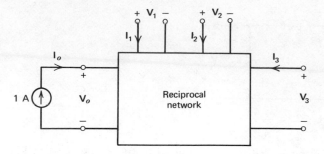

Figure P15.8

9. Express the sensitivities $\partial V_o/\partial R_1$, $\partial V_o/\partial L$, $\partial V_o/\partial C$, and $\partial V_o/\partial R_2$ in terms of the node voltages $\mathbf{E}_1$ and $\mathbf{E}_2$ in Figure P15.9 and in terms of the node voltages $\hat{\mathbf{E}}_1$ and $\hat{\mathbf{E}}_2$ of the adjoint circuit.

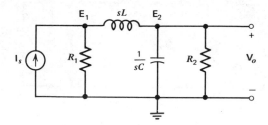

Figure P15.9

10. Verify that the two-port in Figure P15.10b is the adjoint circuit of the circuit in Figure P15.10a, that is, show that the two networks are interreciprocal by showing that $v_o/i_s|_{i_o=0} = \hat{v}_s/\hat{i}_o|_{\hat{i}_s=0}$.

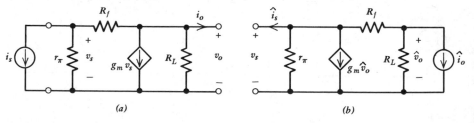

$\qquad$ (a) $\qquad\qquad\qquad\qquad\qquad$ (b)

Figure P15.10

11. Show that the admittance matrix of the circuit in Figure P15.10b is the transpose of the admittance matrix of the circuit in Figure P15.10a.

12. In Figure P15.10a express the sensitivities $\partial v_o/\partial g_\pi$, $\partial v_o/\partial G_f$, $\partial v_o/\partial g_m$, and $\partial v_o/\partial G_L$ in terms of v_s, v_o, $\hat{v}_s$, and $\hat{v}_o$ when $\hat{i}_o = 1$ A.

13. In Figure P15.13 show that $v_o/v_s = -\hat{i}_s/\hat{i}_o$. Why is this true?

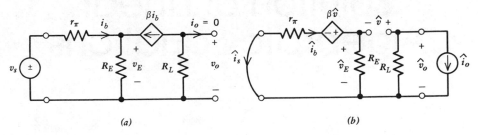

(a)   (b)

Figure P15.13

14. In Figure P15.13a express the sensitivities $\partial v_o/\partial r_\pi$, $\partial v_o/\partial R_E$, $\partial v_o/\partial R_L$, and $\partial v_o/\partial \beta$ in terms of the responses of the circuit and its adjoint in Figure P15.13b with $\hat{i}_o = 1\ A$.

Appendix A
Solution of Linear Algebraic Equations

An equation places an equality constraint on a function of one or more variables. In many physical situations the constraints may be dependent on time or on the spatial coordinates (the independent variables). If the function does not contain any differential values of the variables, then we say that the equation is algebraic. This implies that the variables do not depend on any past history. Examples of algebraic equations are

$$5x_1 + 3x_2 = \sin 5t \tag{1}$$

$$5x_1 + (\cos t)x_2 = 5 \tag{2}$$

$$x_1 x_2 + 5x_2 = \cos t \tag{3}$$

$$x_1^2 + tx_2 = 4 \tag{4}$$

where x_1 and x_2 denote the dependent variables and t denotes the given quantity time (independent variable).

Frequently, equations are prefixed with adjectives, such as linear or nonlinear, and constant or nonconstant (time-invariant or time-varying when the independent variable is time). *An equation is said to be linear if each term in the equation contains at most one dependent variable of the first power. The equation is said to be constant if all the coefficients of the dependent variables are constant.* In the above equations (1) is linear constant, (2) is linear time-varying, (3) is nonlinear constant, and (4) is nonlinear time-varying.

We say that an equation places a constraint on a function. In the next section we examine sets of equations and the independence of their constraints.

476

A.1 INDEPENDENCE OF EQUATIONS

Consider the two equations

$$f_1(x_1, x_2) = 0 \tag{5a}$$
$$f_2(x_1, x_2) = 0 \tag{6a}$$

These two equations are said to be *dependent* if

$$af_1(x_1, x_2) + bf_2(x_1, x_2) \equiv 0 \tag{7}$$

for all x_1, x_2 and for a and b not equal to zero. For example, the equations

$$x_1 + 2x_2 - 4 = 0 \tag{8}$$
$$-3x_1 - 6x_2 + 12 = 0 \tag{9}$$

are dependent since $-3 \times$ Equation 8 = Equation 9. Geometrically this means that both (8) and (9) describe the same line in Figure A.1 and the solutions to (8) and (9) lie on this line. Therefore, since (8) and (9) describe the same line, both of them are not needed and we can discard either (8) or (9), but not both.

Now let us suppose that (5a) and (6a) are independent, and define

$$f_3(x_1, x_2) = af_1(x_1, x_2) + bf_2(x_1, x_2) \tag{10}$$

then the set of equations

$$f_1(x_1, x_2) = 0 \tag{5b}$$
$$f_3(x_1, x_2) = 0 \tag{6b}$$

are independent if $b \neq 0$, and the equations

$$f_3(x_1, x_2) = 0 \tag{5c}$$
$$f_2(x_1, x_2) = 0 \tag{6c}$$

are independent if $a \neq 0$. Furthermore, if $\hat{x}_1, \hat{x}_2$ is a solution to (5a) and (6a) it follows that $\hat{x}_1, \hat{x}_2$ is a solution to (5b) and (6b), and also (5c) and (6c), and vice

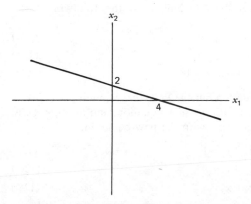

Figure A.1. Constraint imposed by both (8) and (9).

versa. However, (5a), (6a), and (10) form a dependent set of equations so that we have redundant information and we can discard at least one of the equations.

The above results are valid for all types of equations and are useful in the solution of equations. In particular, in this appendix we intend to discuss the solution of linear algebraic equations. The above ideas are essential for this task. The solution of nonlinear algebraic equations will not be discussed, since no general methods exist for this task except for iterative techniques that require the use of a computer.

A.2 ELIMINATION OF VARIABLES AND GAUSSIAN ELIMINATION

The most common technique for the solution of equations is the elimination of variables. This technique is particularly powerful for linear algebraic equations since it always yields the unique solution if one exists. The method is demonstrated by example. Consider the two algebric equations

$$x_1 + x_2 = 1 \tag{11a}$$

$$2x_1 + x_2 = 0 \tag{11b}$$

Let us form a new equation by multiplying (11a) by -2 and adding it to (11b).

or
$$0 \cdot x_1 - x_2 = -2$$
$$x_2 = 2 \tag{11c}$$

Now (11a), (11b), and (11c) form a set of dependent equations, but any combination of the two form an independent set, for example,

$$x_1 + x_2 = 1 \tag{11a}$$

$$x_2 = 2 \tag{11c}$$

The above two equations are easy to solve. Obviously $\hat{x}_2 = 2$ and upon substitution of this solution into (11a) we obtain $\hat{x}_1 = -1$.

Let us apply the elimination of variables technique to the following three algebraic equations.

$$x_1 + x_2 + x_3 = 1 \tag{12a}$$

$$x_1 + 2x_2 + 2x_3 = 1 \tag{12b}$$

$$x_1 \qquad + 2x_3 = -1 \tag{12c}$$

First, let us eliminate x_1 from every equation but the first, that is, multiply (12a) by -1 and add it to (12b) and replace (12b) with the new equation, and multiply (12a) by (-1) and add it to (12c) and replace (12c) with the new equation.

$$x_1 + x_2 + x_3 = 1 \tag{12a}$$

$$+ x_2 + x_3 = 0 \tag{12b'}$$

$$- x_2 + x_3 = -2 \tag{12c'}$$

Next we use the second equation (12b') to eliminate x_2 from every equation, but the first and second, that is, add (12b') to (12c') and replace (12c') with the new equation

$$x_1 + x_2 + x_3 = 1 \tag{12a}$$

$$x_2 + x_3 = 0 \tag{12b'}$$

$$2x_3 = -2 \tag{12c''}$$

Obviously the solution is $\hat{x}_3 = -1$, and upon substitution of $\hat{x}_3$ into (12b') we obtain $\hat{x}_2 = 1$, and upon substitution of $\hat{x}_1$ and $\hat{x}_2$ into (12a) we obtain $\hat{x}_1 = 1$. This process is called back substitution, and the entire method is called *Gaussian elimination*. The object is to eliminate variables from the equations to obtain the triangular form above. This method is very fast and is a very popular computer algorithm.

Let us work one other example.

$$x_1 + x_2 + x_3 = 1 \tag{13a}$$

$$x_1 + x_2 + 2x_3 = 0 \tag{13b}$$

$$x_1 \quad\quad + 2x_3 = 1 \tag{13c}$$

We proceed to eliminate x_1 from (13b) and (13c) to obtain

$$x_1 + x_2 + x_3 = 1 \tag{13a}$$

$$x_3 = -1 \tag{13b'}$$

$$-x_2 + x_3 = 0 \tag{13c'}$$

Now we need to use (13b') to eliminate x_2 from (13c'). However, the coefficient of x_2 in (13b') is zero, which makes this operation impossible. To correct this problem simply interchange (13b') and (13c').

$$x_1 + x_2 + x_3 = 1 \tag{13a}$$

$$-x_2 + x_3 - 0 \tag{13c'}$$

$$x_3 = -1 \tag{13b'}$$

By back substitution the solution is $\hat{x}_3 = -1$, $\hat{x}_2 = -1$, $\hat{x}_1 = 3$.

In the next section we present a convenient notation for the representation of linear algebraic equations.

A.3 MATRICES

Let us express (12) in the form

$$\begin{bmatrix} 1 & 1 & 1 \\ 1 & 2 & 2 \\ 1 & 0 & 2 \end{bmatrix} \begin{bmatrix} x_1 \\ x_2 \\ x_3 \end{bmatrix} = \begin{bmatrix} 1 \\ 1 \\ -1 \end{bmatrix} \tag{14}$$

and define

$$\mathbf{A} = \begin{bmatrix} 1 & 1 & 1 \\ 1 & 2 & 2 \\ 1 & 0 & 2 \end{bmatrix} \quad \mathbf{x} = \begin{bmatrix} x_1 \\ x_2 \\ x_3 \end{bmatrix} \quad \mathbf{b} = \begin{bmatrix} 1 \\ 1 \\ -1 \end{bmatrix}$$

so that we can write (14) as

$$Ax = b \tag{15}$$

The rectangular arrays A, x, and b are called *matrices*. When the number of rows is equal to the number of columns, as in the case of the array, A, the matrix is called a *square matrix*. In (14) a row of A contains the coefficients of the equation corresponding to that row, and the columns of A denote the dependent variable with which that coefficient is associated. For example, the numbers in column 2 are the coefficients of x_2.

In linear algebraic equations the matrix

$$Ag = [A \vdots b] \tag{16}$$

formed by appending the column b to A is called the *augmented matrix*. This matrix is useful in the Gaussian elimination method. For example, for the set of algebraic equations (12)

$$Ag = \begin{bmatrix} 1 & 1 & 1 & \vdots & 1 \\ 1 & 2 & 2 & \vdots & 1 \\ 1 & 0 & 2 & \vdots & -1 \end{bmatrix} \tag{17}$$

Let us now operate on the rows of Ag to obtain zeros below the diagonal of A. Multiply row 1 by -1 and add it to rows 2 and 3 to obtain

$$Ag' = \begin{bmatrix} 1 & 1 & 1 & \cdot & 1 \\ 0 & 1 & 1 & \vdots & 0 \\ 0 & -1 & 1 & \cdot & -2 \end{bmatrix} \tag{18}$$

To eliminate x_2 from row 3, add row 2 to row 3.

$$Ag'' = \begin{bmatrix} 1 & 1 & 1 & \cdot & 1 \\ 0 & 1 & 1 & \vdots & 0 \\ 0 & 0 & 2 & \cdot & -2 \end{bmatrix} \tag{19}$$

so that our modified equations are

$$\begin{aligned} x_1 + x_2 + x_3 &= 1 \\ x_2 + x_3 &= 0 \\ 2x_3 &= -2 \end{aligned} \tag{20}$$

In a computer program it is very easy to form the augmented matrix Ag and to operate on the rows to make the square matrix A triangular (all zeros in the array below or above the diagonal).

To generalize the above discussion consider the algebraic equation

$$\begin{bmatrix} a_{11} & a_{12} & a_{13} \\ a_{21} & a_{22} & a_{23} \\ a_{31} & a_{32} & a_{33} \end{bmatrix} \begin{bmatrix} x_1 \\ x_2 \\ x_3 \end{bmatrix} = \begin{bmatrix} b_1 \\ b_2 \\ b_3 \end{bmatrix} \tag{21}$$

and the augmented matrix

$$\begin{bmatrix} a_{11} & a_{12} & a_{13} & \cdot & b_1 \\ a_{21} & a_{22} & a_{23} & \vdots & b_2 \\ a_{31} & a_{32} & a_{33} & \cdot & b_3 \end{bmatrix}$$

First, to eliminate x_1 in the second and third equations multiply row 1 by $-a_{21}/a_{11}$ and add it to row 2, multiply row 1 by $-a_{31}/a_{11}$ and add it to row 3 to obtain

$$\begin{bmatrix} a_{11} & a_{12} & a_{13} & \cdot & b_1 \\ 0 & a_{22}^{(2)} & a_{23}^{(2)} & \vdots & b_2^{(2)} \\ 0 & a_{32}^{(2)} & a_{33}^{(2)} & \cdot & b_3^{(2)} \end{bmatrix}$$

where, for example, $a_{22}^{(2)} = a_{22} - a_{21}a_{12}/a_{11}$, that is, the superscript 2 denotes that the coefficients below row 1 are changed by this operation. Next we multiply row 2 by $-a_{32}^{(2)}/a_{22}^{(2)}$ and add it to row 3 to obtain

$$\begin{bmatrix} a_{11} & a_{12} & a_{13} & \cdot & b_1 \\ 0 & a_{22}^{(2)} & a_{23}^{(2)} & \vdots & b_2^{(2)} \\ 0 & 0 & a_{33}^{(3)} & \cdot & b_3^{(3)} \end{bmatrix}$$

Note that if a_{11} or $a_{22}^{(2)} = 0$ we should interchange rows to correct the problem. We now have our algebraic equations in the form

$$\begin{aligned} a_{11}x_1 + a_{12}x_2 + a_{13}x_3 &= b_1 \\ a_{22}^{(2)}x_2 + a_{23}^{(2)}x_3 &= b_2^{(2)} \\ a_{33}^{(3)}x_3 &= b_3^{(3)} \end{aligned} \tag{22}$$

which can be solved by back substitution.

A.4 EXISTENCE AND UNIQUENESS

If a unique solution exists, the method Gaussian elimination will find it. However, sometimes the set of linear algebraic equations may have multiple solutions or no solution. In this section we show how the Gaussian elimination method behaves in such cases. We begin with some simple examples.

Example 1

Consider the equations

$$x_1 + x_2 = 1 \tag{23a}$$

$$2x_1 + x_2 = 0 \tag{23b}$$

The augmented matrix is

$$\mathbf{Ag} = [\mathbf{A} \ \vdots \ \mathbf{b}] = \begin{bmatrix} 1 & 1 & \vdots & 1 \\ 2 & 1 & \vdots & 0 \end{bmatrix} \tag{24}$$

which can be reduced to the form

$$\mathbf{Ag'} = [\mathbf{A'} \vdots \mathbf{b'}] = \begin{bmatrix} 1 & 1 & \vdots & 1 \\ & & \vdots & \\ 0 & -1 & \vdots & -2 \end{bmatrix} \tag{25}$$

so that the equivalent constraints are

$$\begin{aligned} x_1 + x_2 &= 1 \\ -x_2 &= -2 \end{aligned} \tag{26}$$

which yields the solution $\hat{x}_2 = 2$, $\hat{x}_1 = -1$. Equations 23 are plotted in Figure A.2. Note that the point of intersection yields values of x_1 and x_2 which simultaneously satisfy (23a) and (23b).

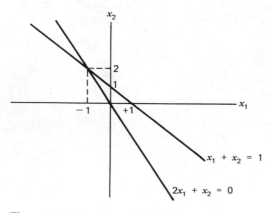

Figure A.2.   A unique solution exists.

Example 2
Next we consider the equations

$$\begin{aligned} x_1 + x_2 &= 1 \\ x_1 + x_2 &= 0 \end{aligned} \tag{27}$$

The augmented matrix is

$$\mathbf{Ag} = \begin{bmatrix} 1 & 1 & \vdots & 1 \\ & & \vdots & \\ 1 & 1 & \vdots & 0 \end{bmatrix} \tag{28}$$

which reduces to the form

$$\mathbf{Ag'} = \begin{bmatrix} 1 & 1 & \vdots & 1 \\ & & \vdots & \\ 0 & 0 & \vdots & -1 \end{bmatrix} \tag{29}$$

upon the elimination of x_1 from row 2. This represents the equations

$$\begin{aligned} x_1 + x_2 &= 1 \\ 0 \cdot x_2 &= -1 \end{aligned} \tag{30}$$

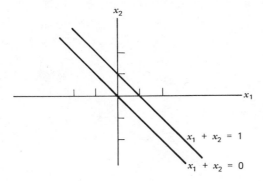

Figure A.3.   No solution exists.

which obviously have no solution. Equations 27 are plotted in Figure A.3. They represent parallel lines that do not intersect and hence no value of x can simultaneously satisfy both equations.

Example 3

Finally, consider the equations

$$2x_1 + x_2 = 1$$
$$8x_1 + 4x_2 = 4$$

(31)

The augmented matrix is

$$\mathbf{Ag} = \begin{bmatrix} 2 & 1 & \vdots & 1 \\ & & \vdots & \\ 8 & 4 & \cdot & 4 \end{bmatrix}$$

(32)

and upon the elimination of x_1 from the second equation

$$\mathbf{Ag'} = \begin{bmatrix} 2 & 1 & \vdots & 1 \\ & & \vdots & \\ 0 & 0 & \cdot & 0 \end{bmatrix}$$

(33)

which represents the equations

$$2x_1 + x_2 = 1$$
$$0 = 0$$

(34)

Thus any point along the line $2x_1 + x_2 = 1$ satisfies (31). In Figure A.4, the equations (31) are plotted and they are parallel lines that lie on top of each other so that there is an infinite number of solutions along this line.

A close look at the rows of $\mathbf{A}$ and $\mathbf{Ag}$ reveals the following important information. In Example 1 there are two independent rows of $\mathbf{A}$ and $\mathbf{Ag}$ and there are two unknowns, and a unique solution exists. In Example 2 row 1 of $\mathbf{A}$ is dependent on row 2 of $\mathbf{A}$, thus we obtain zeros in the last row of $\mathbf{A'}$, but the rows of $\mathbf{Ag}$ are independent because $\mathbf{Ag'}$ has a nonzero element in the last column of the last row, and so no solution exists. In Example 3, row 2 is dependent on row 1 in both $\mathbf{A}$ and

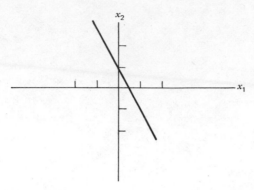

Figure A.4. Many solutions are possible.

Ag thus we obtain zeros in row 2 of **Ag'**, and the number of independent rows is less than the number of unknowns. This observation leads us to the following theorem.

Theorem

Given a set of algebraic equations Ax = b, if the number of independent rows of A is equal to the number of independent rows of the augmented matrix Ag then a solution exists, and that solution is unique (only one set of values satisfy the equations) if the number of independent rows of A and Ag is equal to the number of unknowns. A solution does not exist if the number of independent rows of A does not equal the number of independent rows of Ag.

The number of independent rows of a matrix is called its row rank.

To prove the above theorem, consider the m algebraic equations in n unknowns below.

$$
\begin{aligned}
a_{11}x_1 + a_{12}x_2 + \cdots + a_{1n}x_n &= b_1 \\
a_{21}x_1 + a_{22}x_2 + \cdots + a_{2n}x_n &= b_2 \\
&\ \ \vdots \\
a_{m1}x_1 + a_{m2}x_2 + \cdots + a_{mn}x_n &= b_m
\end{aligned}
\tag{35}
$$

By means of Gaussian elimination let us systematically eliminate the coefficients a_{ij} for $i > j$. If we obtain a row of zeros then we know that row is dependent on the rows above it. For example, upon Gaussian elimination, (35) reduces to

$$
\begin{bmatrix}
a_{11} & a_{12} & \cdots & & a_{1n} \\
0 & a_{22}^{(2)} & \cdots & & a_{2n}^{(2)} \\
& & \ddots & & \\
& & a_{kk}^{(k)} & \cdots & a_{kn}^{(k)} \\
& & 0 & \cdots & 0 \\
& \mathbf{0} & & \ddots & \vdots \\
& & & & 0
\end{bmatrix}
\begin{bmatrix}
x_1 \\ x_2 \\ \vdots \\ \\ \\ \\ x_n
\end{bmatrix}
=
\begin{bmatrix}
b_1 \\ b_2^{(2)} \\ \vdots \\ b_k^{(k)} \\ \vdots \\ \\ b_m^{(m)}
\end{bmatrix}
\tag{36}
$$

where the zeros on the diagonal occur only if dependent rows of $\mathbf{A}$ are present. The last m-k equations have the form

$$\sum_{j=1}^{n} 0 \cdot x_j = b_\alpha^{(\alpha)} \qquad \alpha > k \qquad \text{and} \qquad \alpha = k+1, \ldots, m \qquad (37)$$

Therefore (36) does not have a solution unless $b_\alpha^{(\alpha)} = 0$ for $\alpha > k$. This means that the matrix

$$\begin{bmatrix} a_{11} & a_{12} & \cdots & a_{1n} \\ a_{21} & a_{22} & \cdots & a_{2n} \\ \cdot & & & \cdot \\ \cdot & & & \\ \cdot & & & \\ a_{m1} & a_{m2} & \cdots & a_{mn} \end{bmatrix}$$

must have the same number of independent rows as the matrix

$$\begin{bmatrix} a_{11} & a_{12} & \cdots & a_{1n} & \cdot & b_1 \\ a_{21} & a_{22} & \cdots & a_{2n} & \cdot & b_2 \\ \cdot & & & & & \cdot \\ \cdot & & & & & \cdot \\ \cdot & & & & & \\ a_{m1} & a_{m2} & \cdots & a_{mn} & \cdot & b_m \end{bmatrix}$$

In other words, the row rank of $\mathbf{A}$ is equal to the row rank of $\mathbf{Ag}$. Furthermore, to obtain a unique solution we not only require that the row ranks be equal, but we also need the same number of equations as unknowns ($m = n$). In this case, (36) has the form

$$\begin{bmatrix} a_{11} & a_{12} & \cdots & a_{1n} \\ 0 & a_{22}^{(2)} & \cdots & a_{2n}^{(2)} \\ & & \cdot & \\ & \mathbf{0} & \cdot & \\ & & & a_{nn}^{(n)} \end{bmatrix} \begin{bmatrix} x_1 \\ x_2 \\ \\ \\ x_n \end{bmatrix} = \begin{bmatrix} b_1 \\ b_2^{(2)} \\ \\ \\ b_n^{(n)} \end{bmatrix} \qquad (38)$$

where it is required that $a_{ii}^{(i)} \neq 0$ for $i = 1, 2, \ldots, n$ or else the rows of A are not independent. Once we have reduced our n equations in n unknowns with the row rank of $\mathbf{A}$ equal to n to the form (38) by means of Gaussian elimination, it is a simple matter to obtain the unique solution by means of back substitution.

A.5 MATRIX OPERATIONS

In the previous section we learned that the set of n algebraic equations in n unknowns

$$\mathbf{Ax} = \mathbf{b} \qquad (39)$$

has a unique solution if the $n \times n$ matrix $\mathbf{A}$ has n independent rows. This solution could be found by a method called Gaussian elimination. If we consider $\mathbf{A}$ as an operator, then the solution could also be obtained if we could find the inverse operator such that

$$\mathbf{x} = \mathbf{A}^{-1}\mathbf{b} \qquad (40)$$

We will show that the determinant is one way to find the inverse of a square matrix **A**, but first we must define some simple matrix operations.

A.5.1 MATRIX MULTIPLICATION

Recall that in matrix notation we write (35) as

$$\mathbf{Ax} = \mathbf{b} \tag{41}$$

where

$$\sum_{j=1}^{n} a_{ij}x_j = b_i \qquad i = 1, 2, \ldots, m \tag{42}$$

In order to get a more general definition of the multiplication of two matrices, consider the multiplication of the $r \times m$ matrix

$$\mathbf{B} = \begin{bmatrix} b_{11} & b_{12} & \cdots & b_{1m} \\ b_{21} & b_{22} & \cdots & b_{2m} \\ \cdot & \cdot & & \cdot \\ \cdot & \cdot & & \cdot \\ \cdot & \cdot & & \cdot \\ b_{r1} & b_{r2} & \cdots & b_{rm} \end{bmatrix} \tag{43}$$

with the vector **b**. Then

$$\mathbf{BAx} = \mathbf{Bb} \tag{44}$$

Applying (42) to **Bb** and **B(Ax)** allows us to define a new matrix

$$\mathbf{C}^{r \times n} = \mathbf{B}^{r \times m}\mathbf{A}^{m \times n} \tag{45}$$

where the superscript simply reminds us that the matrix C has r rows and n columns, the matrix **B** has r rows and m columns, and so on. The scalar element c_{ij} in the ith row and the jth column of the matrix **C** is given by the summation

$$c_{ij} = \sum_{k=1}^{m} b_{ik}a_{kj} \tag{46}$$

Note that the number of columns of B must equal the number of rows of A or matrix multiplication cannot be defined. The multiplication of an $r \times m$ matrix with an $m \times n$ matrix results in an $r \times n$ matrix. Also, the element of the $i - j$th position is equal to the product of the first element in the ith row of the matrix on the left with the first element in the jth column of the matrix on the right plus the product of the second element in the ith row of the matrix on the left with the second element in the jth column of the matrix on the right, etc. Let us consider an example.

Example 4

Consider the matrices

$$\mathbf{A} = \begin{bmatrix} 1 & 2 & 0 \\ 3 & -1 & 2 \end{bmatrix} \qquad \mathbf{B} = \begin{bmatrix} 1 & 1 \\ 0 & 2 \\ -2 & 1 \end{bmatrix}$$

The product

$$\mathbf{AB} = \begin{bmatrix} 1 & 5 \\ -1 & 3 \end{bmatrix} = \mathbf{C} = \begin{bmatrix} c_{11} & c_{12} \\ c_{21} & c_{22} \end{bmatrix}$$

To find the scalar c_{11} we take row 1 of the $\mathbf{A}$ matrix, rotate it by $-90°$ and multiply it by column 1 of the $\mathbf{B}$ matrix, term for term, and sum the result. The element c_{12} is found by again rotating row 1 by $-90°$ and multiplying it by column 2 of the $\mathbf{B}$ matrix and summing the result, and so on.

Note that the product

$$\mathbf{BA} = \begin{bmatrix} 4 & -1 & 2 \\ 6 & -2 & 4 \\ 1 & -5 & 2 \end{bmatrix}$$

Thus, $\mathbf{BA} \neq \mathbf{AB}$, or matrix multiplication is not necessarily commutative.

A.5.2 ADDITION AND SUBTRACTION OF MATRICES

Addition and subtraction are defined only when the matrices involved all have the same number of rows and the same number of columns. Let $\mathbf{A}$ be an $m \times n$ matrix and $\mathbf{B}$ be an $m \times n$ matrix; their sum is

$$\mathbf{A} + \mathbf{B} = \mathbf{C} \tag{47}$$

where $c_{ij} = a_{ij} + b_{ij}$.

In subtraction

$$\mathbf{A} - \mathbf{B} = \mathbf{C} \tag{48}$$

where $c_{ij} = a_{ij} - b_{ij}$.

Example 5

$$\mathbf{A} = \begin{bmatrix} 2 & 1 \\ 0 & 1 \\ 4 & 3 \end{bmatrix} \quad \mathbf{B} = \begin{bmatrix} 3 & 0 \\ 1 & 2 \\ 0 & 1 \end{bmatrix}$$

$$\mathbf{A} + \mathbf{B} = \begin{bmatrix} 5 & 1 \\ 1 & 3 \\ 4 & 4 \end{bmatrix} \quad \mathbf{A} - \mathbf{B} = \begin{bmatrix} -1 & 1 \\ -1 & -1 \\ 4 & 2 \end{bmatrix}$$

A.5.3 THE IDENTITY MATRIX AND THE TRANSPOSE OF A MATRIX

Let us define the *identity matrix* $\mathbf{I}_n$. The identity matrix is an $n \times n$ square matrix

$$\mathbf{I}_n = \begin{bmatrix} 1 & 0 & 0 & \dots & 0 \\ 0 & 1 & 0 & \dots & 0 \\ 0 & 0 & 1 & \dots & 0 \\ \cdot & \cdot & \cdot & & \cdot \\ \cdot & \cdot & \cdot & \cdot & \cdot \\ \cdot & \cdot & \cdot & & \cdot \cdot \\ 0 & 0 & 0 & \dots & 1 \end{bmatrix} \tag{49}$$

which has ones on the diagonal and zeros for the off-diagonal terms. It has the property that

$$I_m B^{m \times n} = B^{m \times n} \tag{50}$$

or

$$B^{m \times n} I_n = B^{m \times n} \tag{51}$$

The *transpose* of the matrix A is denoted by A^t and $a^t_{ij} = a_{ji}$, that is, the indices are transposed.

Example 6

$$A = \begin{bmatrix} 1 & 0 \\ 1 & 3 \\ 2 & 1 \end{bmatrix} \qquad A^t = \begin{bmatrix} 1 & 1 & 2 \\ 0 & 3 & 1 \end{bmatrix}$$

Note that $a^t_{ij} = a_{ji}$.

A matrix is said to be symmetric if $a_{ij} = a_{ji}$, then $A^t = A$.

A.5.4 INVERSE OF A MATRIX

The inverse of a scalar α is $1/\alpha = \alpha^{-1}$ and $\alpha^{-1}\alpha = \alpha\alpha^{-1} = 1$. We define the left inverse of a matrix A to be a matrix B such that multiplication by B results in the identity matrix.

$$B^{n \times m} A^{m \times n} = I_n \tag{52}$$

We define the right inverse of A to be the matrix C such that

$$AC = I_m \tag{53}$$

If $m = n$, and A has a left inverse B, then A has a right inverse C and

$$B = C \tag{54}$$

because

$$B = BI_n = B(AC) = (BA)C = I_n C = C \tag{55}$$

In this case we call B the inverse of A and denote it as A^{-1}.

The inverse of the diagonal matrix

$$D = \begin{bmatrix} d_1 & & & \mathbf{0} \\ & d_2 & & \\ & & \ddots & \\ \mathbf{0} & & & d_n \end{bmatrix} \tag{56}$$

is

$$\mathbf{D}^{-1} = \begin{bmatrix} d_1^{-1} & & & \mathbf{0} \\ & d_2^{-1} & & \\ & & \cdot & \\ & & & \cdot \\ \mathbf{0} & & & d_n^{-1} \end{bmatrix} \tag{57}$$

Not all matrices have an inverse as shown by the next example.

Example 7

Let us show that the matrix

$$\mathbf{A} = \begin{bmatrix} 0 & 1 \\ 0 & 0 \end{bmatrix}$$

does not have an inverse. If we multiply $\mathbf{A}$ by the matrix

$$\mathbf{B} = \begin{bmatrix} a & b \\ c & d \end{bmatrix}$$

then

$$\mathbf{BA} = \begin{bmatrix} 0 & a \\ 0 & c \end{bmatrix}$$

No choice of constants a, b, c, and d can make $\mathbf{BA}$ a diagonal matrix. Therefore the matrix $\mathbf{A}$ does not have an inverse.

Obviously if we can determine $\mathbf{A}^{-1}$, then we can find a solution to (41), that is,

$$\mathbf{x} = \mathbf{A}^{-1}\mathbf{b} \tag{58}$$

In the next section we show that the inverse of a square matrix can be found by the method of determinants.

A.6 DETERMINANTS

The determinant of an $n \times n$ matrix $\mathbf{A}$ denoted by det $\mathbf{A}$, $|\mathbf{A}|$, or Δ is a scalar that is assigned to the matrix $\mathbf{A}$ by a definite rule. The rule is as follows:

Select exactly n elements from A such that each row and each column of A is represented and form the product $a_{1j_1}a_{2j_2}\ldots a_{nj_n}$. Note that no row or column can be represented more than once. Thus, the $j1, j2, \ldots, jn$ represent a permutation of distinct integers from 1 to n.

For example, let $n = 3$, then the permutation of integers $j1, j2, j3$ could be 123, 132, 213, 231, 312, or 321, a total of six permutations, or $n!$ permutations. Thus, the above product could be written as $a_{11}a_{22}a_{33}$, $a_{11}a_{23}a_{32}$, $a_{12}a_{21}a_{33}$, $a_{12}a_{23}a_{31}$, $a_{13}a_{21}a_{32}$, or $a_{13}a_{22}a_{31}$. Now the permutation is called even or odd depending on the number of permutations necessary to put the integers into ascending order; for example

$$321 \xrightarrow{(1)} 312 \xrightarrow{(2)} 132 \xrightarrow{(3)} 123$$

is an odd permutation, and

$$231 \xrightarrow{(1)} 213 \xrightarrow{(2)} 123$$

is an even permutation. If the permutation $j_1 j_2 \ldots j_n$ is even we multiply the product $a_{1j_1} \ldots a_{n j_n}$ by $(+1)$; if it is odd we multiply this product by (-1). The determinant of $\mathbf{A}$ is

$$\Delta = \sum (-1)^{\alpha_i} a_{1j_1} a_{2j_2} \ldots a_{n j_n} \tag{59}$$

where α_i is the number of permutations of $j_1 \ldots j_n$ necessary to place the integers in ascending order, and the summation is taken over the $n!$ permutations of $j_1 j_2 \ldots j_n$. We will use the symbol Δ to denote the determinant of $\mathbf{A}$. As an example consider the matrix

$$\mathbf{A} = \begin{bmatrix} a_{11} & a_{12} & a_{13} \\ a_{21} & a_{22} & a_{23} \\ a_{31} & a_{32} & a_{33} \end{bmatrix} \tag{60}$$

The det $\mathbf{A}$ will contain six terms, because $n = 3$ and $n! = 6$.

$$\begin{aligned} \Delta = &(-1)^{\alpha_1} a_{11} a_{22} a_{33} + (-1)^{\alpha_2} a_{11} a_{23} a_{32} \\ &+ (-1)^{\alpha_3} a_{12} a_{21} a_{33} + (-1)^{\alpha_4} a_{13} a_{21} a_{32} \\ &+ (-1)^{\alpha_5} a_{13} a_{22} a_{31} + (-1)^{\alpha_6} a_{12} a_{23} a_{31} \end{aligned} \tag{61}$$

We conclude that $\alpha_1 = 0$, zero permutations necessary to place the column integers in ascending order; $\alpha_2 = 1$, since one permutation is necessary on the column integers. Similarly, one finds that $\alpha_3 = 1$, $\alpha_4 = 2$, $\alpha_5 = 3$, $\alpha_6 = 2$. Thus,

$$\Delta = a_{11} a_{22} a_{33} - a_{11} a_{23} a_{32} - a_{12} a_{21} a_{33} + a_{13} a_{21} a_{32} - a_{13} a_{22} a_{31} + a_{12} a_{23} a_{31} \tag{62}$$

A.6.1 LAPLACE EXPANSION FOR FINDING THE DETERMINANT

Lest one be confused by the above procedure, there is a cookbook approach to the evaluation of the determinant, called the Laplace's expansion. The determinant of the $n \times n$ matrix $\mathbf{A}$ is written as

$$\Delta = \sum_{i=1}^{n} a_{ij} \Delta_{ij} \qquad \text{for } i = 1, 2, \ldots, \text{ or } n \tag{63}$$

where the "*cofactor*" $\Delta_{ij} = (-1)^{i+j} M_{ij}$, and the quantity M_{ij} is the determinant of $\mathbf{A}$ with the ith row and the jth column deleted. M_{ij} is called a minor of the matrix $\mathbf{A}$. Considering the previous example,

$$\Delta = \begin{vmatrix} a_{11} & a_{12} & a_{13} \\ a_{21} & a_{22} & a_{23} \\ a_{31} & a_{32} & a_{33} \end{vmatrix} = a_{11} M_{11} - a_{21} M_{21} + a_{31} M_{31} \tag{64}$$

The minors are

$$M_{11} = \begin{vmatrix} a_{22} & a_{23} \\ a_{32} & a_{33} \end{vmatrix} \qquad M_{21} = \begin{vmatrix} a_{21} & a_{13} \\ a_{32} & a_{33} \end{vmatrix} \qquad M_{31} = \begin{vmatrix} a_{12} & a_{13} \\ a_{22} & a_{23} \end{vmatrix} \tag{65}$$

Hence,

$$\Delta = a_{11}(a_{22} a_{33} - a_{23} a_{32}) - a_{21}(a_{12} a_{33} - a_{32} a_{13}) + a_{31}(a_{12} a_{23} - a_{22} a_{13}) \tag{66}$$

A.6.2 PROPERTIES OF THE DETERMINANT

From the definition of a determinant one can derive a number of interesting properties for determinants. We will state these properties sometimes without proof. For convenience we will denote det $\mathbf{A}$ as $\det(\mathbf{A}_1, \mathbf{A}_2, \ldots, \mathbf{A}_n)$, where the $\mathbf{A}_i$'s represent the rows of the matrix $\mathbf{A}$.

Property 1. If all of the elements in any row of the matrix $\mathbf{A}$ are zero, then det $\mathbf{A} = 0$.

This result follows nicely from the Laplace expansion as do the following two properties.

Property 2. If any row of $\mathbf{A}$ is multiplied by k, then the determinant of $\mathbf{A}$ is multiplied by k.

$$\det(\mathbf{A}_1, \mathbf{A}_2, \ldots, k\mathbf{A}_i, \ldots, \mathbf{A}_n) = k \det(\mathbf{A}_1, \ldots, \mathbf{A}_i, \ldots, \mathbf{A}_n). \tag{67}$$

Property 3.

$$\det(\mathbf{A}_1, \ldots, \mathbf{A}_i + \mathbf{B}_i, \ldots, \mathbf{A}_n) =$$
$$\det(\mathbf{A}_1, \ldots, \mathbf{A}_i, \ldots, \mathbf{A}_n) + \det(\mathbf{A}_1, \ldots, \mathbf{B}_i, \ldots, \mathbf{A}_n) \tag{68}$$

Property 4. If any two adjacent rows of a matrix $\mathbf{A}$ are interchanged, then the determinant of the resulting matrix is $-\det \mathbf{A}$, for example,

$$\begin{vmatrix} a_{21} & a_{22} \\ a_{11} & a_{12} \end{vmatrix} = a_{12}a_{21} - a_{11}a_{22} \tag{69}$$

This can be proved by noting that in interchanging the adjacent rows the index α_j is changed by 1 for each permutation; hence, the negative sign.

Property 5. If row $\mathbf{A}_i$ is interchanged with row $\mathbf{A}_j$ then the determinant of the new matrix is $-\det \mathbf{A}$.

If row $\mathbf{A}_i$ is interchanged r times with its adjacent neighbor until it reaches row j, then $\mathbf{A}_j$ must be interchanged $r - 1$ times with its adjacent neighbor to reach position i; hence, there are $2r - 1$ interchanges. This is an odd number that implies that det $\mathbf{A}$ changes sign (Property 4).

Property 6. If any two rows of a matrix $\mathbf{A}$ are identical, then det $\mathbf{A} = 0$.

This result follows from Property 5. If we interchange the identical rows, then we obtain

$$\det \mathbf{A} = -\det \mathbf{A} \tag{70}$$

which can be satisfied only when det $\mathbf{A} = 0$.

Property 7. If in the matrix $\mathbf{A}$ a constant times any row is added to another row, the determinants of each array are the same.

$$\det(\mathbf{A}_1, \ldots, \mathbf{A}_n) = \det(\mathbf{A}_1, \ldots, \mathbf{A}_i + k\mathbf{A}_j, \ldots, \mathbf{A}_n) \tag{71}$$

This result follows from Properties 2, 3, and 6.

Property 8. If the rows of $\mathbf{A}$ are linearly dependent then det $\mathbf{A} = 0$.

If the rows are dependent, then by linear combinations of the rows one can reduce one row to zero; hence, by Property 1 det $\mathbf{A} = 0$.

At this point we should comment that all of the above properties are true if the word "row" is replaced by the word "column."

Property 9. The determinant of $\mathbf{A}$ and $\mathbf{A}^t$ are identical.

This follows from the basic definition of the determinant. Each product contains only one term from each row and column.

Property 10.   If **A** and **B** are $n \times n$ matrices, then

$$\det(\mathbf{AB}) = (\det \mathbf{A})(\det \mathbf{B}) \tag{72}$$

This property can be very useful and its proof is left as an exercise.

In the next section we will discuss the application of determinants in finding the inverse of a matrix and the resulting Cramer's rule for the solution of linear algebraic equations.

A.6.3 THE ADJOINT MATRIX AND CRAMER'S RULE

The *adjoint matrix* of an $n \times n$ square **A** is defined as

$$\text{Adj } \mathbf{A} = \begin{bmatrix} \Delta_{11} & \Delta_{21} & \cdots & \Delta_{n1} \\ \Delta_{12} & \Delta_{22} & \cdots & \Delta_{n2} \\ \vdots & \vdots & & \vdots \\ \Delta_{1n} & \Delta_{2n} & \cdots & \Delta_{nn} \end{bmatrix} \tag{73}$$

where Δ_{ij} is the $i - j$th cofactor of the **A** matrix. The adjoint matrix has a very important property, which is

$$\mathbf{A}(\text{adj } \mathbf{A}) = \Delta \mathbf{I}_n$$

or

$$\boxed{\mathbf{A}^{-1} = \text{adj } \mathbf{A}/\Delta} \tag{74}$$

To prove this result let the matrix **B** represent the product of **A** with its adjoint.

$$\mathbf{B} = \begin{bmatrix} a_{11} & a_{12} & \cdots & a_{1n} \\ a_{-1} & a_{22} & \cdots & a_{2n} \\ \vdots & \vdots & & \vdots \\ a_{n1} & a_{n2} & \cdots & a_{nn} \end{bmatrix} \begin{bmatrix} \Delta_{11} & \Delta_{21} & \cdots & \Delta_{n1} \\ \Delta_{12} & \Delta_{22} & \cdots & \Delta_{n2} \\ \vdots & \vdots & & \vdots \\ \Delta_{1n} & \Delta_{2n} & \cdots & \Delta_{nn} \end{bmatrix} \tag{75}$$

Now

$$b_{11} = a_{11}\Delta_{11} + a_{12}\Delta_{12} + \cdots + a_{1n}\Delta_{1n} = \det \mathbf{A} \tag{76}$$

However,

$$b_{12} = a_{11}\Delta_{21} + a_{12}\Delta_{22} + \cdots + a_{1n}\Delta_{2n} \tag{77}$$

This is the determinant of the matrix

$$\begin{bmatrix} a_{11} & a_{12} & \cdots & a_{1n} \\ a_{11} & a_{12} & \cdots & a_{1n} \\ a_{31} & a_{32} & \cdots & a_{3n} \\ \vdots & \vdots & & \vdots \\ a_{n1} & a_{n2} & \cdots & a_{nn} \end{bmatrix}$$

Since rows 1 and 2 are identical, it follows that $b_{12} = 0$. If one continues this this reasoning for all b_{ij}, then $b_{ii} = \det \mathbf{A}$, and $b_{ij} = 0$ for $i \neq j$.   Q.E.D.

Recall that $\Delta \neq 0$ if and only if the rows of A are linearly independent. Again, we see that A^{-1} exists if and only if the rows of A are linearly independent, and the equation

$$Ax = b \tag{78}$$

has a unique solution if and only if $\det A \neq 0$. That solution is

$$x = A^{-1}b \tag{79}$$

or, in terms of the adjoint matrix,

$$
\begin{bmatrix} x_1 \\ x_2 \\ \vdots \\ x_n \end{bmatrix} = \frac{1}{\Delta}
\begin{bmatrix}
\Delta_{11} & \Delta_{21} & \cdots & \Delta_{n1} \\
\Delta_{12} & \Delta_{22} & \cdots & \Delta_{n2} \\
\vdots & \vdots & & \vdots \\
\Delta_{1n} & \Delta_{2n} & \cdots & \Delta_{nn}
\end{bmatrix}
\begin{bmatrix} b_1 \\ b_2 \\ \vdots \\ b_n \end{bmatrix}
\tag{80}
$$

Thus,

$$x_i = \frac{1}{\Delta}(b_1 \Delta_{1i} + b_2 \Delta_{2i} + \cdots + b_n \Delta_{ni}) \tag{81}$$

We can write (81) as

$$x_i = (\det A_{bi})/\det A \tag{82}$$

where the square matrix A_{bi} is obtained from A by replacing the ith column of A with the column matrix b. This result is commonly known as *Cramer's rule*.

Let us consider the equations

$$
\begin{bmatrix} a_{11} & a_{12} \\ a_{21} & a_{22} \end{bmatrix}
\begin{bmatrix} x_1 \\ x_2 \end{bmatrix} =
\begin{bmatrix} b_1 \\ b_2 \end{bmatrix}
\tag{83}
$$

as an example. Upon application of Cramer's rule we obtain

$$
x_1 = \frac{\begin{vmatrix} b_1 & a_{12} \\ b_2 & a_{22} \end{vmatrix}}{a_{11}a_{22} - a_{12}a_{21}}
\qquad
x_2 = \frac{\begin{vmatrix} a_{11} & b_1 \\ a_{21} & b_2 \end{vmatrix}}{a_{11}a_{22} - a_{12}a_{21}}
\tag{84}
$$

or

$$
x_1 = \frac{b_1 a_{22}}{\Delta} - \frac{b_2 a_{12}}{\Delta}
\qquad
x_2 = \frac{a_{11}b_2}{\Delta} - \frac{a_{21}b_1}{\Delta}
$$

where $\Delta = a_{11}a_{22} - a_{12}a_{21}$.

Example 8

Consider the algebraic equations

$$
\begin{bmatrix}
2 & -1 & -1 \\
-1 & 3 & -1 \\
-1 & -1 & 3
\end{bmatrix}
\begin{bmatrix} x_1 \\ x_2 \\ x_3 \end{bmatrix} =
\begin{bmatrix} 2 \\ -1 \\ 0 \end{bmatrix}
$$

Let us use Cramer's rule to find the unknown x_1.

$$x_1 = \frac{\begin{vmatrix} 2 & -1 & -1 \\ -1 & 3 & -1 \\ 0 & -1 & 3 \end{vmatrix}}{\begin{vmatrix} 2 & -1 & -1 \\ -1 & 3 & -1 \\ -1 & -1 & 3 \end{vmatrix}}$$

Let us simplify the evaluation of these determinants by linearly combining the rows in a way that will produce all zeros in column 1, but one. If we multiply row 2 by 2 and add it to row 1 in both the numerator and the denominator, we obtain

$$x_1 = \frac{\begin{vmatrix} 0 & 5 & -3 \\ -1 & 3 & -1 \\ 0 & -1 & 3 \end{vmatrix}}{\begin{vmatrix} 0 & 5 & -3 \\ -1 & 3 & -1 \\ 0 & -4 & 4 \end{vmatrix}} = \frac{\begin{vmatrix} 5 & -3 \\ -1 & 3 \end{vmatrix}}{\begin{vmatrix} 5 & -3 \\ -4 & 4 \end{vmatrix}} = \frac{3}{2}$$

A.7 A COMPARISON

Which method should be used to solve linear algebraic equations—Gaussian elimination or determinants? In terms of the total number of operations involved to find all the unknowns $x_1, x_2, \ldots, x_n$ this question is easy to answer. Gaussian elimination requires $\frac{2}{3}n^3 + \frac{3}{2}n^2 - \frac{1}{6}n$ additions, multiplications, and so on, and Cramer's rule requires $n!(n^2 + n) - 1$ operations. Consult Table A.1 and it is obvious why Gaussian elimination is the choice of computer programmers. It should be noted that the above numbers were computed assuming that $a_{ij} \neq 0$ for all i, j. In practical systems the matrix **A** usually has some zero elements. In particular in

Table A.1   Operations Count

n	Gaussian Elimination	Cramer's Rule
2	11	11
3	31	71
5	120	3600
10	815	3.992×10^8
20	5930	1.022×10^{21}

large systems the matrix $\mathbf{A}$ is sparse, that is, a majority of the elements in $\mathbf{A}$ are zero. Recently, special Gaussian elimination algorithms have been written to take into account this sparsity and significantly reduce the number of operations [4, 5].

Finally, in this book our analysis is restricted to very small circuits to simplify the analysis and avoid clouding the concepts with tedious calculations. In such cases we will be able to eliminate a few obvious variables from the equations to reduce the number of equations to two or three, at which point the determinant approach will be perfectly satisfactory to obtain the solution, particularly when we need to determine only one of the responses.

A.8 PROPERTIES OF LINEAR ALGEBRAIC EQUATIONS

Let us conclude this section with the discussion of two properties of linear algebraic equations which are very useful in the analysis of systems.

A.8.1 HOMOGENEITY

Suppose the algebraic equation

$$\mathbf{Ax} = \mathbf{b} \tag{85}$$

has a solution $\hat{\mathbf{x}}$, then the algebraic equation

$$\mathbf{Ax} = k\mathbf{b} \tag{86}$$

where k is a scalar, has a solution $k\hat{\mathbf{x}}$. Note that in circuit analysis $\mathbf{b}$ represents the inputs to the circuit and $\mathbf{x}$ is the response. Thus, if *all* the inputs are scaled by k, the responses are scaled by k. The proof of this property is left as an exercise.

A.8.2 ADDITIVITY

If $\hat{\mathbf{x}}_\alpha$ is a solution to the equation

$$\mathbf{Ax} = \mathbf{b}_\alpha \tag{87}$$

and $\hat{\mathbf{x}}_\beta$ is a solution to the equation

$$\mathbf{Ax} = \mathbf{b}_\beta \tag{88}$$

then $\hat{\mathbf{x}}_\alpha + \hat{\mathbf{x}}_\beta$ is a solution to the equation

$$\mathbf{Ax} = \mathbf{b}_\alpha + \mathbf{b}_\beta. \tag{89}$$

This property is sometimes called supersition, since one can find the response to several inputs by finding the response to each input separately and then adding the responses to obtain the response when all the inputs are applied to the circuit. This property is not true in general for systems described by *nonlinear* equations, that is, systems whose component characteristics are nonlinear.

REFERENCES

1. Forsythe, George E. and Cleve B. Moler. *Computer Solution of Linear Algebraic Systems*, Prentice-Hall, Inc., 1967.
2. Hohn, F. E. *Elementary Matrix Algebra*, Macmillan Co. New York, 1964.
3. Gantmacher, F. R. *The Theory of Matrices*, Vols. I and II, Chelsea Publishing Co., New York, 1959.
4. Tinney, W. F. and J. W. Walker. "Direct Solution of Sparse Network Equations by Optimally Ordered Triangular Factorization," *Proceedings of the IEEE*, Vol. 55, pp. 1801–1809, November 1967.
5. Berry, Robert D. "An Optimal Ordering of Electronic Circuit Equations for a Sparse Matrix Solution," *IEEE Transactions on Circuit Theory*, Vol. CT-18, No. 1, pp. 40–50, January 1971.

PROBLEMS

1. Solve the equations below by means of the elimination of variables.

$$x_1 - 2x_2 = 5$$
$$2x_1 + 3x_2 = 3$$

2. Repeat Problem 1 for the following equations.

$$x_1 + 2x_2 + x_3 = 1$$
$$\frac{1}{2}x_1 - x_2 - 2x_3 = 0$$
$$3x_1 - x_2 + 4x_3 = 0$$

3. Write the augmented matrix for the equations in Problem 1. Show that the rows of **A** and **Ag** are independent.
4. Repeat Problem 3 using the equations in Problem 2.
5. The augmented matrices for several systems of equations are given below. For each augmented matrix find out if a solution to the system of equations exists and if it is unique.

$$\begin{bmatrix} 1 & 0 & 1 & \vdots & 2 \\ 1 & -1 & 0 & \vdots & 0 \end{bmatrix}$$
(a)

$$\begin{bmatrix} 1 & 2 & -1 & \vdots & 1 \\ -2 & -4 & 2 & \vdots & 0 \end{bmatrix}$$
(b)

$$\begin{bmatrix} 1 & -1 & \vdots & -1 \\ 3 & 1 & \vdots & -3 \end{bmatrix}$$
(c)

$$\begin{bmatrix} 1 & -\frac{1}{2} & \vdots & -1 \\ -2 & 1 & \vdots & 2 \end{bmatrix}$$
(d)

$$\begin{bmatrix} 1 & -\frac{1}{2} & \vdots & -1 \\ -2 & 1 & \vdots & 0 \end{bmatrix}$$
(e)

$$\begin{bmatrix} 1 & 1 & \vdots & 1 \\ 1 & -1 & \vdots & 0 \\ 3 & 1 & \vdots & 2 \end{bmatrix}$$
(f)

6. Find the matrix $\mathbf{C} = \mathbf{AB}$ where

(a)

$$\mathbf{A} = \begin{bmatrix} 1 & 1 \\ 0 & 1 \end{bmatrix} \qquad \mathbf{B} = \begin{bmatrix} 3 & -3 \\ 1 & 5 \end{bmatrix}$$

(b)

$$\mathbf{A} = \begin{bmatrix} 3 & -3 \\ 1 & 5 \end{bmatrix} \qquad \mathbf{B} = \begin{bmatrix} 1 & 1 \\ 0 & 1 \end{bmatrix}$$

(c)

$$\mathbf{A} = \begin{bmatrix} 1 & -1 & 3 \\ 2 & 1 & -3 \\ 3 & 2 & 1 \end{bmatrix} \qquad \mathbf{B} = \begin{bmatrix} 0 & 1 & 4 \\ 3 & -1 & -1 \\ -2 & 2 & 1 \end{bmatrix}$$

(d)

$$\mathbf{A} = \begin{bmatrix} 1 & 2 \\ 2 & 0 \\ 1 & 1 \end{bmatrix} \qquad \mathbf{B} = \begin{bmatrix} 1 \\ -1 \end{bmatrix}$$

7. Which of the matrices below have all of their rows linearly independent?

$$\mathbf{A} = \begin{bmatrix} 1 & -1 & 2 \\ 2 & 0 & -3 \\ 3 & 1 & 2 \end{bmatrix} \qquad \mathbf{B} = \begin{bmatrix} 1 & -1 & 2 \\ 2 & 0 & -3 \\ 5 & -3 & 3 \end{bmatrix}$$

$$\mathbf{C} = \begin{bmatrix} 1 & 1 \\ -1 & 1 \\ 3 & -1 \end{bmatrix} \qquad \mathbf{D} = \begin{bmatrix} 1 & 0 & 3 \\ 2 & 1 & 3 \end{bmatrix}$$

8. Find the transpose of the matrices in Problem 7.
9. Find the determinants of the square arrays given below.

$$\begin{bmatrix} 1 & -1 \\ 3 & 1 \end{bmatrix} \qquad \begin{bmatrix} 1 & -\dfrac{1}{2} \\ -2 & 1 \end{bmatrix}$$

$$\text{(a)} \qquad\qquad\qquad \text{(b)}$$

$$\begin{bmatrix} 1 & -1 & 2 \\ 2 & 0 & -3 \\ 3 & 1 & 2 \end{bmatrix} \qquad \begin{bmatrix} 1 & 2 & 1 \\ \dfrac{1}{2} & -1 & -2 \\ 3 & -1 & 4 \end{bmatrix}$$

$$\text{(c)} \qquad\qquad\qquad \text{(d)}$$

10. Solve the equations in Problem 1 by means of Cramer's rule.
11. Repeat Problem 10 for the equations in Problem 2.

12. Find the inverse of the matrix

$$\begin{bmatrix} 1 & 2 & 1 \\ \dfrac{1}{2} & -1 & -2 \\ 3 & -1 & 4 \end{bmatrix}$$

in Problem 2. Use this inverse to compute x_1, x_2, and x_3.

Appendix B
Complex Numbers

The complex number system evolved from the need to take roots of negative real numbers. For example, if $a^2 = -5$, then

$$a = \sqrt{-5} = \sqrt{-1} \cdot \sqrt{5} = j\sqrt{5}$$

a real number preceded by $j = \sqrt{-1}$ is called an imaginary number.[1] A number z of the form

$$z = x + jy \tag{1}$$

is called a complex number, and x is referred to as the *real part* of z, $x = \text{Re}[z]$, and y is called the *imaginary part* of z, $y = \text{Im}[z]$. As in the real number system, certain rules of operation are defined such as addition, multiplication, division, and so on. These rules follow along with a geometric interpretation of z.

B.1 COMPLEX PLANE

A complex number is represented geometrically as illustrated in Figure B.1. The x and y represent coordinates of a point in a set of rectangular cartesian coordinates, and to each point (x, y) there corresponds one and only one complex number (vector) z. Note that we can also specify z uniquely by the angle θ and the magnitude or length of the vector z, that is,

$$z = |z|\underline{/\theta} \tag{2}$$

---

[1] In mathematical literature the symbol i is often used to denote $\sqrt{-1}$, but we use j to avoid confusion since the symbol i denotes current in electrical circuits.

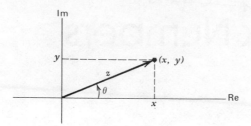

Figure B.1. Complex plane.

where $|z| = \sqrt{x^2 + y^2}$ and $\theta = \tan^{-1} y/x$. Equation 2 is referred to as the polar form of z. Note that

$$x = |z| \cos \theta \tag{3}$$

and

$$y = |z| \sin \theta \tag{4}$$

Therefore,

$$z = |z|(\cos \theta + j \sin \theta) \tag{5}$$

Next we show that the complex number $\cos \theta + j \sin \theta$ has an interesting exponential representation.

B.2 THE EXPONENTIAL FORM

In the complex plane the x coordinate of the complex number $\cos \theta + j \sin \theta$ is $\cos \theta$, and its y coordinate is $\sin \theta$. Therefore, since

$$|\cos \theta + j \sin \theta| = \sqrt{\cos^2 \theta + \sin^2 \theta} = 1 \tag{6}$$

it follows that this complex number has a magnitude of unity regardless of the angle θ. In addition we note that if

$$z = \cos \theta + j \sin \theta$$

then

$$\frac{dz}{d\theta} = -\sin \theta + j \cos \theta = j(\cos \theta + j \sin \theta)$$

We use the fact that $j \cdot j = \sqrt{-1} \cdot \sqrt{-1} = -1$. Thus,

$$\frac{dz}{d\theta} = jz \tag{7}$$

the solution of this equation is

$$\frac{d}{d\theta} \ln z = j$$

and

$$\ln z = j\theta + c$$

Therefore,

$$z = k\, e^{j\theta} \tag{8}$$

where $k = e^c$. Since

$$z(0) = \cos 0 = 1$$

we conclude that $k = 1$ and

$$z = e^{j\theta} = \cos \theta + j \sin \theta \tag{9}$$

Equation 9 is called the *Euler formula.*

With this background the rules of addition, multiplication, and so on, for complex numbers follow.

B.3 ADDITION OF COMPLEX NUMBERS

Given two complex numbers

$$z_1 = a + jb \tag{10}$$

and

$$z_2 = c + jd \tag{11}$$

we define their sum as

$$z_a = z_1 + z_2 = (a + c) + j(b + d) \tag{12}$$

and we define their difference as

$$z_d = z_1 - z_2 = (a - c) + j(b - d) \tag{13}$$

These operations result in another complex number. The first closed parenthesis contains the real part of the resultant complex number, and the second closed parenthesis, prefixed by j, contains the imaginary part. The operations are illustrated graphically in Figure B.2.

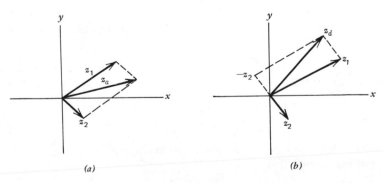

Figure B.2.   Addition and substraction of complex numbers. (*a*) Addition. (*b*) Subtraction.

B.4 COMPLEX CONJUGATE OF A COMPLEX NUMBER

The complex conjugate of a complex number $z = x + jy$ is defined as

$$z^* = x - jy \tag{14}$$

Note that the imaginary part is multiplied by -1.

B.5 MULTIPLICATION AND DIVISION OF COMPLEX NUMBERS

In polar form the *product* of $z_1 = |z_1| \, e^{j\theta_1}$ with $z_2 = |z_2| \, e^{j\theta_2}$ is defined as

$$z_1 z_2 = |z_1| |z_2| \, e^{j(\theta_1 + \theta_2)} \tag{15}$$

that is, the new vector has a magnitude equal to the product of the magnitudes of z_1 and z_2 and its angle is the sum of the angles of z_1 and z_2.

In rectangular coordinates this is equivalent to

$$z_1 z_2 = (x_1 x_2 - y_1 y_2) + j(x_1 y_2 + x_2 y_1) \tag{16}$$

The *division* of two complex numbers in polar form is defined as

$$\frac{z_1}{z_2} = \frac{|z_1| \, e^{j\theta_1}}{|z_2| \, e^{j\theta_2}} = \frac{|z_1|}{|z_2|} \, e^{j(\theta_1 - \theta_2)} \tag{17}$$

Note that the magnitude of divisor is divided into the magnitude of the dividend resulting in the magnitude of the quotient. The angle of the quotient is the difference between the angle of the dividend and the angle of the divisor.

In rectangular coordinates this is equivalent to

$$\frac{z_1}{z_2} = \frac{x_1 + jy_1}{x_2 + jy_2} \tag{18}$$

If we multiply the numerator and denominator of (18) by z_2^* we obtain

$$\frac{z_1}{z_2} = \frac{(x_1 x_2 + y_1 y_2) + j(y_1 x_2 - y_2 x_1)}{x_2^2 + y_2^2} \tag{19}$$

It should be clear that these operations are easier to perform in the polar coordinate system.

B.6 POWERS AND ROOTS

The complex number raised to the nth power is defined by

$$z^n = |z|^n \, e^{jn\theta} \tag{20}$$

Thus, again we note the usefulness of the polar coordinate representation.

It is also interesting to note that

$$e^{j\theta} \equiv e^{j(\theta + 2\pi k)} = e^{j\theta} \, e^{j2\pi k}, \quad k \text{ an integer} \tag{21}$$

since $\cos(2\pi k) = 1$ and $\sin(2\pi k) = 0$. With this information we can now define the nth root of z. For example, consider the equation

$$z^3 + 1 = 0 \tag{22}$$

We can write

$$z^3 = -1 = e^{j180°} \tag{23}$$

therefore

$$z = (e^{j180°})^{1/3} = e^{j60°} = \frac{1}{2} + j\frac{\sqrt{3}}{2} \tag{24}$$

Note that $z = -1$ is also a solution to (22). Hence, Equation 23 does not locate all of the roots. Therefore, let us use identity (21). Then

$$z = (e^{j(180°+2\pi k)})^{1/3}, \quad k \text{ an integer}$$

This is the same as

$$z = e^{j(60°+120°k)} \tag{25}$$

For $k = 0$ we obtain

$$z_1 = e^{j60°} = \frac{1}{2}(1 + j\sqrt{3}) \tag{26}$$

For $k = 1$,

$$z_2 = e^{j180°} = -1 \tag{27}$$

and for $k = 2$

$$z_3 = e^{j300°} = \frac{1}{2}(1 - j\sqrt{3}) \tag{28}$$

These three roots lie on the unit circle in Figure B.3. Note that the third root is the complex conjugate of the first. In any polynomial with real coefficients the complex roots must appear in conjugate pairs. Convince yourselves that any other integer values of k result in the same roots.

Thus we define the nth root of a complex number as

$$z^{1/n} = |z|^{1/n} e^{j(\theta+2\pi k/n)}, \quad k = 0, 1, 2, \dots, n - 1 \tag{29}$$

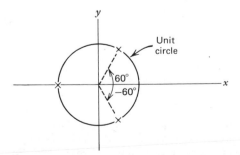

Figure B.3. Roots of $z^3 + 1 = 0$ in the complex plane.

PROBLEMS

1.   Show that $zz^* = |z|^2$.
2.   Prove the following identities:

(a) $\text{Re}[e^{x+jy}] = e^x \cos y$
(b) $\text{Im}[e^{x+jy}] = e^x \sin y$
(c) $e^{jx} + e^{-jx} = 2 \cos x$
(d) $e^{jx} - e^{-jx} = 2j \sin x$

3.   Prove that $(\cos \theta + j \sin \theta)^n = \cos n\theta + j \sin n\theta$.
4.   Convert to polar form

(a) $1 + j2$ (b) $-5 + j3$
(c) $-3 - j5$ (d) $2 - j1$
(e) $0.1 + j10$ (f) $5.6 - j0.1$

5.   Convert to rectangular form
(a) $10\underline{/30°}$ (b) $5\underline{/205°}$
(c) $1\underline{/60°}$ (d) $8.2\underline{/-45°}$

6.   Evaluate the following expression and give your answer in rectangular form.

(a) $z = \dfrac{10\underline{/70°} - 5\underline{/10°}}{1 + j2}$

(b) $z = \ln(5\underline{/90°})$

(c) $z = \dfrac{(1 - j2)(-3 + j1)}{3 + j4}$

(d) $z = 4 \, e^{1+j}$

7.   Find the roots of the following equations:

(a) $z^2 + 1 = 0$ (b) $z^2 - 1 = 0$
(c) $z^3 - 1 = 0$ (d) $z^5 + 1 = 0$

8.   Solve the following equations for the unknown constants

(a) $(10 + j20)2\underline{/\theta} = K\underline{/30°}$

(b) $2\underline{/-120°} + 3.5 + jb = a + j5$

9.   Show that Equations 15 and 16 are equivalent.
10. Show that Equations 17 and 19 are equivalent.
11. Show that

$$\text{Re}[z] = \frac{1}{2}(z + z^*)$$

$$\text{Im}[z] = \frac{1}{2j}(z - z^*)$$

12. Any polynomial with real coefficients a_i can be factored as

$$x^n + a_{n-1}x^{n-1} + \cdots + a_1 x + a_o = (x - z_1)(x - z_2) \cdots (x - z_n)$$

where the z_i's are the roots or zeros of the polynomial. Show that if z_i is a complex root (imaginary part not zero) then there must be another root $z_j = z_i^*$ for some $j \neq i$, that is, the complex roots of a polynomial with all real coefficients must occur in conjugate pairs.

INDEX

Physical Constants

Constant	Symbol	Value
Charge of an electron	$-e$	-1.602×10^{-19} C
Mass of an electron	m	9.1085×10^{-31} kg
Electron volt	eV	1.602×10^{-19} J
Dielectric permittivity of free space	ϵ_o	8.854×10^{-12} F/m
Magnetic permittivity of free space	μ_o	1.257×10^{-6} H/m
Velocity of light in free space	c	2.998×10^{8} m/s
Boltzmann's constant	k	1.380×10^{-23} J/K
Avogadro's number	N_o	6.025×10^{23} molecules/gram-mole